Comprehensive

MECHANICS

B.A./B.Sc. IInd Year

Comprehensive

MECHANICS

(B.A./B.Sc. IInd Year)

Paer III

(Strictly as per MDU, Rohtak KU, Kurukshetra and Himachal University Syllabus)

By

Dr. A.K. GUPTA

M.Sc. (Maths, Stat.), Ph.D.

Head, Deptt. of Mathematics

G.M.N. College

Ambala City

Haryana

UNIVERSITY SCIENCE PRESS

(An Imprint of Laxmi Publications Pvt. Ltd.)

An ISO 9001:2015 Company

BENGALURU • CHENNAI • GUWAHATI • HYDERABAD

KOCHI • KOLKATA • LUCKNOW • RANCHI

NEW DELHI

Printed and bound in India

Fourth Edition : 2008 Reprint : 2009, 2012 *Edition* : 2025

ISBN : 978-81-3180-235-9

Branches

✆	Bengaluru	080-26 75 69 30
✆	Chennai	044-24 34 47 26
✆	Guwahati	0361-254 36 69
✆	Hyderabad	040-27 55 53 83
✆	Kochi	0484-405 13 03
✆	Kolkata	033-40 04 77 79
✆	Lucknow	0522-430 36 13
✆	Ranchi	0651-224 24 64

PUBLISHED IN INDIA BY

UNIVERSITY SCIENCE PRESS

(An Imprint of Laxmi Publications Pvt. Ltd.)

An ISO 9001:2015 Company

113, GOLDEN HOUSE, GURUDWARA ROAD, DARYAGANJ,

NEW DELHI - 110002, INDIA

Corporate Office : B-33, Sector-5, Noida,

Gautambuddha Nagar (U.P.) - 201 301, INDIA

Telephone : +91-120-44 46 900

www.laxmipublications.com

info@laxmipublications.com

PREFACE

The present book is an effort to meet the requirement of the students of B.A./B.Sc. II Paper III of K.U. Kurukshetra, M.D.U. Rohtak and other Indian Universities which have accepted the new syllabus of UGC. The objective behind the book is to make the subject clear to the students. For this purpose a number of books have been consulted and the articles and examples have been solved in detail.

Throughout the book anticlockwise system of co-ordinate axes has been taken and the units used in problems are either C.G.S. or M.K.S. system of units.

Keeping in view of the difficulty faced by the students in solving the problems, some more examples have been added to make the topics more understandable.

Though every care has been taken to keep the book error free, yet some of them might have been left. Every error pointed out by the fellow teachers or by the students will be thankfully acknowledged.

I can never forget the inspiration and motivation provided by my family, my friends and above all Dr. V.P. Kaushik, Chairman, Deptt. of Maths, K.U. Kurukshetra for valuable suggestions given by him from time to time.

Every suggestion to improve the book will be gratefully acknowledged.

—AUTHOR

SYLLABUS

BM 203 MECHANICS

(K.U., Kurukshetra)

	Maximum Marks	*Time*
B.A.	: 30	3 hours
B.Sc.	: 45	3 hours

Section I (3 Questions)

Analytical conditions of equilibrium of Coplanar forces. Virtual work. Catenary.

Section II (2 Questions)

Forces in three dimensions. Poinset's central axis. Wrenches. Null lines and planes. Stable and unstable equilibrium.

Section III (3 Questions)

Velocities and accelerations along radial and transverse directions and along tangential and normal directions. Simple harmonic motion. Elastic strings.

Motion on smooth and rough plane curves. Motion in a resisting medium. Motion of particles of varying mass.

Section IV (2 Questions)

Central Orbits. Kepler's laws of motion.

Motion on a particle in three dimensions. Acceleration in terms of different coordinate systems.

Note. The examiner is requested to set **ten** questions in all selecting questions sectionwise as indicated in the syllabus. The candidate is required to attempt **five** questions selecting at least one question from each section.

SYLLABUS

BM 203 MECHANICS

(M.D.U., Rohtak)

	Maximum Marks	*Time*
B.A.	: 30	3 hours
B.Sc.	: 45	3 hours

Section I (3 Questions)

Conditions of equilibrium of Coplanar forces. Virtual work in two dimensions. Proof of principle of virtual work and its converse for coplanar forces. Forces which may be omitted in forming the equation of virtual work.

Section II (2 Questions)

Forces in three dimensions. Stable and unstable equilibrium.

Section III (3 Questions)

Velocities and accelerations along radial and transverse directions, and along tangential and normal directions. Simple harmonic motion. Elastic strings. Motion on smooth and rough plane curves. Motion in a resisting medium. Motion of particles of varying mass.

Section IV (2 Questions)

Central Orbits. Kepler's laws of motion. Motion of a particle in three dimensions. Acceleration in terms of different coordinate systems.

Note. The examiner is requested to set **ten** questions in all selecting questions sectionwise as indicated in the syllabus. The candidate is required to attempt **five** questions from each section.

CONTENTS

***Note.** Chapters 3 and 6 are not included in the syllabus of M.D.U., Rohtak.

Preliminary Concepts

In order to study the forces in three dimension we need the knowledge of the properties of forces in two dimension. Some results of statics are reproduced below :

(*i*) **Parallelogram Law of Forces.** If two forces acting at a point are represented in magnitude and direction by the sides of a parallelogram drawn from that point, then their resultant is represented by the diagonal of the parallelogram drawn from that point.

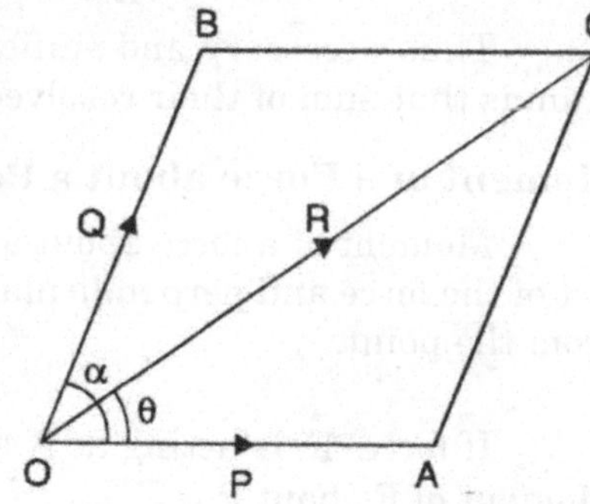

Thus if P and Q are the forces acting along OA and OB of sides of parallelogram then the resultant R is along OC and is given by

$$R^2 = P^2 + Q^2 + 2PQ \cos \alpha$$

$$\tan \theta = \frac{Q \sin \alpha}{P + Q \cos \alpha}$$

where α is the angle between OA and OB.

(*ii*) Component of a force F in two given directions inclined to α and β are $\dfrac{F \sin \beta}{\sin (\alpha + \beta)}$ and $\dfrac{F \sin \alpha}{\sin (\alpha + \beta)}$ respectively. [See Fig. (*i*)]

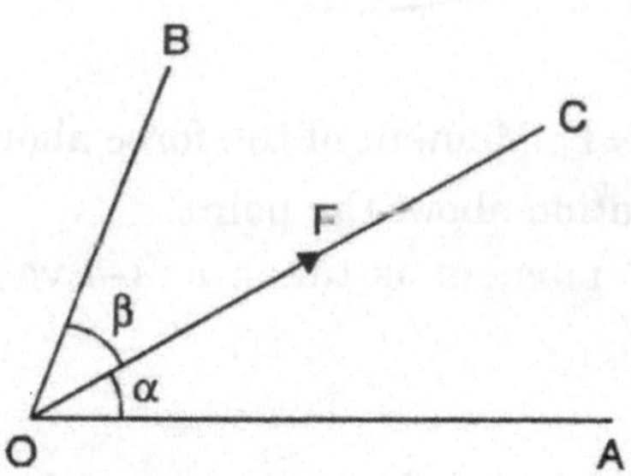

(*i*)

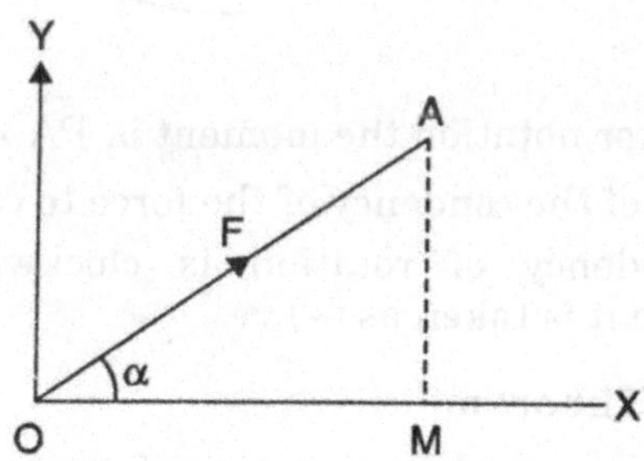

(*ii*)

(*iii*) Resolved parts of a force F in a direction OX inclined at an angle α with F is F cos α and F sin α in a direction perpendicular to OX. [See Fig. (*ii*)]

(*iv*) **Triangle Law of Forces.** If three coplaner forces acting at a point can be represented in magnitude and direction by the sides of a triangle taken in order the forces are in equilibrium.

Converse of Triangle Law of Forces

If three coplaner forces acting at a point are in equilibrium and a triangle be drawn with its sides parallel to the directions of the forces, then the forces will be proportional to sides of a triangle in magnitude and will be taken in order (clockwise or anticlockwise).

(*v*) **Lami's Theorem.** If three coplaner forces acting at a point are in equilibrium, then each force is proportional to the sine of the angle between the other two.

(*vi*) **Condition of Equilibrium of Forces Acting at a Point.** Forces acting at a point are in equilibrium when their resultant is zero.

If the forces are resolved along rectangular axes in one plane and sum of their resolved parts along the axes are X and Y then

$$R^2 = X^2 + Y^2$$

$$R = 0 \quad \Rightarrow \quad X = 0 \quad \text{and} \quad Y = 0$$

Thus necessary and sufficient condition that the forces acting at a point be in equilibrium is that sum of their resolved parts in two perpendicular direction must be zero.

Moment of a Force about a Point

Moment of a force about a point is defined as the product of the force and perpendicular distance of its line of action from the point.

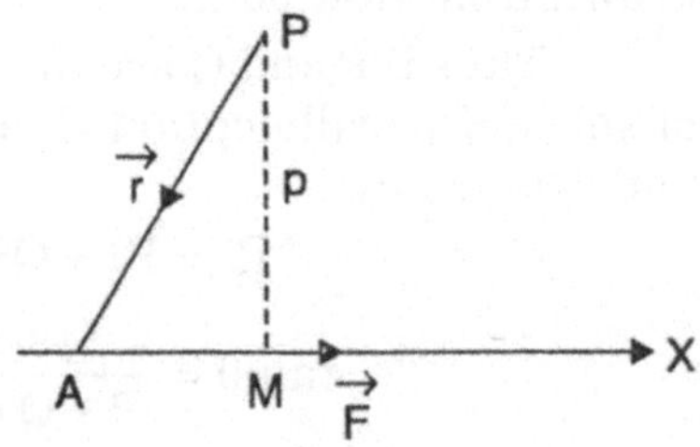

If force $\vec{F}$ is acting at A along AX and P is any point. Moment of F about P is F. p.

Where p is length of perpendicular from P on F.

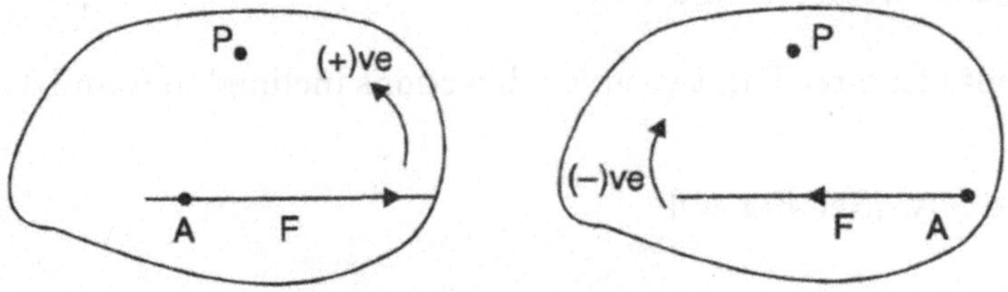

In vector notation the moment is $\vec{PA} \times \vec{F} = \vec{r} \times \vec{F}$. Moment of the force about a point is the measure of the tendency of the force to cause rotation about the point.

If tendency of rotation is clockwise the moment is taken as (–) ve and if it is anticlockwise it is taken as (+) ve.

Varignon's Theorem

When any number of coplaner forces act at a point, the algebraic sum of their moments about any point in their plane is equal to the moment of their resultant about the same point.

Parallel forces. When the lines of action of two forces are parallel the forces are called parallel forces. Where the directions are in same sense they are like parallel forces (*i*) and when they are in opposite direction forces are unlike parallel (*ii*).

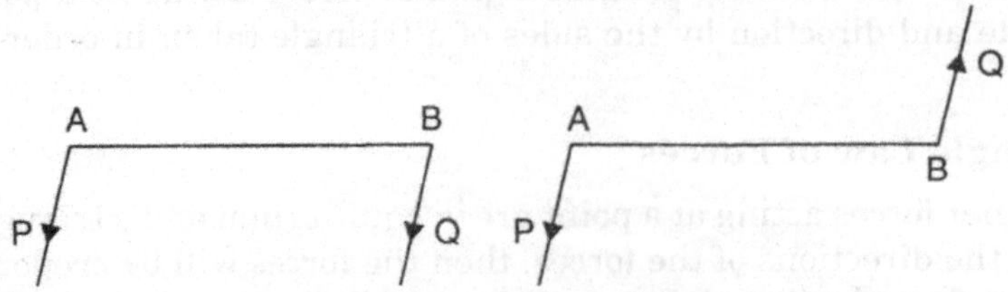

(i) (ii)

Resultant of two like parallel forces P and Q acting at A and B respectively is P + Q and acts at a point C between A and B such that P. AC = Q . BC.

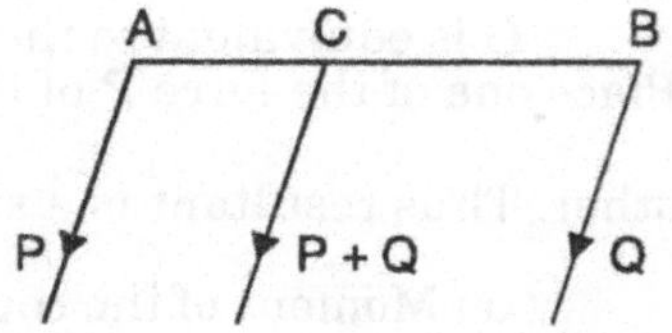

Resultant of two unlike parallel forces P and Q (P > Q) acting at A and B is P – Q and acts at C such that P. AC = Q. BC.

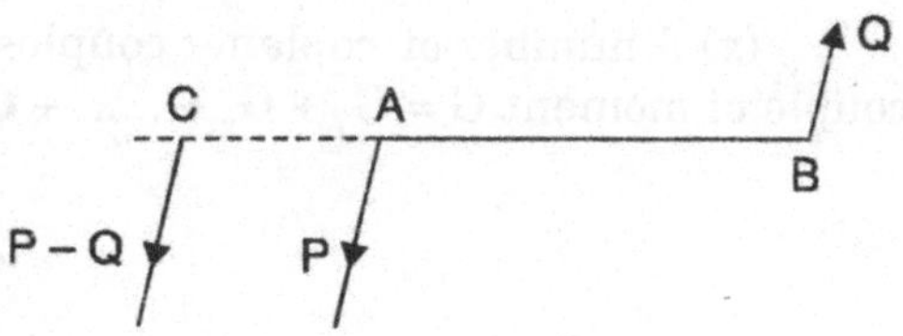

Varignon theorem of moments also holds true on parallel forces *i.e.,* **the algebraic sum of moments of two parallel forces about any point in their plane is equal to the moment of their resultant about the same point.**

Couples. Two equal unlike parallel forces whose lines of action are not the same, from a couple.

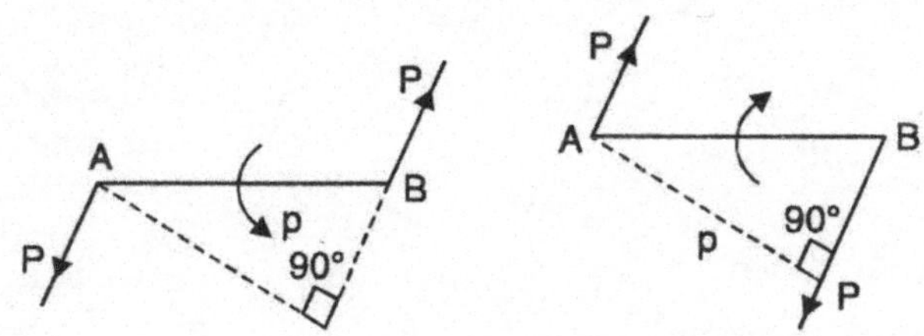

Moment of the couple = P. p where p is the perpendicular distance between the two forces.

Sign of the moment is (+) ve if tendency of rotation is anticlockwise and (–) ve if it is clockwise.

Properties of Couples

(*i*) Since a couple consists of two equal and opposite parallel forces, the algebraic sum of resolved parts of the forces in every direction is zero.

(*ii*) A couple can not be replaced by a single force.

(*iii*) Tendency of a couple is to produce rotation.

(*iv*) Algebraic sum of moments of the couple is independent of the choice of the point.

Moment of the couple is P.p where P is force and p is $\perp$ distance between the forces.

(*v*) Couples of equal moments in the same plane are equivalent.

(*vi*) Couples of equal and opposite moments in the same plane balance one another.

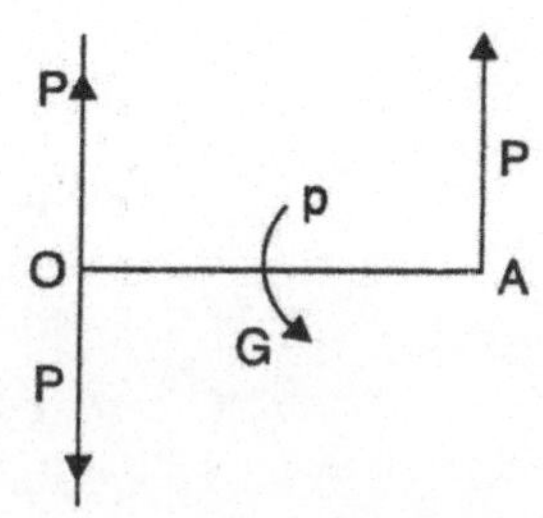

(*vii*) A couple and a force can not be in equilibrium.

(*viii*) A force and a couple in the same plane can always be reduced to a single force.

Let P be the force acting at O. G is the moment of an anticlockwise couple.

G is equivalent to the couple with forces of magnitude P at distance p such that $P.p = G$. Place one of the force P of the couple at O opposite to given force P so that they balance each other. Thus resultant is a single force P at distance $\frac{G}{P}$ from O.

(*ix*) Moment of the couple is a vector whose direction is perpendicular to the plane of the forces.

(*x*) A number of coplaner couples of moments $G_1, G_2, \ldots\ldots G_n$ can be combined to a single couple of moment $G = G_1 + G_2 + \ldots\ldots + G_n$.

SECTION I

1

Analytical Conditions of Equilibrium of Coplanar Forces

1. Introduction

In this chapter we will study the equilibrium of a rigid body acted upon by a number of coplanar forces. Forces are said to be coplanar if the lines of action of the forces and their points of applications are in the same plane. If a rigid body under the action of a number of forces is not in equilibrium its motion is of two types *i.e.,* translation and rotation.

In case of translation the motion of every particle of the rigid body is in the same direction and is of equal magnitude. In case of rotation the body rotates about some point of itself and the particles of the rigid body have different displacement and direction depending on the position of the particle relative to the point about which body rotates. For the equilibrium both translation and rotation should vanish.

Theorem 1. *A system of coplanar forces acting on a rigid body can be reduced to a single force through any arbitrary chosen point and a couple.* (K.U. 2004, 2007)

Proof. Let $\vec{F}_1, \vec{F}_2, \ldots, \vec{F}_n$ be the coplanar forces acting at the points $A_1, A_2, \ldots A_n$. Choose an arbitrary point O on the body. Let $\vec{a}_1, \vec{a}_2, \ldots, \vec{a}_n$ be the position vector of $A_1, A_2, \ldots A_n$ w.r.t. O. Now consider the force $\vec{F}_r$ acting at A_r. Apply the forces $\vec{F}_r$ and $-\vec{F}_r$ at O. The resultant of these two forces is zero. But if we consider $\vec{F}_r$ at A_r and $-\vec{F}_r$ at O they form a couple of moment $\vec{a}_r \times \vec{F}_r$ and we are left with a single force $\vec{F}_r$ at O. Putting $r = 1, 2, \ldots n$ we can say that $\vec{F}_1, \vec{F}_2, \ldots, \vec{F}_n$ reduces to the couples of moment $\vec{a}_1 \times \vec{F}_1, \vec{a}_2 \times \vec{F}_2, \ldots, \vec{a}_n \times \vec{F}_n$ and forces $\vec{F}_1, \vec{F}_2, \ldots, \vec{F}_n$ at O. Thus the resultant of the whole system is force, $\vec{R} = \vec{F}_1 + \vec{F}_2 + \ldots, \vec{F}_n$ at O and couple of moment $\vec{G} = \vec{a}_1 \times \vec{F}_1 + \vec{a}_2 \times \vec{F}_2 + \ldots + \vec{a}_n \times \vec{F}_n$.

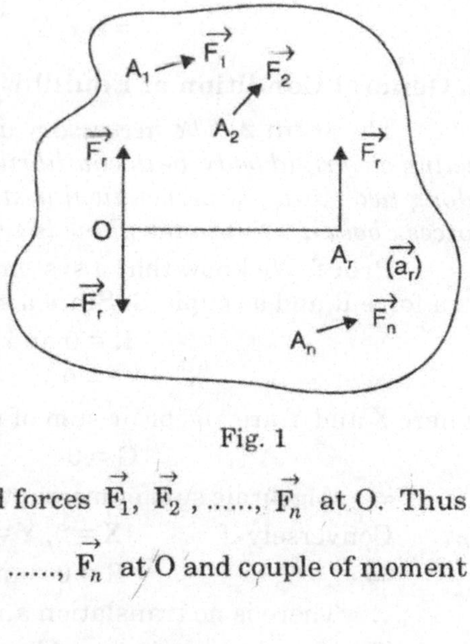

Fig. 1

Hence the system of coplanar forces on a rigid body reduces to a single force R at arbitrary taken point O and a couple of moment G.

2. Reduction from Vector to Cartesian Coordinates

If coordinates of $A_1, A_2, \ldots\ldots A_n$ are $(x_1, y_1), (x_2, y_2), \ldots\ldots (x_n, y_n)$ respectively and resolved parts of forces along X and Y axes are $X_1, X_2, \ldots\ldots X_n$ and $Y_1, Y_2, \ldots\ldots Y_n$ then by the theorem of resolved parts sum of resolved part of the forces in any direction is equal to resolved part of the resultant in same direction, we get

$$X = R\cos\theta = X_1 + X_2 + \ldots\ldots + X_n$$

$$Y = R\sin\theta = Y_1 + Y_2 + \ldots\ldots + Y_n.$$

$$\therefore \qquad R = \sqrt{X^2 + Y^2} \text{ and } \theta = \tan^{-1}\frac{Y}{X}$$

$$\vec{a}_r \times \vec{F}_r = \begin{vmatrix} \hat{i} & \hat{j} & \hat{k} \\ x_r & y_r & 0 \\ X_r & Y_r & 0 \end{vmatrix} = \hat{k}\,(x_r Y_r - y_r X_r)$$

$$\therefore \qquad \vec{G} = \sum_{r=1}^{n} (x_r Y_r - y_r X_r)\,\hat{k}$$

Hence magnitude of the couple is equal to

$$G = (x_1 Y_1 - y_1 X_1) + (x_2 Y_2 - y_2 X_2) + \ldots\ldots + (x_n Y_n - y_n X_n).$$

In case we shift the point O to some other point $O'(h, k)$ then the moment of the couple is G' where

$$G' = \Sigma\,(x_r - h)\,Y_r - (y_r - k)\,X_r$$

$$= \Sigma\, x_r Y_r - y_r X_r - h\,\Sigma\, Y_r + k\,\Sigma\, X_r = G - hY + kX.$$

3. General Condition of Equilibrium *(K.U. 2007)*

Theorem 2. *The neccessary and sufficient condition that a system of coplanar forces acting on a rigid body, be in equilibrium, is that algebraic sum of resolved parts of all the forces along two mutually perpendicular straight lines be seperately zero and the moment of all the forces about a point in the plane is zero.*

Proof. We know that a system of coplanar forces acting on a rigid body can be reduced to a force R and a couple G. Since a single force can not balance a couple G ; for equilibrium

$$R = 0 \text{ and } G = 0 \text{ separately.}$$

$$\Rightarrow \qquad X^2 + Y^2 = 0 \qquad \Rightarrow \quad X = 0,\ Y = 0$$

where X and Y are algebraic sum of resolved parts along two perpendicular directions.

$$G = 0$$

$\Rightarrow$ Algebraic sum of moments is zero.

Conversely if $\quad X = 0,\ Y = 0,\ G = 0 \ \Rightarrow\ R^2 = X^2 + Y^2 = 0$

$\Rightarrow \quad R = 0$ and $G = 0$.

$\therefore$ There is no translation and no rotation and hence system is in equilibrium.

To solve the practical problems on equilibrium we resolve the forces along two perpendicular chosen direction and equate to zero and take moment of the forces about some suitable point and equate it to zero *i.e.*, $X = 0$, $Y = 0$ and $G = 0$.

Theorem 3. *If the algebraic sum of moments of all the coplanar forces acting on a rigid body about three non collinear points is separately zero, then the body is in equilibrium.*

Proof. Let the three non collinear points be A, B and C. Choose A as origin and AB as x-axis. Let the system of forces be reduced to force R at A and a couple of moment G. Let coordinate of A, B, C be (0, 0), $(a, 0)$ and (b, c) respectively. We know that moment of couple about some point is $G' = G - xY + yX$.

Taking the moments about A, B, C we have

$$G - 0Y + 0X = 0 \quad ...(1)$$

$$G - aY + 0X = 0 \quad ...(2)$$

$$G - bY + cX = 0 \quad ...(3)$$

Solving (1), (2) we have $G = 0, Y = 0$

Putting G and Y in (3), we have $cX = 0$

$\because$ If $c = 0$ then C lies on AB. But A, B and C are non collinear

$\therefore$ $c \neq 0$

$\therefore$ $X = 0$

$X = 0, Y = 0, G = 0$

$\Rightarrow$ System is in equilibrium.

Theorem 4. *If the algebraic sum of moments of all the coplanar forces acting on a rigid body about two points in the plane is zero and sum of resolved part of the forces in any direction not perpendicular to the straight line joining these two points is zero then the system is in equilibrium.* (K.U. 2005)

Proof. Let A and B be two given points. Take A as origin and the given direction as x-axis. Let coordinate of A and B be (0, 0) and (a, b).

$\because$ Moment about A and B is zero

$\therefore$ $G - 0Y + 0X = 0 \Rightarrow G = 0$

$G - aY + bX = 0$

$X = 0$ (given), we have $aY = 0$

$a \neq 0$ otherwise AB will be perpendicular to x-axis.

Hence $Y = 0 \Rightarrow$ The system is in equilibrium.

Theorem 5. *A system of coplanar forces acting on a rigid body (not in equilibrium) can be reduced to a single force or to a couple.*

Proof. Since the body is not in equilibrium $R = 0, G = 0$ can not hold simultaneously.

$\therefore$ Three cases arise

(*i*) $R \neq 0, G = 0$

(*ii*) $R = 0, G \neq 0$

(*iii*) $R \neq 0, G \neq 0$.

In first two cases the system is either a single force or a single couple. Consider the third case $R \neq 0, G \neq 0$. Replace the couple G by two

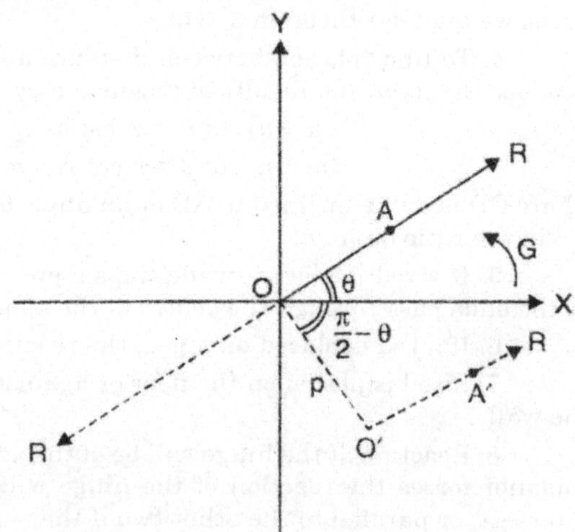

Fig. 2

equal and opposite force one acting at O opposite to the direction of R and another at O′ parallel to it such that

$$R.OO' = G \quad \text{or} \quad OO' = \frac{G}{R}$$

Now equal and opposite forces R acting at O balance each other and we are left with a single force R at O′ along O′A′ where O′A′ || R.

4. Line of Action of the Single Resultant

We know that equation of straight line is $x \cos \alpha + y \sin \alpha = p$

Refer to Fig. 2 we have $\alpha = -\left(\frac{\pi}{2} - \theta\right), \quad p = OO'$

The equation of line of action of R at O′ is

$$x \cos\left(\theta - \frac{\pi}{2}\right) + y \sin\left(\theta - \frac{\pi}{2}\right) = OO'$$

or

$$x \sin\theta - y \cos\theta = \frac{G}{R}$$

$$\because \quad \tan\theta = \frac{Y}{X}, \; \sin\theta = \frac{Y}{\sqrt{X^2 + Y^2}} = \frac{Y}{R}, \; \cos\theta = \frac{X}{\sqrt{X^2+Y^2}} = \frac{X}{R}$$

Thus line of action of R is

$$x.\frac{Y}{R} - y\frac{X}{R} = \frac{G}{R}$$

Hence, the equation, the line of action of single resultant is $xY - yX = G$.

Note. In solving the problems of equilibrium we keep the following points in mind.

1. If three coplanar forces keep a body in equilibrium, they are either concurrent or parallel.

2. In case the forces are concurrent we can use Lami's Theorem or triangle law of forces.

3. If equilibrium is under three or more coplanar forces we may use theorem 2, 3 or 4.

4. To find relation between distance and angles we may use the following results of trigonometry.

$$(m + n) \cot\theta = m \cot\alpha - n \cot\beta$$
$$(m + n) \cot\theta = n \cot A - m \cot B$$

where CD is a line inclined to AB at an angle θ and divides AB in the ratio of $m : n$.

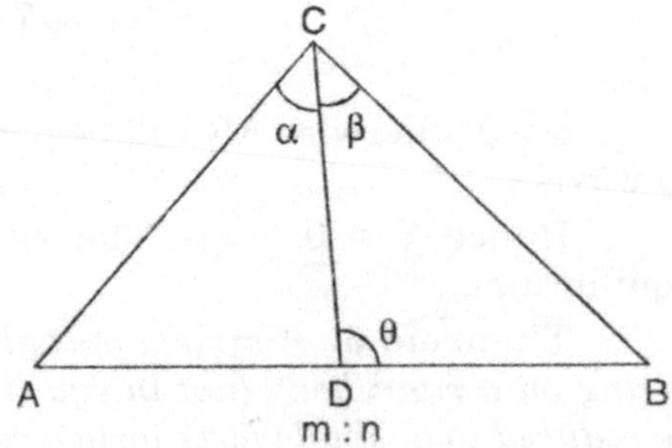

Fig. 3

5. If a rod is placed inside the sphere the reactions at the ends pass through the centre of the sphere.

6. If a rod is placed on a peg, the reaction will be perpendicular to the rod.

7. If rod is placed on the floor or against a wall, the reaction will be perpendicular to the floor or the wall.

8. Reaction at the hinge will be of the adjusting type *i.e.*, if the body is in equilibrium under three coplaner forces the reaction of the hinge will pass through the intersection of the other two if they intersect, or parallel to the other two if they are parallel.

9. Tension in the string will be towards the fixed point.

Example 1. *A heavy uniform rod AB of weight W and length 2a is hinged at A and the lower end is kept at a distance 2b from the vertical through A by a horizontal force P applied at B. Find the reaction at the hinge and the force at B.*

Sol. The rod AB is in equilibrium under three coplanar forces

(*i*) Weight W acting at mid point of rod AB, vertically downward.

(*ii*) Force P acting horizontally at B.

(*iii*) Reaction R at A will pass through the point of intersection of W and P.

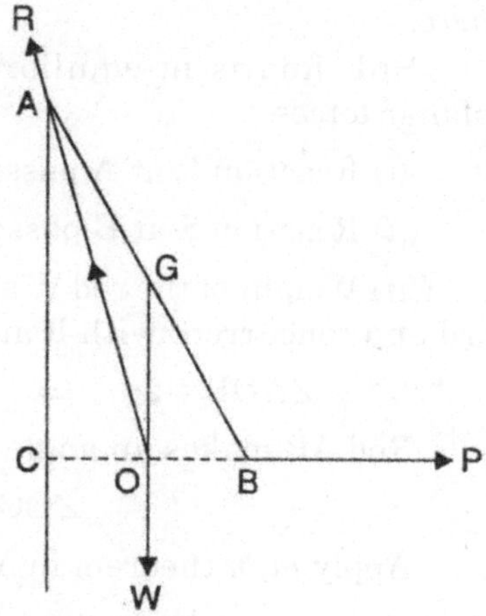

Fig. 4

$$AB = 2a,\ BC = 2b$$

$\because$ G is middle point of AB hence AG = GB = a

$$OG \parallel AC \quad \therefore \quad OB = OC = b$$

From $\Delta ABC\ AC^2 = AB^2 - BC^2 = 4a^2 - 4b^2$

$$\therefore \quad AC = 2\sqrt{a^2 - b^2}$$

From $\Delta OAC,\ OA^2 = OC^2 + AC^2 = b^2 + 4a^2 - 4b^2$

$$\therefore \quad OA = \sqrt{4a^2 - 3b^2}$$

Now ΔOAC can be taken as the triangle of force.

$\because$ R acting along OA, RP along CO and W can be represented along AC.

$\therefore$ By triangle law of forces

$$\frac{P}{OC} = \frac{R}{OA} = \frac{W}{AC}$$

$$\therefore \quad P = \frac{W.OC}{AC} = \frac{Wb}{2\sqrt{a^2 - b^2}}$$

$$R = W.\frac{OA}{AC} = \frac{W.\sqrt{4a^2 - 3b^2}}{2\sqrt{a^2 - b^2}}.$$

Example 2. *A triangular lamina ABC is suspended from a point O by light strings attached to points A and B and hangs so that BC is vertical. If α and β be the angles which the strings AO, BO make with the vertical, prove that 2 cot α – cot β = 3 cot B.*

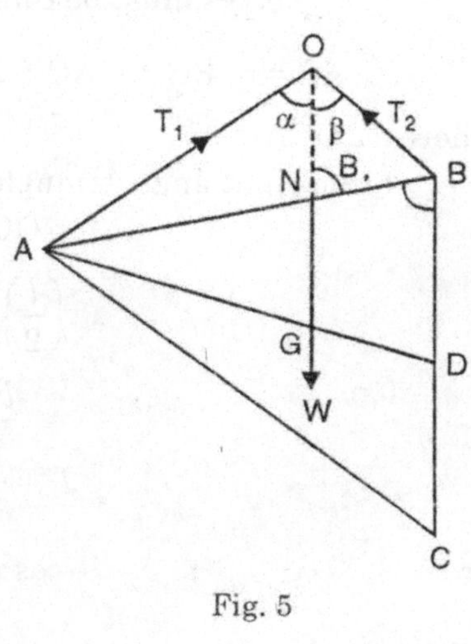

Fig. 5

Sol. ΔABC is suspended by strings AO and BO.

Forces are Tensions T_1 along AO and T_2 along BO and weight W acting at G and passing through O.

$$\angle AOG = \alpha,\ \angle BOG = \beta$$

$$AG : GD = AN : NB = 2 : 1.$$

Applying $m : n$ theorem

$$(2 + 1) \cot B = 2 \cot \alpha - 1 \cot \beta$$

$$3 \cot B = 2 \cot \alpha - \cot \beta.$$

Example 3. *A rod whose C.G. divides it in two portions of lengths a and b is placed in a smooth sphere. The inclination of the rod to the horizontal θ in the position of equilibrium is given by* $\tan\theta = \dfrac{(b-a)\tan\alpha}{b+a}$ *where 2α is the angle which rod subtends at the centre of the sphere.*

Sol. Rod is in equilibrium under the following coplanar forces :

(*i*) Reaction R at A passing through the centre

(*ii*) Reaction S at B passing through the centre

(*iii*) Weight of the rod W acting at G vertically downward and concurrent with R and S

$$\angle AOB = 2\alpha \quad \Rightarrow \quad \angle ABO = \angle OAB = 90 - \alpha$$

Rod AB makes an angle θ with horizontal

$$\therefore \qquad \angle OGB = 90 - \theta$$

Apply *m*, *n* theorem in ΔOAB

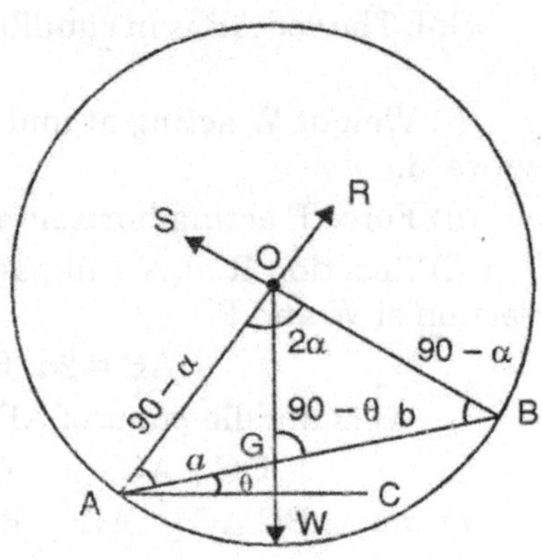

Fig. 6

$$(a+b)\cot(90-\theta) = b\cot(90-\alpha) - a\cot(90-\alpha)$$

or $$(a+b)\tan\theta = (b-a)\tan\alpha$$

$$\therefore \qquad \tan\theta = \frac{b-a}{b+a}\tan\alpha.$$

Example 4. *A uniform rod AB of length a rests against a vertical wall at A and the other end B is supported by a string of length l (l > a) which is tied at a point C vertically above A. Show that inclination of the rod with the wall is given by* $\cos^2\theta = \dfrac{l^2-a^2}{3a^2}$.

Sol. Forces on the rod in equilibrium are :

(*i*) Weight W acting at G vertically downwards.

(*ii*) Reaction R perpendicular to wall at A.

(*iii*) Tension T along BC.

Forces must be concurrent at O

$$AB = a,\ BC = l,\ AC = 2OG = 2\cdot\frac{a}{2}\cos\theta$$

where $\angle OGA = \theta$.

From right angle triangle OAC

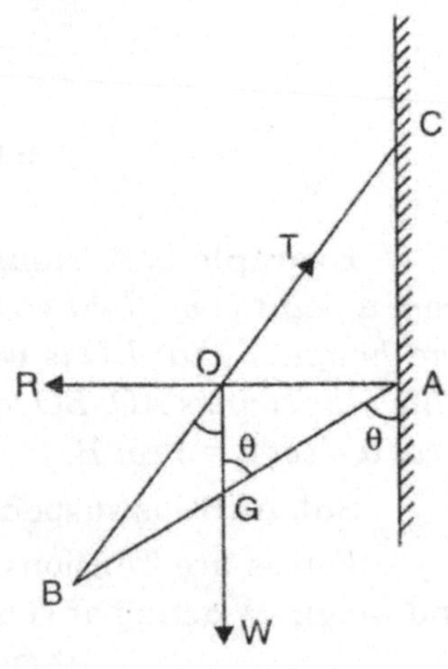

Fig. 7

$$OC^2 = OA^2 + AC^2$$

$$\left(\frac{l}{2}\right)^2 = \left(\frac{a}{2}\sin\theta\right)^2 + (a\cos\theta)^2$$

$$\frac{l^2}{4} = \frac{a^2}{4}(1-\cos^2\theta) + a^2\cos^2\theta$$

$$l^2 - a^2 = 3a^2\cos^2\theta$$

or $$\cos^2\theta = \frac{l^2-a^2}{3a^2}.$$

Example 5. *A solid cone of height h and semi vertical angle α is placed against a smooth vertical wall and is supported by a string attached to its vertex and to a point in the wall. Show that the greatest possible length of the string is*

$$h\sqrt{1+\frac{16}{9}\tan^2\alpha}\,. \qquad \text{(K.U. 2005)}$$

Sol. The cone OAB is in equilibrium under the forces :

(*i*) Weight W acting at G vertically downward

(*ii*) Reaction at A horizontally

(*iii*) Tension in the string OC passing through the intersection of W and R.

Let OD = h then OG = $\frac{3h}{4}$

ΔOO′G is similar to ΔOCD

$$\therefore \quad \frac{OC}{OO'} = \frac{OD}{OG} = \frac{h}{\frac{3}{4}h} = \frac{4}{3}$$

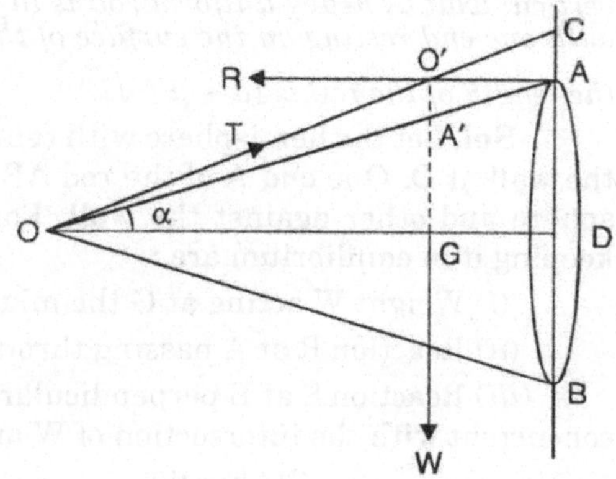

Fig. 8

or
$$OC = \frac{4}{3}\,OO' \quad \text{...(1)}$$

From ΔOAD, we have AD = OD tan α = h tan α. O′G = AD = h tan α.

$$\therefore \quad OO' = \sqrt{OG^2 + O'G^2}$$

$$= \sqrt{\frac{9h^2}{16} + h^2\tan^2\alpha}\,.$$

$$\therefore \quad OC = \frac{4}{3}\,OO' = \frac{4}{3}\times h\sqrt{\frac{9}{16}+\tan^2\alpha}$$

Length of string = OC = $h\sqrt{1+\frac{16}{9}\tan^2\alpha}$.

Example 6. *A rigid wire without weight in the form of the arc of a circle subtending an angle α at its centre having two weights P and Q at its extremities, rests with its convex surface in contact with the horizontal plane. If θ is the angle with the vertical which the radius to the end at which P is suspended then*

$$\tan\theta = \frac{Q\sin\alpha}{P+Q\cos\alpha}\,.$$

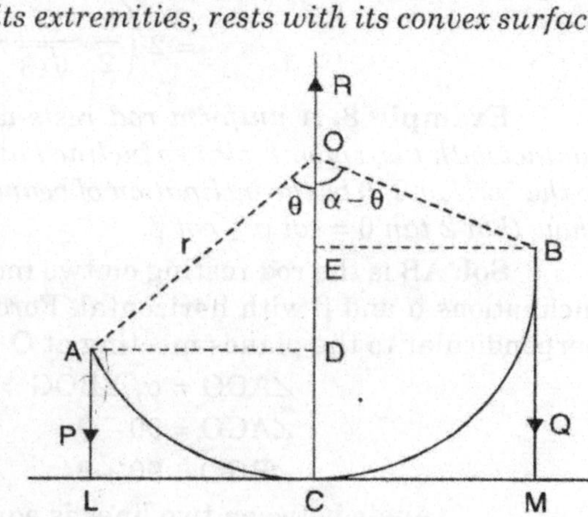

Fig. 9

Sol. The wire ACB is an arc of a circle of radius r and centre O. Weights P and Q are acting vertically downward at A and B. For equilibrium the reaction at C is a parallel force passing through the centre. Let ∠AOC = θ, ∠AOB = α (given)

∵ System of forces is in equilibrium hence algebraic sum of moments about a point is zero. Taking moments of the forces about C

$$P.CL - Q.CM + R.0 = 0$$

or $$P.CL = Q.CM. \quad ...(1)$$

But $$CL = AD = r \sin\theta$$
$$CM = BE = r\sin(\alpha - \theta)$$

Putting CL and CM in (1), we get

$\therefore$ $P.r\sin\theta = Qr.\{\sin\alpha\cos\theta - \cos\alpha\sin\theta\}$

or $$\sin\theta(P + Q\cos\alpha) = Q\sin\alpha\cos\theta$$

or $$\tan\theta = \frac{Q\sin\alpha}{P + Q\cos\alpha}.$$

Example 7. *A smooth hemisphere of diameter a is placed with its edge touching a smooth vertical wall. A heavy uniform rod is in equilibrium inclined at an angle 60° to the horizontal with one end resting on the surface of the bowl and other against the smooth wall. Show that the length of the rod is* $(a + a/\sqrt{13})$.

Sol. Let the hemisphere with centre at C touches the wall at D. One end A of the rod AB is in the hemisphere and other against the wall. Forces on the rod keeping it in equilibrium are :

(*i*) Weight W acting at G the middle point of AB

(*ii*) Reaction R at A passing through the centre

(*iii*) Reaction S at B perpendicular to the wall and concurrent with the intersection of W and R.

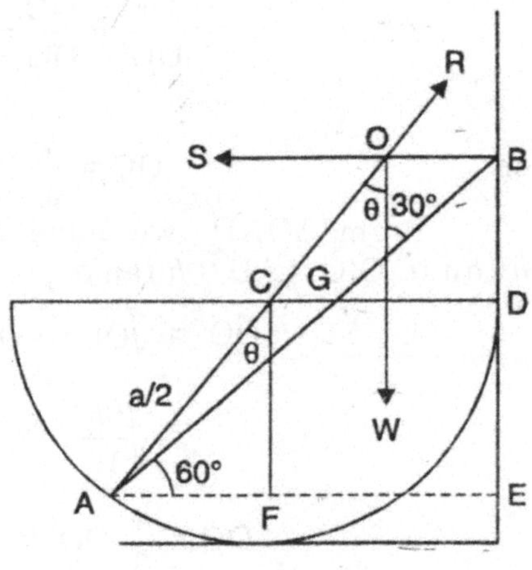

Fig. 10

$$\angle BAE = 60°$$

Let $\angle AOG = \theta$ and Let $AB = 2l$.

Applying $m - n$ theorem on ΔAOB

$$(l + l)\cot 30 = l\cot\theta - l\cot 90°$$

or $$\cot\theta = 2\sqrt{3} \Rightarrow \sin\theta = \frac{1}{\sqrt{13}}$$

Now length of rod $AB = 2l = AE\sec 60°$

or $$AB = (AF + FE)\,2 = 2\,(AC\sin\theta + CD)$$

$$= 2\left(\frac{a}{2}\cdot\frac{1}{\sqrt{13}} + \frac{a}{2}\right) = a + \frac{a}{\sqrt{13}}.$$

Example 8. *A uniform rod rests with its ends in contact with two smooth planes inclined at angles α and β to the horizon. If θ be the inclination of beam to the horizon ; show that 2 tan θ = cot α ~ cot β.*

Sol. AB is the rod resting on two inclined planes of inclinations α and β with horizontal. Forces are R and S perpendicular to the planes meeting at O

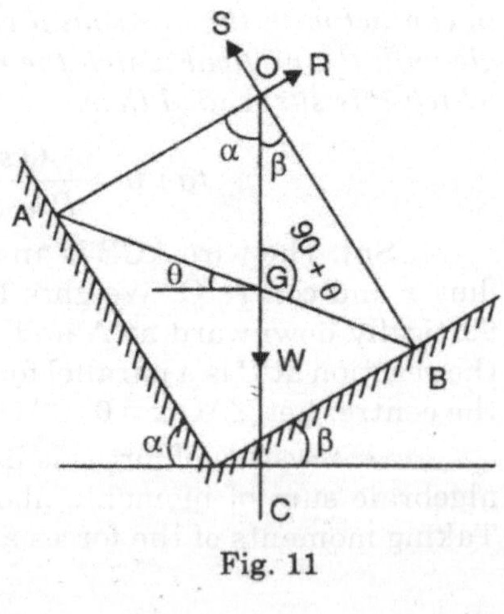

Fig. 11

$$\angle AOG = \alpha,\ \angle BOG = \beta$$
$$\angle AGO = 90 - \theta$$
$$\angle BGO = 90 + \theta$$

(Angle between two lines is equal to the angle between their perpendiculars)

Applying $m : n$ theorem

$$(1 + 1) \cot (90 + \theta) = 1 \cot \alpha - 1 \cot \beta$$

$$-2 \tan \theta = \cot \alpha - \cot \beta$$

or $$2 \tan \theta = \cot \beta \sim \cot \alpha.$$

Example 9. *A square of side 2a is placed with its plane vertical between two smooth pegs which are in the same horizontal line at a distance c apart. Show that the square is in equilibrium when inclination of one of its edge to the horizontal is either*

$$\frac{\pi}{4} \text{ or } \frac{1}{2} \sin^{-1}\left(\frac{a^2 - c^2}{c^2}\right)$$

Sol. Let ABCD be the square of weight W resting on pegs P and Q (PQ = c). Forces acting on the square are

(*i*) Reactions R and S perpendicular to the edges meeting at O.

(*ii*) Weight W acting at G passing through O.

Resolving the forces horizontally

$$R \sin \theta - S \cos \theta = 0$$

or $$R \sin \theta = S \cos \theta \qquad ...(1)$$

Taking moment of the forces about G we have

$$R . GL = S . GM$$

or $$R . (GK - LK) = S . (GH - MH)$$

or $$R . (a - OQ) = S . (a - OP)$$

or $$R . (a - c \cos \theta) = S . (a - c \sin \theta) \qquad ...(2)$$

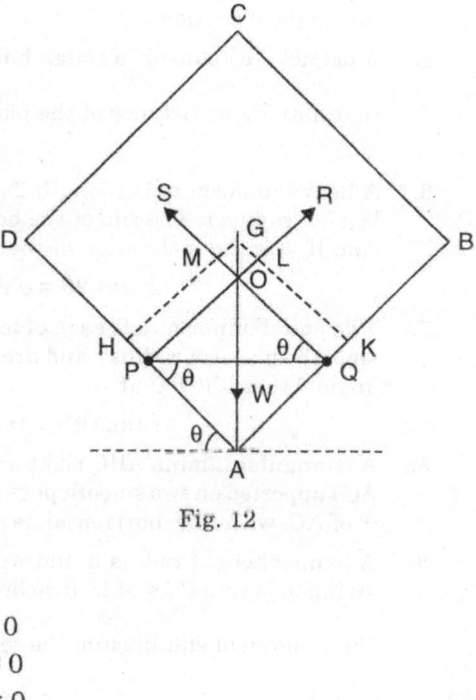

Fig. 12

Dividing (2) by (1), we have

$$\frac{a - c \cos \theta}{\sin \theta} = \frac{a - c \sin \theta}{\cos \theta}$$

or $$a \cos \theta - c \cos^2 \theta = a \sin \theta - c \sin^2 \theta.$$

$$a (\cos \theta - \sin \theta) - c (\cos^2 \theta - \sin^2 \theta) = 0$$

$$(\cos \theta - \sin \theta) \{a - c (\cos \theta + \sin \theta)\} = 0$$

$$\Rightarrow \quad \cos \theta - \sin \theta = 0 \quad \text{or} \quad a - c (\cos \theta + \sin \theta) = 0$$

$$\Rightarrow \quad \theta = \frac{\pi}{4} \quad \text{or} \quad \sin \theta + \cos \theta = \frac{a}{c}$$

Squaring $\sin^2 \theta + \cos^2 \theta + 2 \sin \theta \cos \theta = \dfrac{a^2}{c^2}$

$$\sin 2\theta = \frac{a^2}{c^2} - 1$$

$$\theta = \frac{1}{2} \sin^{-1}\left(\frac{a^2 - c^2}{c^2}\right).$$

EXERCISE 1

1. A beam whose C.G. divides it in the ratio 1 : 2, placed in a hemispherical bowl subtends an angle 90° at the centre. Find the inclination of the rod with the horizontal.

2. A uniform rod of length $2m$. rests against a vertical wall. The lower end is supported by a string of length $3m$. attached to a point in the vertically wall. Find the inclination of the rod with the wall.

3. A uniform rod of length $2a$ rests in equilibrium with one end resting against a smooth vertical wall and resting upon a peg at a distance b from the wall. Prove that inclination of the rod to the vertical is $\sin^{-1}\left(\frac{b}{a}\right)^{1/3}$.

4. Prove that least force which will move a weight W along a horizontal plane is W sin λ, where λ is the angle of friction.

5. A particle rests inside a rough hollow sphere of radius a. If μ be the coefficient of friction ; show that maximum distance of the particle from a vertical diameter is $\frac{\mu a}{\sqrt{1+\mu^2}}$.

6. A heavy uniform rod of length $2a$ rests partly within and partly outside a fixed hemispherical bowl of radius r. The rim of the bowl is horizontal and one point of the rod is in contact with the rim. If θ is the inclination of the rod to the horizontal show that

$$2r \cos 2\theta = a \cos \theta. \qquad \textit{(K.U. 2007)}$$

7. Two equal uniform rods each of length $2a$ are hinged together at one end. They are placed over a smooth sphere of radius r and are in equilibrium. If each rod is inclined with the horizontal at an angle θ then show that

$$r \tan^3 \theta + r \tan \theta = a.$$

8. A triangular lamina ABC right angle at A rests with its plane vertical and with its sides AB and AC supported on two smooth pegs at distance d apart in same horizontal line. Show that inclination θ of AC with the horizontal is given by $b \cos \theta - c \sin \theta = 3d \cos 2\theta$, where AB = c and AC = b.

9. A hemisphere of radius a and weight W is placed with its curved surface on a smooth table. A string of length l ($l < a$) is attached to a point on its rim and to a point on the table. Prove that in the position of equilibrium the tension in the string is $\frac{3\,W}{8}\frac{(a-l)}{\sqrt{2al-l^2}}$.

10. Inside a fixed hollow cylinder of radius R whose generators are horizontal, are placed two equal cylinders each of radius r. A third equal cylinder is placed symmetrically on them. Show that equilibrium can not exist unless $R < r(1 + 2\sqrt{7})$.

11. A square lamina of side $2a$ rests in a vertical plane on two smooth pegs in a horizontal line. Show that if the sum of the distances of the pegs from the lowest corner is equal to a, there is equilibrium.

12. The algebraic sum of moments of a system of coplaner forces about the points (1, 0), (0, 2) and (2, 3) referred to rectangular axes are G_1, G_2, G_3.

Show that $\tan\theta = \frac{G_1 - 3G_2 + 2G_3}{2G_1 - G_2 - G_3}$

where θ is the angle which the resultant makes with x-axis.

13. Equal weights P and P are attached to two strings ACP and BCP passing over a smooth peg C. AB is a beam of weight W whose C.G. is at the distance a m from A and b m from B. Show that AB is inclined to horizon at an angle

$$\tan^{-1}\left[\frac{a-b}{a+b}\tan\left(\sin^{-1}\frac{W}{2P}\right)\right].$$

14. A cup is formed from a hollow sphere whose radius is a. It is so placed that the radius joining the point on the rim to the centre makes an angle α with the vertical and radius from a point A on the cup makes an angle β with the vertical. If a smooth uniform rod with one end at A in the cup and if the rod touches the rim show that its length must be $4a \sin\beta \sec\frac{\alpha-\beta}{2}$.

ANSWERS

1. $\tan^{-1}1/3$. **2.** $\cos^{-1}\sqrt{5/12}$.

SOME MORE ILLUSTRATIVE EXAMPLES

Example 1. *Two equal beams AB and AC are hinged at A and are placed in a vertical plane with the ends B and C resting on a horizontal plane. Two strings are attached at B and C joining the middle points of opposite beams. Show that in the position of equilibrium the ratio of tension in the string and weight of a beam is* $\frac{1}{8}\sqrt{9\cot^2\theta+1}$, *where* θ *is the inclination of each beam with the horizontal.*

Sol. The forces acting are as shown in the Fig. 13

(*i*) Tensions in the strings

(*ii*) Reactions at B and C

(*iii*) Weight of rods at middle points D and E

(*iv*) Reaction of the hinge at A in unknown direction.

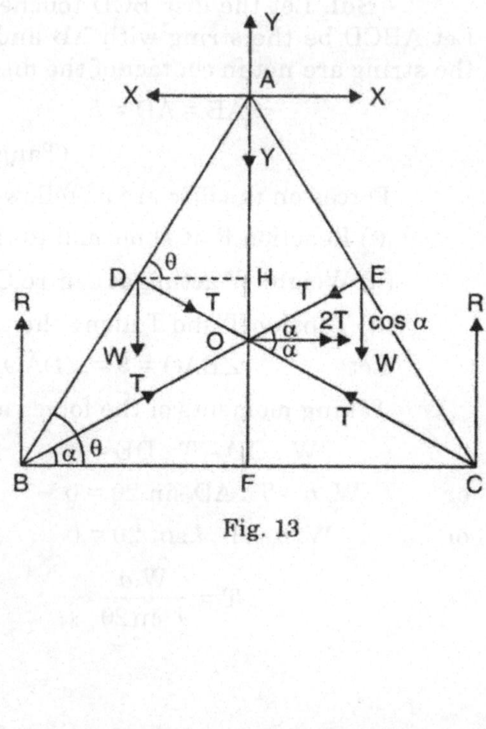

Fig. 13

Resolving the forces in vertical direction

$$2R = 2W \Rightarrow R = W \qquad ...(1)$$

Resolved parts of T in vertical direction balance each other.

Taking moments of all the forces on the rod AB about A (A is chosen to avoid the unknown reaction at A)

$$W.DH - R.BF + 2T\cos\alpha \,.\, AO = 0 \qquad ...(2)$$

Let $AB = AC = 2a$

$DH = a\cos\theta,\ BF = 2a\cos\theta$

$$AO = \frac{2}{3}AF \qquad (\because \text{ O is point of intersection of medians})$$

$$= \frac{2}{3}.\, 2a\sin\theta.$$

$$\therefore\quad W.\, a\cos\theta - W.\, 2a\cos\theta + 2T\cos\alpha\,.\,\frac{4}{3}a\sin\theta = 0$$

$$W a\cos\theta = \frac{8}{3}Ta\sin\theta\cos\alpha.$$

$$\therefore\quad \frac{T}{W} = \frac{3}{8}\cot\theta\sec\alpha = \frac{3}{8}\cot\theta\sqrt{1+\tan^2\alpha}$$

But $$\tan\alpha = \frac{OF}{BF} = \frac{1}{3}\frac{AF}{BF} = \frac{1}{3}\tan\theta$$

Hence, $$\frac{T}{W} = \frac{3}{8}\cot\theta\sqrt{1+\frac{1}{9}\tan^2\theta} = \frac{1}{8}\sqrt{9\cot^2\theta+1}\,.$$

Example 2. *A circular disc BCD of radius a and weight W is hanging by means of thin light string touching the arc of the disc and the ends are at point A of the vertical wall. The disc is normal to the wall. Part AD of the string touches the wall. If 2l be the length of portion of the string not touching the disc then show that tension in the string is* $W\left(\frac{a^2+l^2}{2l^2}\right)$.

Sol. Let the disc BCD touches the wall at D. Let ABCD be the string with AB and AD portions of the string are not in contact of the disc.

$$AB = AD = l.$$

(Tangents to the disc)

Forces on the disc are as follows :

(*i*) Reaction R at D normal to the wall.

(*ii*) Weight W acting at centre O.

(*iii*) Tension T and T along the tangents.

Let $\angle BAO = \theta = \angle DAO$

Taking moments of the forces about D

$$W.\, OD - T.\, DE = 0$$

or $$W.\, a - T.\, AD\sin 2\theta = 0$$

or $$W.\, a - T.\, l.\sin 2\theta = 0$$

$$T = \frac{W a}{l\sin 2\theta}$$

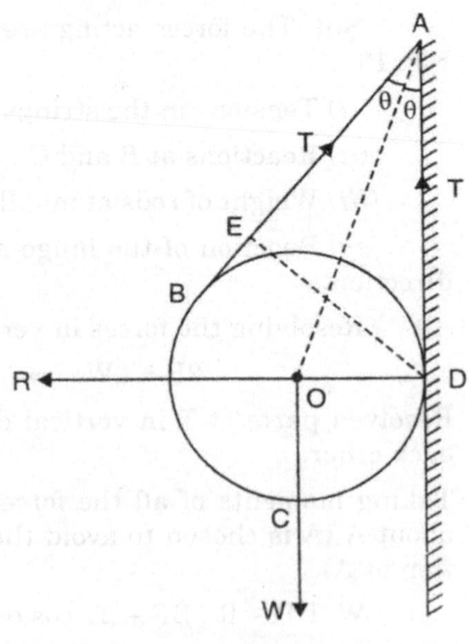

Fig. 14

$$= \frac{W\,a\,(1+\tan^2\theta)}{l\,.\,2\tan\theta} \qquad \tan\theta = \frac{a}{l}$$

$$= \frac{W\,.\,a\,.\left(1+\dfrac{a^2}{l^2}\right)}{2\dfrac{a}{l}\,l} = \frac{W\,(a^2+l^2)}{2\,l^2}.$$

Example 3. *Two equal uniform rods AB and AC each of weight W are freely joined at A and rest with the extremities B and C on the inside of a smooth circular loop, whose radius is greater than the length of either rod, the whole being in a vertical plane and the middle points of the rods being joined by a light string ; show that if the string is stretched its tension is W (tan α – 2 tan β), where 2α is the angle between the rods, and β the angle either rod subtends at the centre.*

Sol. Let O be the centre of the circular loop of radius *r*. Let AB and AC be the rods and DE be the strings joining the middle points of the rods.

Let $\angle BAF = \alpha$

$\angle BOF = \beta$

Forces keeping the rods in equilibrium are :

(*i*) Reaction R and R at B and C passing through the centre of the hoop.

(*ii*) Weights W and W vertically downward at D and E.

(*iii*) Tension in the string DE.

Resolving the forces vertically for the whole system

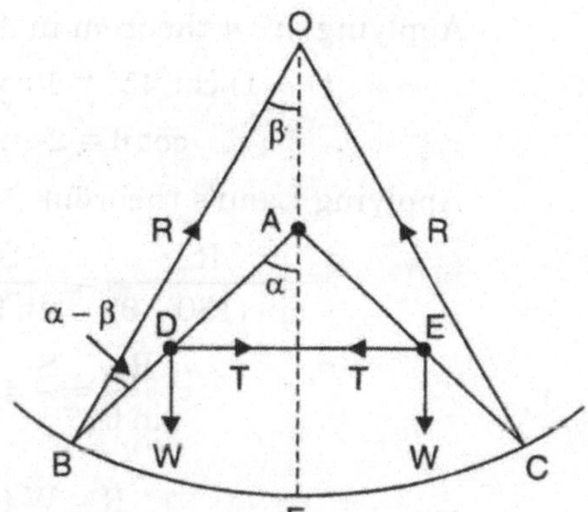

Fig. 15

$$W + W = R\cos\beta + R\cos\beta$$

or $$W = R\cos\beta \qquad ...(1)$$

Taking moments for the forces on rod AC about A, we have

$$R\,.\,AC\sin(\alpha-\beta) - T.\,AE\cos\alpha - WAE\sin\alpha = 0$$

$$R\,.\,2\,AE\sin(\alpha-\beta) - T.\,AE\cos\alpha - WAE\sin\alpha = 0 \qquad (\because\ AC = 2\,AE)$$

$$\therefore\quad T\cos\alpha = 2R\sin(\alpha-\beta) - W\sin\alpha$$

Putting the value of R from (1)

$$T = \frac{2\,.\dfrac{W}{\cos\beta}\sin(\alpha-\beta) - W\sin\alpha}{\cos\alpha}$$

$$= \frac{W\,[2(\sin\alpha\cos\beta - \cos\alpha\sin\beta) - \sin\alpha\cos\beta]}{\cos\alpha\cos\beta}$$

$$T = W\,[\tan\alpha - 2\tan\beta].$$

Example 4. *A cylinder of radius r whose axis is fixed horizontally touches a vertical wall along a generating line. A flat beam of uniform material of length 2l and weight W rests with its extremities in contact with the wall and the cylinder, making an angle 45° with the vertical. Show that in the absence of friction,* $\frac{l}{r} = \frac{\sqrt{5}-1}{\sqrt{10}}$, *pressure on the wall is W/2 and the reaction of the cylinder is* $\frac{1}{2}W\sqrt{5}$.

Sol. Rod AB is in equilibrium on the cylinder and against the wall under the forces.

(*i*) Reaction R ⊥ wall at B.

(*ii*) Reaction S at A passing through the centre D.

(*iii*) Weight W acting at middle point G of the rod vertically downwards.

For equilibrium forces must be concurrent at O

$\angle ABL = \angle OGB = 45°$ (given)

Let $\angle AOG = \theta$

Applying $m : n$ theorem in ΔAOB

$$(1 + 1) \cot 45° = 1 \cot \theta - 1 \cot 90$$

$$\therefore \quad \cot \theta = 2 \quad \ldots(1)$$

Applying Lami's theorem

$$\frac{R}{\sin(180-\theta)} = \frac{S}{\sin 90°} = \frac{W}{\sin(90+\theta)}$$

or

$$\frac{R}{\sin\theta} = \frac{S}{1} = \frac{W}{\cos\theta}$$

$$\therefore \quad R = W\tan\theta = \frac{W}{2} \quad \text{(using 1)}$$

$$S = W\sec\theta = W\sqrt{1+\tan^2\theta}$$

$$= W\sqrt{1+\frac{1}{4}} = \frac{1}{2}W\sqrt{5}$$

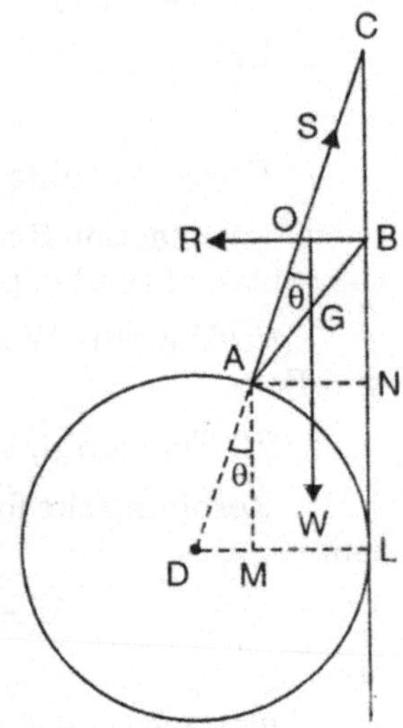

Fig. 16

Further $DL = DM + ML = DM + AN = r\sin\theta + 2l\sin 45°$

$$\therefore \quad r = r\cdot\frac{1}{\sqrt{5}} + 2l\cdot\frac{1}{\sqrt{2}} \quad \Rightarrow \quad \frac{r(\sqrt{5}-1)}{\sqrt{5}} = l\sqrt{2}$$

$$\Rightarrow \quad \frac{l}{r} = \frac{(\sqrt{5}-1)}{\sqrt{10}}.$$

Example 5. *A ladder of weight W rests with one end against a smooth vertical wall and other end on a smooth floor. If inclination of ladder to the horizon is 60°, find the horizontal force that must be applied at lower end to prevent it from slipping.* (M.D.U. 2005)

Sol. Let AB be the ladder placed on horizontal floor AO against vertical wall OB. Forces on the ladder are

(*i*) Reaction R at A $\perp$ AO

(*ii*) Reaction S at B $\perp$ OB

(*iii*) Weight W acting vertically downwards at G (Mid point of AB)

(*iv*) Let P be the horizontal force at the lower end A towards AO to avoid slipping. Resolving the forces horizontally and vertically

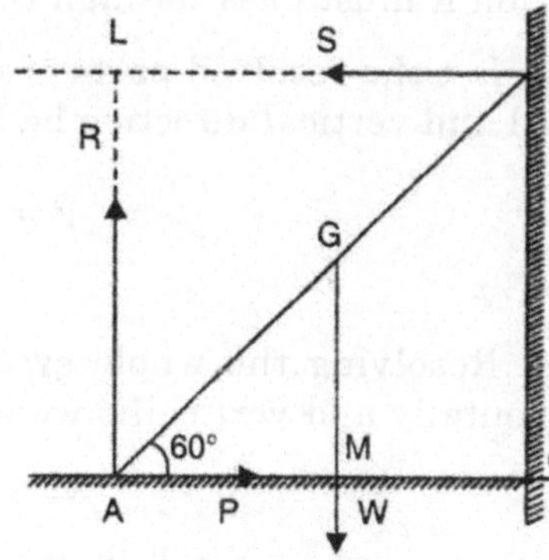

Fig. 17

$$R = W \quad \text{...(1)}$$

$$P = S \quad \text{...(2)}$$

$$\angle BAO = 60° \text{ (Given)}$$

Taking moment of the forces about A

$$-W \,.\, AM + S \,.\, AL = 0 \quad \text{...(3)}$$

Let $AB = 2a \quad \therefore \quad AG = a$

$$\therefore \quad AM = AG \cos 60° = a \cos 60° = \frac{a}{2}$$

$$AL = OB = AB \sin 60° = 2a\frac{\sqrt{3}}{2} = a\sqrt{3}$$

Putting S = P from (2) and values of AL and AM in (3), we get

$$-W.\frac{a}{2} + P\,.\,a\sqrt{3} = 0 \;\Rightarrow\; P = \frac{W}{2\sqrt{3}}$$

which is the required force.

Example 6. *One end of a uniform rod is attached to a hinge and the other end is supported by a string attached to the extremity of the rod, the rod and the string are inclined at the same angle θ to the horizontal. If w be the weight of the rod, show that the reaction at the hinge is* $\frac{w}{4}\sqrt{8 + \operatorname{cosec}^2\theta}$ *or* $\frac{w}{4}\sqrt{9 + \cot^2\theta}$. *Also find the tension in the string.*

Sol. Let AB and BC be the rod and the string each inclined at an angle θ with the horizontal AC

$$\therefore \quad AB = BC = l$$

One end A of the rod AB is hinged and the other end B is connected by a string BC.

The system is in equilibrium under the forces (*i*) weight *w* of the rod acting vertically downwards at middle point G of the rod AB.

(*ii*) Tension T in the string along BC.

(*iii*) Reaction R at the hinge. Since the system is in equilibrium under three coplanar forces they must be concurrent or parallel. Since lines of action of T and W meet at O, the reaction R must pass through O.

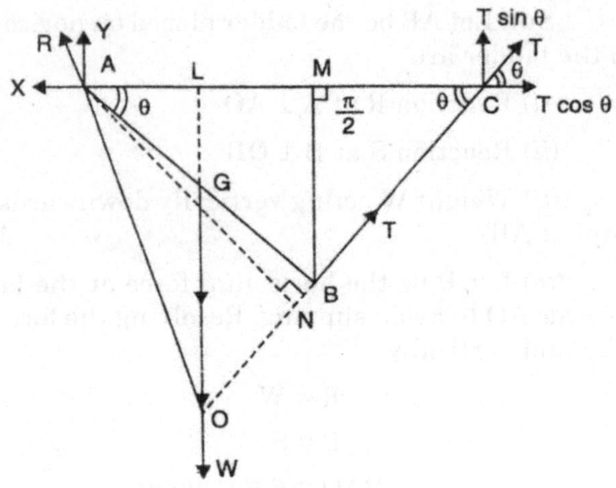

Fig. 18

Let the resolved parts of R along horizontal and vertical direction be X and Y.

$$\therefore \quad R = \sqrt{X^2 + Y^2} \quad ...(1)$$

Resolving the whole system of forces horizontally and vertically, we have

$$X = T \cos \theta$$

$$Y + T \sin \theta = W \quad ...(2)$$

Taking moment of the forces about A, we have

$$T \,.\, AN - W \,.\, AL = 0 \quad ...(3)$$

From Δ ACN, $\quad AN = AC \sin \theta$

$$= 2\ AM \sin \theta$$

$$= 2 \,.\, AB \cos \theta \sin \theta$$

$$= l \sin 2\theta$$

From Δ AGL, $AL = AG \cos \theta = \frac{l}{2} \cos \theta$

Putting AL and AN in (3), we get

$$T \,.\, l \,.\, \sin 2\theta = W \,.\, \frac{l}{2} \cos \theta$$

or $$T = \frac{W \cos \theta}{2 \,.\, 2 \sin \theta \cos \theta} = \frac{W}{4} \operatorname{cosec} \theta \quad ...(4)$$

$\therefore$ Tension in the string $T = \frac{W}{4} \operatorname{cosec} \theta$.

Putting the value of T in (2), we get

$$X = \frac{W}{4} \,.\, \operatorname{cosec} \theta \,.\, \cos \theta = \frac{W}{4} \cot \theta$$

and $$Y = W - \frac{W}{4} \operatorname{cosec} \theta \,.\, \sin \theta = \frac{3W}{4}$$

From (1) $\therefore$ $$R = \sqrt{X^2 + W^2} = \sqrt{\frac{W^2}{16}\cot^2\theta + \frac{9W^2}{16}}$$

$$= \frac{W}{4}\sqrt{9+\cot^2\theta}$$

$$= \frac{W}{4}\sqrt{9+\operatorname{cosec}^2\theta - 1} = \frac{W}{4}\sqrt{8+\operatorname{cosec}^2\theta}$$

$\therefore$ $$R = \frac{W}{4}\sqrt{8+\operatorname{cosec}^2\theta} = \frac{W}{4}\sqrt{9+\cot^2\theta}.$$

EXERCISE 2

1. **A uniform rod of weight W is movable in a vertical plane about a smooth hinge at one end and the other end is fastened a weight $\frac{W}{2}$. This end is fastened by a string of length l to a point at a height c vertically above the hinge. Find the tension in the string.**

2. A beam of weight W is divided by its C.G. into two portions of lengths a and b. The beam rests in a vertical plane on a smooth floor and a vertical wall. A string is attached to a point on the rod and the corner of wall and floor. If θ and ϕ be the inclination of the beam and string to the horizontal then tension in the string is

$$\frac{Wa\cos\theta}{(a+b)\sin(\theta-\phi)}.$$

3. The algebraic sum of moments of a system of coplaner forces about the point A_r (x_r, y_r) in their plane are G_r, $r = 1, 2, 3, 4$, show that

$$\begin{vmatrix} G_1 & x_1 & y_1 & 1 \\ G_2 & x_2 & y_2 & 1 \\ G_3 & x_3 & y_3 & 1 \\ G_4 & x_4 & y_4 & 1 \end{vmatrix} = 0.$$

4. Three equal uniform rods each of weight W, are smoothly joined so as to form an equilateral triangle. If the system be supported at the middle point of one of the rods, show that action at the lowest angle is $\frac{1}{6}W\sqrt{3}$ and that at each of the other is $W\sqrt{\frac{13}{12}}$.

5. Two equal uniform rods AB and AC are freely joined at A. The ends B and C are connected by a fine string. The rods are suspended from B by a string. If in the position of equilibrium angle between the rods is 2α then tension in the string is $\frac{3\omega\sin\alpha}{\sqrt{1+3\sin^2\alpha}}$ where 2ω is the weight of either rod.

6. A picture frame, rectangular in shape rests against a smooth vertical wall from points in which it is suspended by parallel strings attached to two points in the upper edge of the back of the frame. If the C.G. of the frame coincides with the centre of the figure, the picture will hang against the wall at an angle $\tan^{-1}\left(\frac{b}{3a}\right)$ to the vertical, where a is the height and b is the thickness of the frame.

7. Two smooth spheres of equal weight rest inside a fixed smooth right circular cone which has its vertex downwards and its axis inclined at an angle β with the vertical, each sphere touching the cone at one point only. Prove that the common normal to the spheres makes an angle

$$\cot^{-1}\left(\frac{\sin 2\beta}{\cos 2\alpha + \cos 2\beta}\right).$$

8. Two spheres each of radius a and weight W lie in contact in a spherical bowl of radius na. Show that in the absence of friction the pressure between them is $W/\sqrt{n^2 - 2n}$.

9. Two equal rods of weight w and length $2a$ are freely joined . Their free ends are attached by strings of length $2a$ each to a fixed point. A circular disc of radius r weight W rests in the angle between the rods. The whole system hangs in a vertical plane. If 2θ is the angle between the rods prove that

$$r = 2a\sin^2\theta\tan\theta\left(\frac{3\omega + 2W}{W}\right).$$

10. A particle P is acted on by forces $\frac{\mu}{r_1}$ and $\frac{\mu}{r_2}$ respectively towards two fixed points O_1 and O_2. Show that the particle will be at rest at any point of a smooth groove whose equation is $r_1 r_2$ = constant.

11. A rectangular lamina ABCD of weight W is hung from A. If 2W weight is hung at B, prove that the inclination of diagonal AC to the vertical is $\tan^{-1}\frac{4ab}{5a^2 + b^2}$.

12. A ladder of weight W and length $2a$ rests with one end against a smooth vertical wall and the other end resting on a smooth floor. The middle point of the ladder is attached to a string of length b which is tied to the wall. The C.G. of ladder is at $\frac{3}{4}$ of its length from the end on the floor. Prove that tension in the string is $\frac{3bW}{2\sqrt{a^2 - b^2}}$.

13. Forces P, Q, R and S are acting along the sides AB, BC, CD and DA of the cyclic quadrilateral ABCD. AB is diameter of the circumscribed circle. If forces are in equilibrium then prove that $R^2 = P^2 + Q^2 + S^2 + 2PQS/R$.

14. In a smooth hemispherical bowl is placed a heavy rod of length equal to radius of the bowl. C.G. of the rod divides it in the ratio 1 : 2. Show that the angle made by the rod with the vertical is $\tan^{-1}(3\sqrt{3})$.

15. **A solid hemisphere is supported by a string fixed to a point on its rim and to a point on a smooth vertical wall with which the curved surface of the hemisphere is in contact. If θ and φ are the inclinations of the string and the plane base of the hemisphere to the vertical then prove that**

$$\tan \phi = \frac{3}{8} + \tan \theta. \qquad \textit{(K.U. 2005)}$$

ANSWER

1. *lW/c.*

2

Virtual Work

1. Work

If the point of application of the force is displaced from one position to the other, then it is said that force has done some work. The amount of work done is the product of the forces and the projection of the displacement on the line of action of the force.

Let the particle A be acted upon by a force P acting parallel to AB and displaces the particle to C. Let CB ⊥ AB, then work done by P

$$= P \times AB = P \,.\, AC \cos \theta$$

or $$= (P \cos \theta)\, AC.$$

i.e., work done by a force is the product of displacement of the point of application and the resolved part of force in the direction of the displacement.

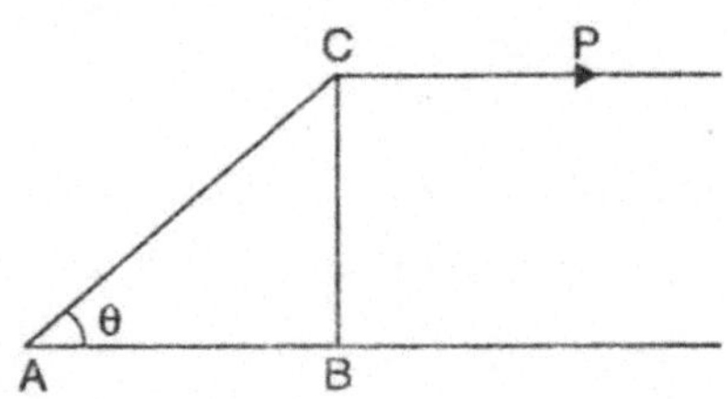

Fig. 1

In vector notation, let position vector of point of application of force F be $\vec{r}$ and the position vector of the displaced position Q be $\overrightarrow{r + \delta r}$ then work done by P = $\vec{F} \,.\, \vec{PQ}$

$$= \vec{F} \,.\, \vec{\delta r} = |\,F\,|\ |\delta r|\ \cos \theta = (F \cos \theta)\ (\delta r)$$

If work done by the force is zero then

$$F \cos \theta \,.\, \delta r = 0 \quad \Rightarrow \quad \delta r = 0$$

or $$F \cos \theta = 0$$

$$\Rightarrow \quad \delta r = 0 \quad \text{or} \quad \theta = \pi/2$$

i.e., a force does no work if it does not have any displacement or has a displacement in the direction perpendicular to the direction of the force.

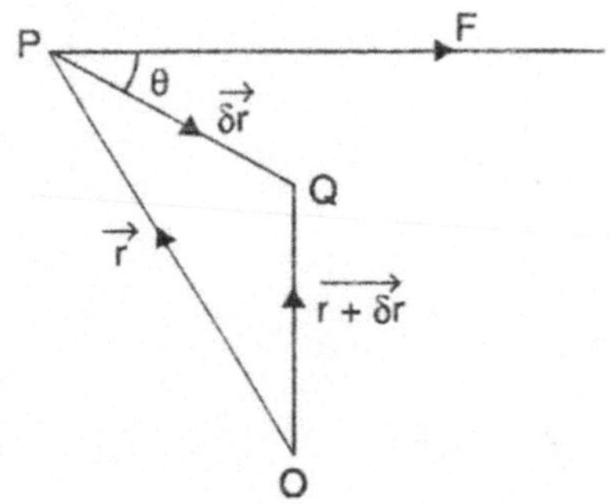

Fig. 2

Units of Work. In C.G.S. system, the unit of work is **Erg.** *i.e.*, the work done by a force of one dyne in displacing its point of application through one cm. in the direction of the force. In M.K.S. system the unit of work is **Joule** *i.e.*, work done by a force of 1 Newton in displacing its point of application through one metre in the direction of force

1 Joule = 1 Newton × 1 metre = 100,000 dynes × 100 cm = 10^7 ergs

In F.P.S. system unit of work is foot – poundal *i.e.*, the work done by a force of one poundal in displacing point of application through one foot in the direction of force

Theorem 1. *The work done by a system of forces in moving a particle is equal to the work done by the resultant of the system of forces.*

Proof. Let $\vec{F_1}, \vec{F_2}, ..., \vec{F_n}$ be the set of forces acting on a particle. Let the particle have the displacement $\vec{dr}$.

Sum of work done by the forces

$$= \vec{F_1} \cdot \vec{d_r} + \vec{F_2} \cdot \vec{d_r} + ... + \vec{F_n} \cdot \vec{d_r}$$

$$= (\vec{F_1} + \vec{F_2} + ... + \vec{F_n}) \cdot \vec{d_r} = \vec{F} \cdot \vec{d_r}$$

where $\vec{F} = \vec{F_1} + \vec{F_2} + ... + \vec{F_n}$ = Resultant force = Work done by the resultant.

Theorem 2. *The work done by a force in moving a particle from one position to another is equal to the sum of the work done by the resolved parts of the forces.*

Proof. Let a force $\vec{F}$ be acting at O along $\vec{OA}$ and displaces it to position O′. Let $\angle AOO' = \theta$. Let the force be inclined by angle α with x-axis.

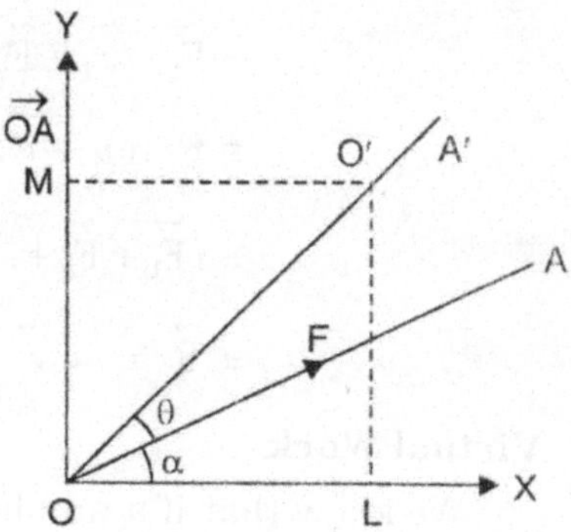

Fig. 3

Work done by the force = $\vec{F} \cdot \vec{OO'} = F . OO' \cos\theta$

Resolving the force along the axes we have,

$X = F\cos\alpha,\ Y = F\sin\alpha.$

Sum of work done by the resolved parts

$= X . OL + Y . OM$

$= X . OO'\cos(\theta + \alpha) + Y . OO'\sin(\theta + \alpha)$

$= OO'[F\cos\alpha\cos(\theta + \alpha) + F\sin\alpha\sin(\theta + \alpha)]$

$= OO'. F\cos(\theta + \alpha - \alpha)$

$= F. OO'\cos\theta = \vec{F} \cdot \vec{OO'}$.

2. Work done by a Couple

Let the forces $\vec{F}$ and $-\vec{F}$ be acting at the points with position vector $\vec{r_1}$ and $\vec{r_2}$ and constitute a couple of moment $\vec{G}$

$$\therefore \qquad \vec{G} = (\vec{r_1} - \vec{r_2}) \times \vec{F}$$

If the body is displaced by a small quantity which is equivalent to translation $\vec{u}$ and rotation $\vec{e}$ then

$$\vec{dr_1} = \vec{u} + \vec{e} \times \vec{r_1} \quad \text{and} \quad \vec{dr_2} = \vec{u} + \vec{e} \times \vec{r_2}$$

$\therefore$ Work done by the couple is $\vec{F} \cdot \vec{dr_1} + (-\vec{F}) \cdot \vec{dr_2}$

$$= \vec{F} \cdot (\vec{u} + \vec{e} \times \vec{r_1}) - \vec{F} \cdot (\vec{u} + \vec{e} \times \vec{r_2})$$

$$= \vec{F} . \vec{e} \times \vec{r_1} - \vec{F} . \vec{e} \times \vec{r_2} = \vec{F} . \vec{e} \times (\vec{r_1} - \vec{r_2})$$

$$= \vec{e} . (\vec{r_1} - \vec{r_2}) \times \vec{F} = \vec{e} . \vec{G}.$$

3. Work done by a System of Forces

Let the forces $\vec{F_1}, \vec{F_2}, ..., \vec{F_n}$ be acting at points with position $\vec{r_1}, \vec{r_2}, ..., \vec{r_n}$ respectively. We know that the system of forces can be reduced to a single force $\vec{F} = \vec{F_1} + \vec{F_2} + ... + \vec{F_n}$ at 0 and a couple of moment $\vec{G} = \vec{r_1} \times \vec{F_1} + \vec{r_2} \times \vec{F_2} + ... + \vec{r_n} \times \vec{F_n}$. If the body undergoes a small displacement which is sum of translation $\vec{u}$ and rotation $\vec{e}$ then $d\vec{r_n} = \vec{u} + \vec{e} \times \vec{r_n}, n = 1, 2, ... n$ ...(1)

$\therefore$ Work done by the system

$$= \vec{F_1} . d\vec{r_1} + \vec{F_2} . d\vec{r_2} + ... + \vec{F_n} . d\vec{r_n}$$

$$= \vec{F_1} . (\vec{u} + \vec{e} \times \vec{r_1}) + \vec{F_2} . (\vec{u} + \vec{e} \times \vec{r_2}) + ... + \vec{F_n} . (\vec{u} + \vec{e} \times \vec{r_n})$$

$$= (\vec{F_1} + \vec{F_2} + ... + \vec{F_n}) . \vec{u} + \vec{e} . (\vec{r_1} \times \vec{F_1} + \vec{r_2} \times \vec{F_2} + ... + \vec{r_n} \times \vec{F_n})$$

$$= \vec{F} . \vec{u} + \vec{e} . \vec{G} \qquad ...(2)$$

4. Virtual Work

We know that if a number of forces act on a rigid body and body undergoes some displacement then some actual work is done by these forces. But if the body is in equilibrium under the action of the forces, there is no displacement and hence no work is done. In some statical problems if we suppose some imaginary displacement or virtual displacement in the body then the imaginary work done by the forces is called **virtual work**.

5. Principle of Virtual Work for a System of Coplaner Forces Acting on a Particle

Statement. *The necessary and sufficient condition that a particle acted upon by a number of coplaner forces be in equilibrium, then the sum of the virtual work done by the forces in any small virtual displacement consistent with the geometric conditions of the system is zero.*

Proof. Necessary Condition. Let $\vec{F_1}, \vec{F_2}$, ..., $\vec{F_n}$ be number of forces acting at O. Let the forces be in equilibrium. If virtual displacement OO′ be given. Then

Some of the virtual work done by the forces is

$$\vec{F_1} . \vec{OO'} + \vec{F_2} . \vec{OO'} + ... + \vec{F_n} . \vec{OO'}$$

$$= (\vec{F_1} + \vec{F_2} + ... + \vec{F_n}) . \vec{OO'}$$

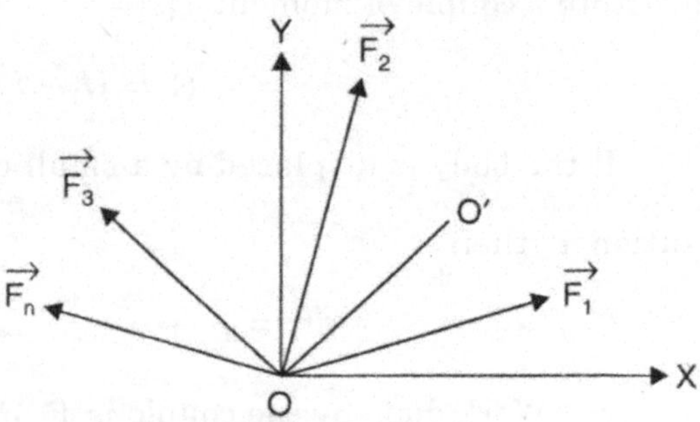

Fig. 4

$$= \vec{F} . \vec{OO'}$$

where $\vec{F} = \vec{F}_1 + \vec{F}_2 + ... + \vec{F}_n$

$= 0$

is the resultant of the forces. (Since the system is in equilibrium)

Conversely, if the algebraic sum of the virtual work done by the forces is zero, then

$$\vec{F} . \vec{OO'} = 0$$

$\because$ $OO' \neq 0 \Rightarrow \vec{F} = 0$

Hence the forces are in equilibrium.

6. Principle of Virtual Work done for a System of Coplaner Forces Acting on a Rigid Body

Statement. *If a system of coplaner forces acting upon a rigid body keep the body in equilibrium and the body is given a small virtual displacement, consistent with the geometrical condition of the system, then the algebraic sum of virtual works done by the forces is zero.*

Proof. Let the forces $\vec{F}_1, \vec{F}_2, ..., \vec{F}_n$ acting at $\vec{r}_1, \vec{r}_2, ..., \vec{r}_n$ on a rigid body keep the body in equilibrium. If the body is given a slight virtual displacement $\vec{dr}_n$ consistent with the geometric conditions between different parts of the rigid body, then we have to prove that

$$\Sigma \vec{F}_n . \vec{dr}_n = 0$$

Now if the body undergoes a small displacement which is sum of constant translation $\vec{u}$ and a small rotation $\vec{e}$ about some arbitrary chosen point O, then $\vec{r}_n$ is displaced to $\vec{r}_n + \vec{dr}_n$

we have, $\vec{dr}_n = \vec{u} + \vec{e} \times \vec{r}_n , \quad n = 1, 2, ..., n.$

$\therefore$ Work done by the forces is

$$\Sigma \vec{F}_n . \vec{dr}_n = \vec{F}_1 . \vec{dr}_1 + \vec{F}_2 . \vec{dr}_2 + ... + \vec{F}_n . \vec{dr}_n$$

$$= \vec{F}_1 . (\vec{u} + \vec{e} \times \vec{r}_1) + \vec{F}_2 . (\vec{u} + \vec{e} \times \vec{r}_2) + ... + \vec{F}_n . (\vec{u} + \vec{e} \times \vec{r}_n)$$

$$= (\vec{F}_1 + \vec{F}_2 + ... + \vec{F}_n) . \vec{u} + \vec{F}_1 . \vec{e} \times \vec{r}_1 + \vec{F}_2 . \vec{e} \times \vec{r}_2 + ... + \vec{F}_n . \vec{e} \times \vec{r}_n$$

$$\because (\vec{a} . \vec{b} \times \vec{c} = \vec{b} . \vec{c} \times \vec{a})$$

$$= \vec{F} . \vec{u} + \vec{e} . (\vec{r}_1 \times \vec{F}_1 + \vec{r}_2 \times \vec{F}_2 + ... + \vec{r}_n \times \vec{F}_n) = \vec{F} . \vec{u} + \vec{e} . \vec{G} \quad ...(1)$$

where $\vec{F}$ is resultant force acting at O and $\vec{G}$ is resultant couple.

$\because$ The system is in equilibrium

$\therefore$ $\vec{F} = 0$ and $\vec{G} = 0$

Hence, $\Sigma \vec{F}_n . \vec{dr}_n = 0 \Rightarrow$ Algebraic sum of virtual work is zero.

Converse of the Theorem. *If the virtual work done by the applied system of forces is zero for any displacement, then the system of forces is in equilibrium.*

Proof. We know that $\Sigma \vec{F}_n \cdot \vec{dr}_n = \vec{u} \cdot \vec{F} + \vec{e} \cdot \vec{G}$ From (1)

$\because$ Virtual work done by the forces is zero.

Therefore, $\vec{u} \cdot \vec{F} + \vec{e} \cdot \vec{G} = 0$ ...(1)

for any displacement.

Let the body undergoes the translation only hence by putting $\vec{e} = 0$ in equation (1), we have,

$$\vec{u} \cdot \vec{F} = 0 \quad \text{or} \quad \vec{F} = 0 \qquad ...(2)$$

Again let the body takes rotation only

$$\therefore \quad \vec{u} = 0 \text{ we have, } \quad \vec{e} \cdot \vec{G} = 0 \quad \text{or} \quad \vec{G} = 0 \qquad ...(3)$$

By (2) and (3), we have, $\vec{F} = 0$ and $\vec{G} = 0$

Hence, the system of forces acting on the rigid body is in equilibrium.

7. Forces Which can be Omitted in Forming the Equation of Virtual Work

(M.D.U. 2005)

If a body undergoes a small displacement under the action of number of forces, the work done by certain forces is zero. Therefore, these forces can be avoided in writing the equation of virtual work. Such forces are given below :

(*i*) **Tension in an inextensible string.** Let the string AB having tension T be virtually displaced to position A′ B′ inclined at an angle θ to AB.

$\because$ String is inextensible

$\therefore$ A′B′ = AB

θ is very small. Draw A′L and B′M perpendicular from A′ and B′ on AB. Draw A′N ⊥ B′N.

Now tension T in the string is equivalent to a force T at A along AB and a force T at B along BA.

$\therefore$ Virtual work done by these forces

$$= T \cdot AL - T \cdot BM$$
$$= T \cdot (AB - LB) - T\,(LM - LB)$$
$$= T \cdot (AB - LM)$$
$$= T \cdot (AB - A'N)$$
$$= T(AB - A'B' \cos\theta)$$
$$= T \cdot AB\,(1 - \cos\theta) \qquad A'B' = AB$$
$$= T.\,AB \left\{1 - \left(1 - \frac{\theta^2}{2!} + \frac{\theta^4}{4!} \ldots\ldots\right)\right\} = 0$$

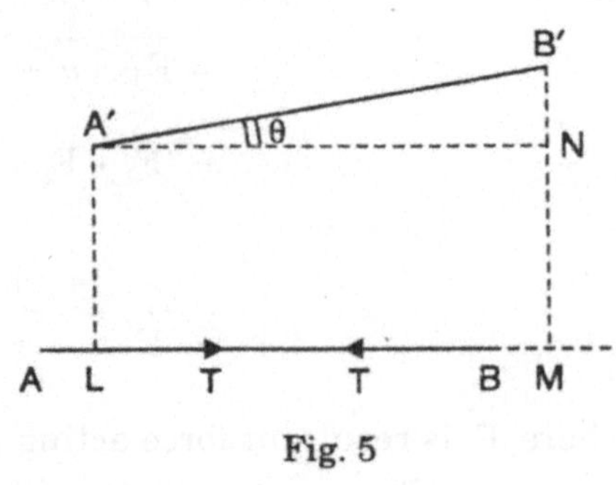

Fig. 5

$\because$ θ is very small, its higher powers can be neglected.

(*ii*) **Thrust in the light rod.** The case is similar to the case of inextensible string. The only difference is that the force T at A is along BA and at B is along AB, while in the case of string it was in opposite direction.

(*iii*) **The reaction R of any smooth surface with which the body is in contact.** Let $\vec{r}$ be the position vector of the point of contact of the body with the smooth surface, then reaction R is perpendicular to the surface. Now if $\vec{r}$ moves to $\vec{r} + \vec{dr}$ and the contact is preserved during displacement then $\vec{dr} \perp \vec{R}$. Hence work done $\vec{R} . \vec{dr} = 0$.

(*iv*) **Mutual pressure between two bodies in contact.** Since action and reaction are equal and opposite, therefore the virtual work done by one force is balance by other.

(*v*) **The reaction at a point of contact with fixed surface on which the body rolls without sliding.** The point of contact of the body with the surface is momentarilly at rest and so it does not go any displacement, hence work done by normal reaction is zero.

8. To Find Tension in a String or Thrust in a Rod

Refer to Fig. (5)

Work done by the tension

$$= T\,[AB - A'B' \cos \theta] \qquad \theta \text{ is very small}$$
$$= T\,[l - (l + \delta l)] \qquad AB = l$$
$$= -T.\,\delta l \qquad A'B' = l + \delta l$$

where δl is the increase in the length of the string.

Similarly, in the case of the rod AB, the work done by the thrust T is $+\,T.\,\delta l$.

In solving the problems on virtual work we measure the distances of the points of application of the forces from a fixed point or a fixed line. If Z be the distance of point of application of force F and small displacement δ_z be given the work done is F. δ_z. The sign of work done is (+) ve if Z is measured in the direction of F and is (–) ve if measured in a direction opposite of F. Then by applying the principle of virtual work we can write the equation of virtual work.

Method of solving the problems on virtual work

Step I. Make the figure according to the given problems.

II. Show the forces which can do virtual work in virtual displacement.

III. Choose some fixed point in the system, which will have no displacement

IV. Measure distances of the forces from the fixed point and write the equation of virtual work, keeping in view the points explained in article (7) and (8) above.

Example 1. *Three light rods AB, AC of length a each and BC of length* $\frac{3a}{2}$ *are freely joined to form a triangle ABC. It rests with BC horizontal, A below BC and the rods AB and AC resting against the smooth pegs E and F in the same horizontal line distant 2b apart. A weight W is suspended from A. In equilibrium position find thrust in BC.*

Sol. Let T be the thrust in the rod BC. Let θ be the angle which AB and AC make with the vertical. Let the system be given a small virtual displacement such that θ changes to $\theta + d\theta$.

Consider the middle point of EF as origin O.

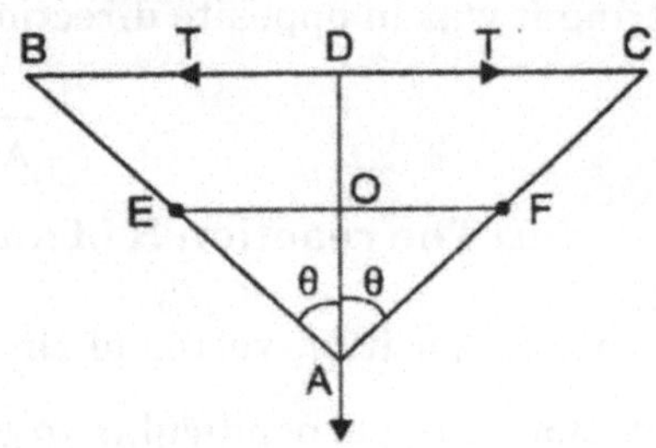

Fig. 6

$$BC = 2BD = 2a \sin\theta$$

$$AO = EO \cot\theta = b \cot\theta$$

$\therefore$ Equation of virtual work is

$$T \,.\, d\,(BC) + W \,.\, d\,(AO) = 0$$

$$T \,.\, d\,(2a \sin\theta) + W \,.\, d\,(b \cot\theta) = 0$$

or $$T.\, 2a \cos\theta \, d\theta - W.b \operatorname{cosec}^2\theta \, d\theta = 0$$

or $$T = \frac{Wb \operatorname{cosec}^2\theta}{2a \cos\theta} \qquad d\theta \neq 0 \qquad \ldots(1)$$

From Δ ABE

$$\sin\theta = \frac{BE}{AB} = \frac{3a/4}{a} = \frac{3}{4}$$

$$\cos\theta = \sqrt{1 - \frac{9}{16}} = \frac{\sqrt{7}}{4}$$

Putting in (1), $$T = \frac{Wb}{2a} \times \frac{16}{9} \times \frac{4}{\sqrt{7}} = \frac{32b}{9\sqrt{7}\,a} W$$

$$T = \frac{32}{9\sqrt{7}} \frac{b}{a} W.$$

Example 2. *A square framework formed of four uniform rods of equal weight W joined together, is hung up by one corner. A weight W is suspended from each of the three lower corners and the shape of the square is preserved by a light rod along the horizontal diagonal. Find the thrust of the light rod.*

Sol. Let ABCD be the square framework suspended from A. The total weight of the rods 4W is acting at the centre O.

The weights W at B and D each is equivalent to 2W at O. Thus the total weight at O is 6 W and W is suspended at C. Thrust T in the rod is replaced by T at B along DB and T at D along BT.

Let the distances be measured from A, the point of suspension of the framework. Let θ be the angle which each rod makes with vertical. Let length of each rod be $2a$.

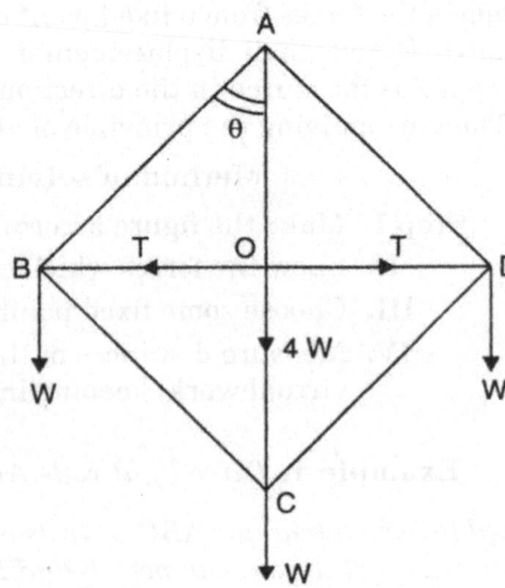

Fig. 7

$$AO = 2a \cos\theta,\ BO = OD = 2a \sin\theta$$

$$AC = 4a \cos\theta,\ BD = 4a \sin\theta = l \text{ (say)}$$

Let the system be given a small virtual displacement such that θ changes to $\theta + \delta\theta$.

The equation of virtual work is

$$6Wd\,(AO) + Wd\,(AC) + T \,.\, d\,(l) = 0$$

$$6\text{W}d\,(2a\cos\theta) + \text{W}d\,(4a\cos\theta) + \text{T}d\,(4a\sin\theta) = 0$$

$$-12a\,\text{W}\sin\theta\,.\,d\theta - 4a\,\text{W}\sin\theta\,d\theta + 4a\,\text{T}\cos\theta\,d\theta = 0 \qquad (\because d\theta \neq 0)$$

$$\text{T} = 4\,\text{W}\tan\theta$$

In the position of equilibrium $\theta = 45°$

$$\therefore \qquad \text{T} = 4\text{W}.$$

Example 3. *A smoothly jointed framework of light rods form a quadrilateral ABCD. The middle points P, Q of an opposite pair of rods are connected by a string having tension T and middle points R, S of the other pair by a light rod under a thrust X. Prove that T/PQ = X/RS.*

Sol. Let ABCD be the framework P, Q are middle points of AD and BC respectively. String PQ has tension T, RS is the rod joining the middle points of AB and CD and has thrust T.

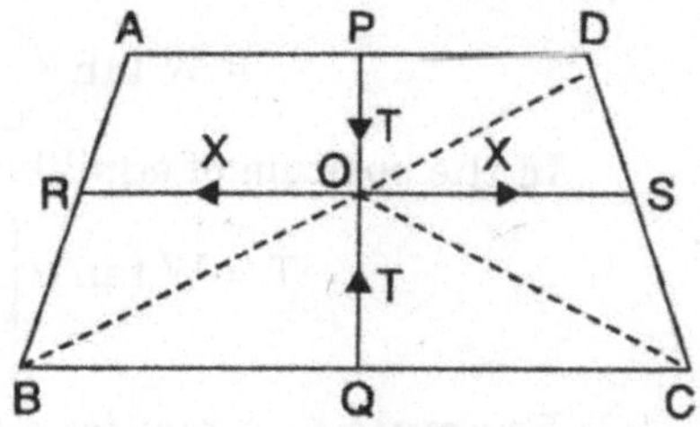

Fig. 8

If the small virtual displacement is given to the framework it changes the angles and not the length of the sides.

Equation of virtual work is

$$-\text{T}d(\text{PQ}) + \text{X}d(\text{RS}) = 0$$

or
$$\text{T}d\,(\text{PQ}) = \text{X}d(\text{RS}) \qquad ...(1)$$

In ΔOBC, OQ is the median

$$\text{OB}^2 + \text{OC}^2 = 2(\text{OQ}^2 + \text{BQ}^2) = 2\left[\left(\frac{1}{2}\text{PQ}\right)^2 + \left(\frac{1}{2}\text{BC}\right)^2\right]$$

$$\therefore \qquad 2[\text{OB}^2 + \text{OC}^2] = \text{PQ}^2 + \text{BC}^2 \qquad ...(2)$$

Similarly from ΔOCD $2[\text{OC}^2 + \text{OD}^2] = \text{RS}^2 + \text{CD}^2$ ...(3)

Subtracting (3) from (2),

we have, $2[\text{OB}^2 - \text{OD}^2] = \text{PQ}^2 - \text{RS}^2 + \text{BC}^2 - \text{CD}^2$ ...(4)

Similarly $2[\text{OB}^2 - \text{OD}^2] = \text{RS}^2 - \text{PQ}^2 + \text{AB}^2 - \text{DA}^2$ ...(5)

Subtracting (5) from (4),

we have, $2[\text{PQ}^2 - \text{RS}^2] = \text{AB}^2 - \text{BC}^2 + \text{CD}^2 - \text{AD}^2$

$\because$ Length of the sides are unaltered, therefore R.H.S. is a constant differentiating

$$\therefore \qquad 4\,\text{PQ}\,.\,d(\text{PQ}) - 4\,\text{RS}\,.\,d(\text{RS}) = 0$$

or
$$\text{PQ}\,.\,d(\text{PQ}) = \text{RS}.\,d(\text{RS}) \qquad ...(6)$$

From (1) and (6), we have, $\dfrac{\text{T}}{\text{PQ}} = \dfrac{\text{X}}{\text{RS}}$.

Example 4. *ABCD is a rhombus formed with four weightless rods each of length l and A weight W is attached to the lowest hinge C and frame rests on two smooth pegs in a horizontal line in contact with the rods AB and AD. B and D are in a horizontal line and are joined by a string. If the distance between pegs is 2c and angle A is 2α, show that the tension in the string is*

$$W\tan\alpha\left(\frac{c}{2l}\,\text{cosec}^3\,\alpha - 1\right).$$

Sol. Proceeding exactly as in the example 2 except the work done by 4ω the equation of virtual work is

$$-Td(BD) + Wd(LC) = 0$$

$$-Td(2l \sin\theta) + Wd(2l\cos\theta - c\cot\theta) = 0$$

$$-2l\,T\cos\theta - 2l\,W\sin\theta + Wc\,\text{cosec}^2\theta = 0$$

$$T = \frac{Wc\,\text{cosec}^2\theta - 2l\,W\sin\theta}{2l\cos\theta}$$

or
$$T = W\left[\frac{c}{2l}\,\text{cosec}^3\theta \,.\, \frac{\sin\theta}{\cos\theta} - \tan\theta\right]$$

$$= W\tan\theta\left[\frac{c}{2l}\,\text{cosec}^3\theta - 1\right]$$

In the position of equilibrium $\theta = \alpha$

$$\therefore \quad T = W\tan\alpha\left[\frac{c}{2l}\,\text{cosec}^3\alpha - 1\right].$$

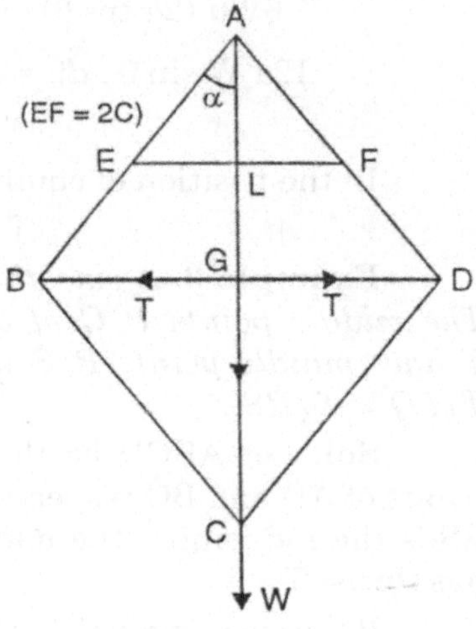

Fig. 9

Example 5. *A regular pentagon ABCDE is formed of five uniform heavy rods, each of weight W smoothly joined at their extremities. It is freely suspended from A and is kept as a regular pentagon by a light rod joining B and E. Show that the stress in the rod is W cot 18°.*

(K.U. 2004)

Sol. Let the thrust T in the rod BE be replaced by forces T at B along EB and T at E along BE. Let $2a$ be the length of each rod. Let upper and the lower slant rods make angles θ and φ with the vertical. The weights W of AB and AE each can be replaced by 2W at L which is middle point of the line joining the middle points of the rod AB and AE. Similarly weights of BC and DE is replaced at N.

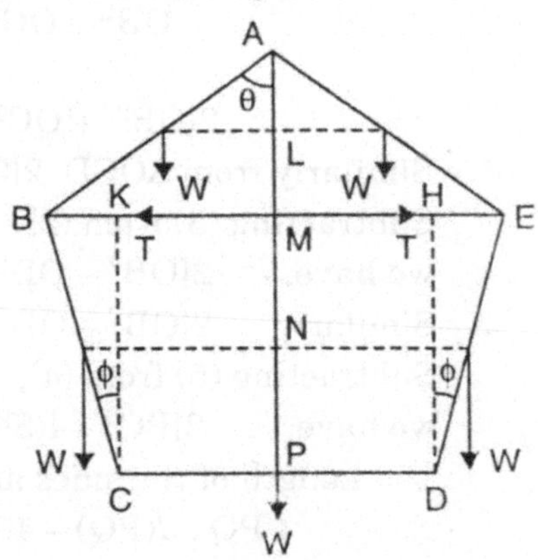

Fig. 10

$$AL = a\cos\theta$$

$$AM = 2a\cos\theta$$

$$CK = 2a\cos\phi$$

$$PN = a\cos\phi = MN$$

$$\therefore \quad AN = AM + MN = 2a\cos\theta + a\cos\phi$$

$$BE = 2BM = 2 \times 2a\sin\theta$$

$$AP = 2a\cos\theta + 2a\cos\phi$$

Let the system be given a small virtual displacement such that θ changes to θ + $d\theta$, and φ to φ + $d\phi$

Taking A as the fixed point, equation of virtual work is

$$2W\,d(AL) + 2W\,d(AN) + W\,d(AP) + T\,d(BE) = 0$$

or
$$2W\,d(a\cos\theta) + 2W\,d(2a\cos\theta + a\cos\phi) + W\,d(2a\cos\theta + 2a\cos\phi) + Td(4a\sin\theta) = 0$$

$$-2a\,W\sin\theta\,d\theta - 4a\,W\sin\phi\,d\phi - 2a\,W\sin\phi\,d\phi$$
$$-2a\,W\sin\theta\,d\theta - 2a\,W\sin\phi\,d\phi + 4a\,T\cos\theta\,d\theta = 0$$

or
$$(T\cos\theta - 2W\sin\theta)\,d\theta = W\sin\phi\,d\phi \qquad ...(1)$$

But $$BM = BK + KM = BK + CP$$

or $$2a \sin\theta = 2a \sin\phi + a$$

Differentiating (2), we have

$$2a \cos\theta \, d\theta = 2a \cos\phi \, d\phi \qquad ...(2)$$

From (1) and (2), we have,

$$\frac{T\cos\theta - 2W\sin\theta}{\cos\theta} = \frac{W\sin\phi}{\cos\phi}$$

or $$T = W(2\tan\theta + \tan\phi) \qquad ...(3)$$

But in the position of equilibrium, the interior angle of pentagon is 108°

$\therefore$ $\theta = 54°$

$\phi + 90° = 108° \Rightarrow \phi = 18°$

Putting θ and ϕ in (3)

$$T = W(2\tan 54° + \tan 18°) = W(2\cot 36° + \tan 18°)$$

$$= W\left[2.\frac{(1-\tan^2 18°)}{2\tan 18°} + \tan 18°\right]$$

$$= W\left[\frac{1}{\tan 18°} - \tan 18° + \tan 18°\right]$$

$$T = W \cot 18°.$$

Example 6. *ABCDEF is a regular hexagon formed of light rods smoothly joined at their ends with a diagonal rod AD. Four equal forces P act inwards at the middle points of the rods AB, CD, DE, FA and perpendicular to the sides. Find the stress in the diagonal rod AD and state whether it is a tension or a thrust.*

Sol. Let K, L, M, N be the middle points of AB, CD, DE and FA respectively and O be the middle point of rod AD. Consider OD as x-axis and a line ⊥ AD through O as y-axis.

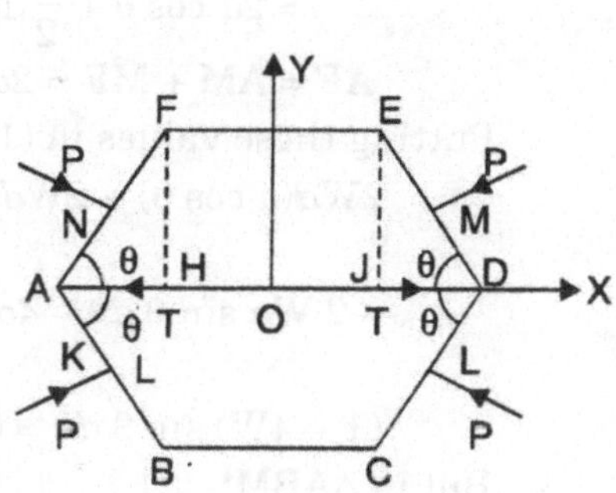

Fig. 11

Let $\angle BAD = \angle FAD = \theta$.

Let T be the stress in AB assumed to be a thrust.

Let $2a$ be the length of each side.

Let us assume that small virtual displacement be so given that it does not alter the position of O and the length of the sides but may alter the length of AD.

Taking O as the fixed point

$$AD = 2(2a\cos\theta + a)$$

Coordinates of M are $(a + a\cos\theta, a\sin\theta)$

Resolved parts of P at M are $X = -P\sin\theta$, $Y = -P\cos\theta$

Virtual work done by P is

$$X\,dx + Y dy = -P\sin\theta \, d(a + a\cos\theta) - P\cos\theta \, d(a\sin\theta)$$

$$= -Pa[-\sin^2\theta + \cos^2\theta]\, d\theta = -Pa\cos 2\theta \, d\theta$$

Virtual work done by T is

$$Td(2a + 4a\cos\theta) = -4a\,T\sin\theta\, d\theta$$

Hence, the equation of virtual work for the whole system is

$$-4a\,T\sin\theta\,d\theta - 4a\,P\cos 2\theta\,d\theta = 0$$

$$T = -\frac{P\cos 2\theta}{\sin\theta}$$

But in the position of equilibrium $\theta = \pi/3$

$$\therefore \qquad T = \frac{P}{\sqrt{3}}$$

$\because$ T is positive hence the stress is a thrust.

Example 7. *A freely joined framework is formed of five equal uniform rods each of weight W. The framework is suspended from one corner which is also joined to the middle of the opposite sides by an inextensible string. If the two upper and lower rods make angle θ and φ respectively with the vertical, the ratio of tension of the string is to the weight of the rod is (4 tan θ + 2 tan φ) : (tan θ + tan φ).*

Sol. Let T be the tension in the string AF. The weights of upper two rods may considered on L and two lower slanted rods at N and of rod CD at F.

Let a small virtual displacement changes θ to θ + δθ and φ to φ + δφ. Measuring distances from the fixed point A, the equation of virtual work is

$$2Wd(AL) + 2Wd(AN) + Wd(AF) - Td(AF) = 0 \qquad ...(1)$$

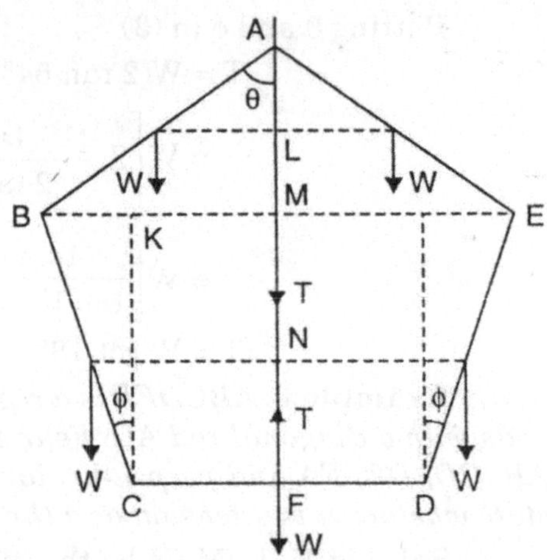

Fig. 12

Let $AB = 2a$

$AL = a\cos\theta$

$AN = AM + MN$

$= 2a\cos\theta + \frac{1}{2}MF = 2a\cos\theta + a\cos\phi$

$AF = AM + MF = 2a\cos\theta + 2a\cos\phi$

Putting these values in (1),

$$2Wd(a\cos\theta) + 2Wd(2a\cos\theta + a\cos\phi) + Wd(2a\cos\theta + 2a\cos\phi) - Td(2a\cos\theta + 2a\cos\phi) = 0$$

$$-2Wa\sin\theta\,d\theta - 4aW\sin\theta\,d\theta - 2aW\sin\phi\,d\phi + W(-2a\sin\theta\,d\theta - 2a\sin\phi\,d\phi) - T(-2a\sin\theta\,d\theta - 2a\sin\phi\,d\phi) = 0$$

$$(T - 4W)\sin\theta\,d\theta = (2W - T)\sin\phi\,d\phi \qquad ...(2)$$

But in ΔABM

$BM = 2a\sin\theta$

Also $\quad BM = BK + KM = 2a\sin\phi + a$

$\therefore \quad 2a\sin\theta = 2a\sin\phi + a$

or $\quad 2a\cos\theta\,d\theta = 2a\cos\phi\,d\phi$

or $\quad \cos\theta\,d\theta = \cos\phi\,d\phi \qquad ...(3)$

Dividing (2) by (3)

$(T - 4W)\tan\theta = (2W - T)\tan\phi.$

$T(\tan\theta + \tan\phi) = W(2\tan\phi + 4\tan\theta)$

or $\quad T/W = (4\tan\theta + 2\tan\phi)/(\tan\theta + \tan\phi).$

Example 8. *A string of length a forms the shorter diagonal of a rhombus of four uniform rods, each of length b and weight w, which are hinged together. If one of the rod is supported in a horizontal position, prove that tension of the string is* $\dfrac{2W(2b^2 - a^2)}{b\sqrt{4b^2 - a^2}}$. (K.U. 2007)

Sol. Let ABCD be the rhombus with side AB horizontal. Weight of each rod is given W. Total weight 4W acts at the intersection of the diagonals at G.

BD is the shorter diagonal and is connected by a string. Let the tension in the string be T. Let $\angle$ BDC $= \theta$. Now suppose that the system is given a small virtual displacements such that the angle θ changes to $\theta + \delta\theta$.

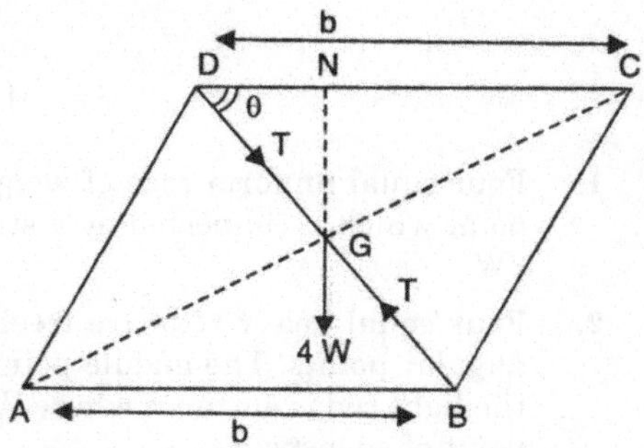

Fig. 13

Draw GN $\perp$ CD

From Δ GCD, $\quad GD = CD \cos\theta = b\cos\theta \quad$ ...(1)

Now depth of C.G. below CD is GN

$$GN = GD \sin\theta$$

$$= b\cos\theta \sin\theta = \frac{b}{2}\sin 2\theta.$$

Now the equation of virtual work is given by

$$-T\,\delta(BD) + 4W\,\delta(GN) = 0$$

$$-T\,\delta(2b\cos\theta) + 4W\,\delta\left(\frac{b}{2}\sin 2\theta\right) = 0$$

or $$2b\sin\theta\,\delta\theta . T + 4\,W\,b\cos 2\theta .\,\delta\theta = 0$$

$\because \quad \delta\theta \neq 0, \quad T = -\dfrac{2W\cos 2\theta}{\sin\theta}$ ...(2)

In the position of equilibrium

$$GD = \frac{1}{2}AC = \frac{a}{2}$$

$\therefore$ From (1), $\quad b\cos\theta = \dfrac{a}{2}$

or $$\cos\theta = \frac{a}{2b}$$

$$\sin\theta = \sqrt{1 - \frac{a^2}{4b^2}} = \frac{\sqrt{4b^2 - a^2}}{2b}$$

Putting the values of cos θ and sin θ in (2), we get

$$T = -2W\frac{(2\cos^2\theta - 1)}{\sin\theta}$$

$$= 2W \frac{\left[1 - 2 . \frac{a^2}{4b^2}\right]}{\frac{\sqrt{4b^2 - a^2}}{2b}} = \frac{W(2b^2 - a^2)}{b\sqrt{4b^2 - a^2}}.$$

EXERCISE 1

1. Four equal uniform rods of weight W each joined to form a square, are hung from an angular point which is connected by a string with the opposite point. Prove that tension in the string is 2W.

2. Four equal heavy rods are freely joined to form a rhombus and is suspended from one of the angular points. The middle points of upper rods are joined by a light rod. Prove that tension in the light rod is $4\omega \tan \alpha$ where W is weight of each rod and 2α is the angle of the rhombus at the point of suspension.

3. Four rods of equal weight ω joined to form a rhombus. The frame is suspended from A and weight W is attached to C. A rod of negligible weight joins the middle points of AB and AD, keeping them inclined at an angle α to AC. Show that the thrust in the light rod is $(2W + 4\omega) \tan \alpha$.

4. A frame consists of five bars forming the sides of a rhombus ABCD with diagonal AC. If four equal forces P act all inwards at the middle point of the sides and perpendicular to the sides then thrust in AC is $\frac{P \cos 2\theta}{\sin \theta}$, where $\theta = \angle BAC$.

5. Four rods are joined to form a parallelogram. The opposite points are joined by strings forming diagonals and the whole system is placed on a smooth horizontal table. Show that their tensions are in the ratio of their lengths.

6. Six equal rods each of weight W form a hexogon ABCDEF. Rod AB is fixed in a horizontal position and the middle points of AB and DE are joined by a string. Find the tension in the string.

7. Six equal heavy rods, freely hinged at the ends form a regular hexagon ABCDEF. It is suspended from A and is kept in shape by two light rods BF and CE. Prove that the thrusts in the rods are in the ratio of 5 : 1.

8. Four equal uniform rods of weight w are freely joined to form a rhombus ABCD. The framework is suspended freely from A and weight W is attached to each of the joints B, C and D. If horizontal forces of magnitude P acting at B and D keep the angle BAD equal to 120°, then prove that $P = (W + w)\, 2\sqrt{3}$.

9. A string of length a forms the shorter diagonal of a rhombus formed of four uniform rods each of length b and weight W by smooth joints. If one of the rod be supported in a horizontal position find the tension in the string.

10. Weight W_1 and W_2 are fastined to a light inextensible string ABC at the points B, C and A being fixed. If a horizontal force P is applied at C and AB and BC are inclined at angles θ and ϕ to the vertical then $P = (W_1 + W_2) \tan \theta = W_2 \tan \phi$.

11. A rod AB of weight W is movable about a point at A, and to B it attached a string whose other end is tied to a ring. The ring slides along a smooth horizontal wire passing through A. Prove that the horizontal force necessary to keep the ring at rest is $\frac{W \cos \alpha \cos \beta}{2 \sin (\alpha + \beta)}$, where α and β are the inclination of the rod and the string to the horizontal.

12. ABCD is a rhombus formed with four equal rods each of length l and weight ω, joined by smooth hinges. A weight W is attached to the lowest hinge C and the frame rests on two smooth pegs in a horizontal line in contact with the rods AB and AD. B and D are in horizontal line and joined by a string. If distance between the pegs is $2c$ and the angle A is 2α, show that the tension in the string is

$$\tan\alpha\left[\frac{c}{2l}(W+4w)\operatorname{cosec}^3\alpha-(W+2w)\right].$$

13. A square of side $2a$ is placed with its plane vertical between two smooth pegs which are in same horizontal line at a distance c apart. Show that it will be in equilibrium when the inclination of one of its edges to the horizontal is either $\frac{\pi}{4}$ or $\frac{1}{2}\sin^{-1}\left(\frac{a^2-c^2}{c^2}\right)$.

14. Six equal rods each of weight W are joined to form a regular hexagon. One rod is fixed in horizontal position and the system lies in vertical plane. The middle points of upper slant rods are connected by a string. Show that tension in the string is 6W cot θ, where θ is the inclination of the slant rods to the horizon.

ANSWERS

6. 3 W

9. $\dfrac{2W(2b^2-a^2)}{b\sqrt{4b^2-a^2}}$.

SOME MORE ILLUSTRATIVE EXAMPLES

Example 1. *Two rods each of weight Wl and length l, are hinged together and placed astride a smooth horizontal cylindrical peg of radius r. The lower ends of the rods are tied together by a string and the rods are inclined at angle ϕ each to the horizontal. Find the tension in the string. If the string be slack then ϕ satisfies the equation $\tan^3\phi+\tan\phi=l/2r$.*

(K.U. 2004)

Sol. Let D and E be the middle points of the rods AB and AC and the G be the middle point of DE. We can assume the weights of the rod acting at G.

Let O be the centre. Draw OL $\perp$ AC.

From Δ AOL AO $= r\sec\phi$

we have,

From Δ ADG $AG=\frac{l}{2}\sin\phi$

we have, $DG=\frac{l}{2}\cos\phi$

$\therefore$ $OG = AO - AG = r\sec\phi - \frac{l}{2}\sin\phi$

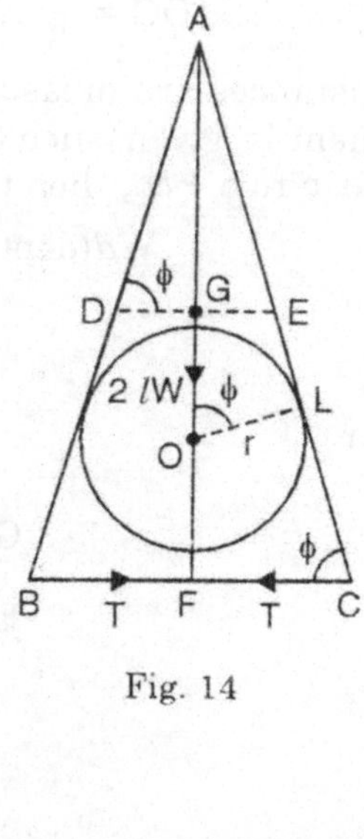

Fig. 14

Let T be the tension in the string BC. Replace it by the forces T at B along BC and T at C along CB respectively.

Since the cylindrical peg is fixed, therefore O is taken as the fixed point. The distances are measured from O

Let the small virtual displacement be given to the system such that ϕ changes to $\phi + \delta\phi$, then the equation of virtual work is

$$-2lW\, d(OG) - Td(BC) = 0$$

or
$$-2lW\, d\left(r\sec\phi - \frac{l}{2}\sin\phi\right) - T\, d(2l\cos\phi) = 0$$

or
$$\left[-2lW\left(r\sec\phi\tan\phi - \frac{l}{2}\cos\phi\right) + 2lT\sin\phi\right] d\phi = 0 \qquad (\because d\phi \neq 0)$$

or
$$T = \frac{W\left(r\sec\phi\tan\phi - \frac{l}{2}\cos\phi\right)}{\sin\phi}$$

$$T = W\left(r\sec^2\phi - \frac{l}{2}\cot\phi\right)$$

If string becomes slack, then T = 0

We have,
$$r\sec^2\phi = \frac{l}{2}\cot\phi$$

or
$$r(1 + \tan^2\phi) = \frac{l}{2}\frac{1}{\tan\phi} \quad \text{or} \quad \tan^3\phi + \tan\phi = \frac{l}{2r}.$$

Example 2. *A solid hemisphere is supported by a string fixed to a point on its rim and to a point on a smooth vertical wall with which the curved surface of the hemisphere is in contact. If θ and ϕ are the inclination of the string and the plane base of the hemisphere to the vertical then prove that* $\tan\phi = \frac{3}{8} + \tan\theta$. (K.U. 2005)

Sol. Let AC be the string of length l attached at A to the hemisphere of radius r and centre O. Let G be the C.G. of the hemisphere on OD such that OG : GD = 3 : 5

$\therefore$ $OG = \frac{3}{8}r$. String is attached to a point c on the wall. If the distances are measured from the fixed point C and virtual displacement is given such that its length remains unaltered θ changes to $\theta + \delta\theta$, ϕ to $\phi + \delta\phi$, then the equation of virtual work is

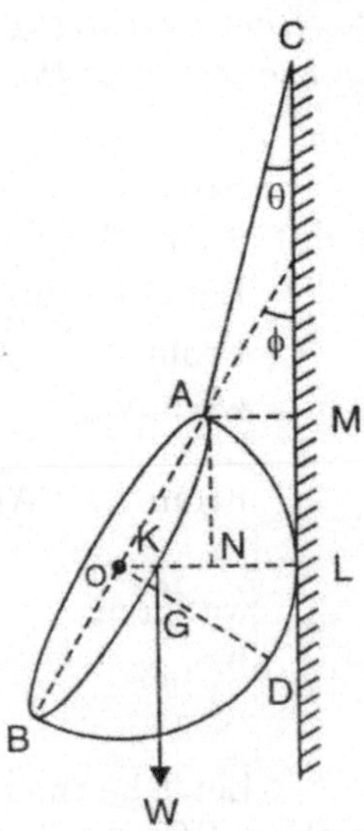

Fig. 15

$$Wd(\text{depth of G below C}) = 0$$

$$Wd\,(GK + LC) = 0$$

(Tension will not do any work)

or
$$d(GK + LC) = 0 \qquad \ldots(1)$$

$$GK = OG\sin\phi = \frac{3}{8}r\sin\phi$$

$$LC = LM + MC = AN + MC$$
$$= r\cos\phi + l\cos\theta \qquad \ldots(2)$$

Putting GK and LC in equation (1)

$$d\left(\frac{3}{8}r\sin\phi + r\cos\phi + l\cos\theta\right) = 0$$

or $$\frac{3}{8}r\cos\phi\, d\phi - r\sin\phi\, d\phi - l\sin\theta\, d\theta = 0$$

or $$l\sin\theta\, d\theta = \left(\frac{3}{8}\cos\phi - \sin\phi\right) r\, d\phi \qquad ...(3)$$

Again, $$r = OL = ON + NL = ON + AM$$

$$r = r\sin\phi + l\sin\theta \qquad ...(4)$$

Differentiating (4), we have,

$$0 = r\cos\phi\, d\phi + l\cos\theta\, d\theta$$

or $$l\cos\theta\, d\theta = -r\cos\phi\, d\phi \qquad ...(5)$$

Dividing (3) by (5),

we have, $$\tan\theta = \tan\phi - \frac{3}{8}$$

or $$\tan\phi = \frac{3}{8} + \tan\theta.$$

Example 3. *A smooth hemisphere of radius r and weight W rests in a horizontal circular hole of radius a. A string is wrapped one round the sphere just above the hole and then pulled tight. What must be the tension in the string so that it will just raise the sphere.*

Sol. Let O be the centre of the sphere, where the weight W of the sphere act. Let G be the centre of the hole of diameter AB. String is held tight around the sphere just above the hole. Let x be the radius of circular band around the hole. Length of the string is $2\pi x$. Let the distance be measured from the level of fixed diameter AB. Let the small virtual displacement be given such that x changes to $x + \delta x$ and sphere rises slightly.

Then the equation of virtual work is

$$-W.d(OG) - T.\, d(2\pi x) = 0$$

or $$W.d\sqrt{r^2 - x^2} + Td(2\pi x) = 0$$

or $$\frac{-W\, x\, dx}{\sqrt{r^2 - x^2}} + 2\pi\, Tdx = 0$$

or $$T = \frac{W\, x}{2\pi\sqrt{r^2 - x^2}}$$

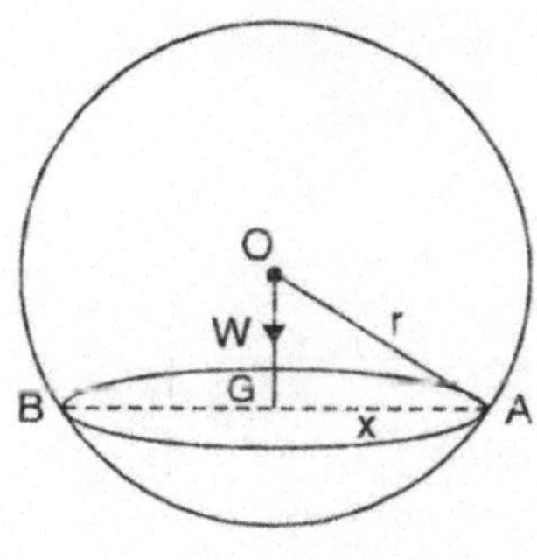

Fig. 16

But in the position of equilibrium $x = a$

Hence, $$T = \frac{W\, a}{2\pi\sqrt{r^2 - a^2}}.$$

Example 4. *A smooth parabolic wire is fixed with its axis vertical and vertex downwards. A uniform rod of length 2l is placed in it with its ends resting on the wire. Show that in the position of equilibrium, the rod is either horizontal or makes with the horizontal an angle θ given by $\cos^2\theta = 2a/l$, where 4a is latus rectum of the parabola.*

Sol. Let the rod AB be resting in the parabola $x^2 = 4ay$. Let the weight of rod AB be acting at G. Draw perpendicular AL, BM and GK on x-axis. Let AN $\perp$ BN.

Let coordinates of A be $(2at, at^2)$.

If $\angle BAN = \theta$ then $AN = 2l \cos\theta$, $BN = 2l \sin\theta$.

The coordinates of B are $(2at + 2l\cos\theta, at^2 + 2l \sin\theta)$

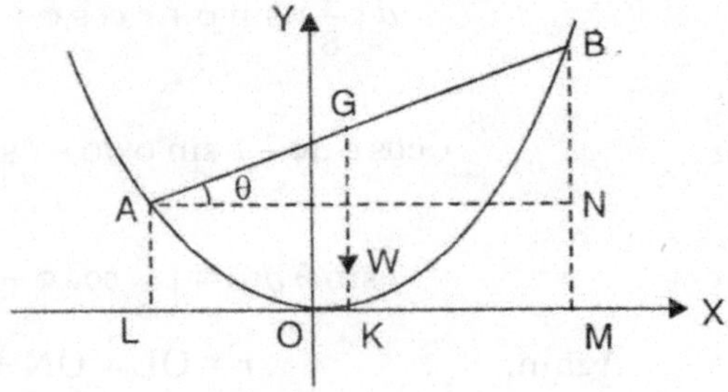

Fig. 17

Putting in the equation of parabola

$$(2at + 2l\cos\theta)^2 = 4a\,(at^2 + 2l\sin\theta)$$

or $$4a^2t^2 + 8\,alt\cos\theta + 4l^2\cos^2\theta = 4a^2t^2 + 8al\sin\theta$$

or $$t = \frac{8al\sin\theta - 4l^2\cos^2\theta}{8al\cos\theta}$$

or $$t = \tan\theta - \frac{l\cos\theta}{2a} \quad ...(1)$$

If small virtual displacement be given such that θ changes to $\theta + d\theta$, then the equation of virtual work is

$$-\,Wd(GK) = 0 \quad ...(2)$$

But $$GK = \frac{1}{2}\,(AL + BM)$$

$$= \frac{1}{2}[at^2 + at^2 + 2l\sin\theta] = at^2 + l\sin\theta$$

$$= a\left[\tan\theta - \frac{l\cos\theta}{2a}\right]^2 + l\sin\theta \quad \text{using (1)}$$

$$GK = a\tan^2\theta + \frac{l^2}{4a}\cos^2\theta$$

Putting the value of GL in (2)

$$-\,Wd\left[a\tan^2\theta + \frac{l^2}{4a}\cos^2\theta\right] = 0$$

or $$\left[2a\tan\theta\sec^2\theta - \frac{l^2}{2a}\cos\theta\tan\theta\right]d\theta = 0 \quad (\because d\theta \neq 0)$$

$\therefore$ $$\frac{\sin\theta}{\cos^3\theta} - \frac{l^2}{4a^2}\sin\theta\cos\theta = 0$$

or $$\sin\theta\left[\frac{4a^2}{l^2} - \cos^4\theta\right] = 0$$

or $$\cos^4\theta = \frac{4a^2}{l^2}$$

or $\sin\theta = 0 \Rightarrow \theta = 0$ *i.e.*, rod is horizontal.

or $$\cos^2\theta = \frac{2a}{l}.$$

Example 5. *A smooth rod passes through a smooth ring at the focus of an ellipse whose major axis is horizontal and rests with its lower end on the quadrant of the curve which is farthest removed from the focus. Find the position of equilibrium and show that its length must at least be* $\frac{3}{4}a + \frac{1}{4}a\sqrt{1+8e^2}$ *, where 2a is the major axis and e is its ecentricity.*

Sol. Let the rod AB of weight W passes through the ring at S.

Let weight of the rod W be acting at G, the middle point of the rod.

The equation of the ellipse referred to S as the pole is

$$\frac{l}{r} = 1 - e\cos\theta$$

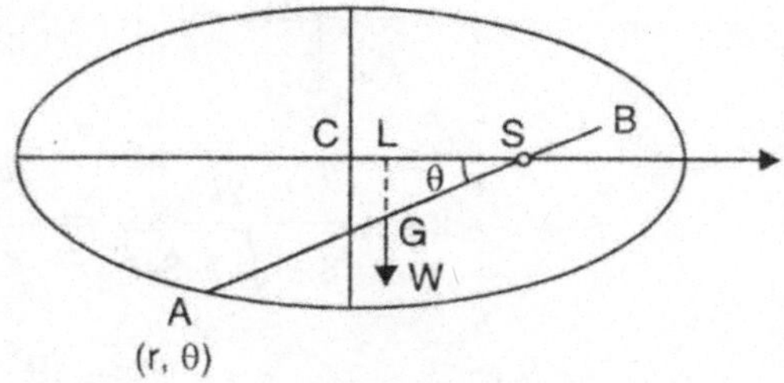

Fig. 18

where l is semi latus rectum of the ellipse. Let polar coordinate of A be (r, θ). Let AB = $2d$

$$\therefore \quad SG = SA - AG = r - d$$
$$GL = SG \sin\theta = (r - d)\sin\theta$$

If distances are measured downwards from the major axis and the system be given a small displacement such that θ changes to $\theta + \delta\theta$ then the equation of virtual work is

$$W d(GL) = 0 \quad \text{or} \quad d(GL) = 0 \qquad ...(1)$$

or $$d[(r - d)\sin\theta] = 0$$

$$d\left[\left\{\frac{l}{1 - e\cos\theta} - d\right\}\sin\theta\right] = 0$$

or $$d\left(\frac{l\sin\theta}{1 - e\cos\theta}\right) - d\,(d\sin\theta) = 0$$

or $$\frac{l\cos\theta\,(1 - e\cos\theta) - l\sin\theta \,.\, e\sin\theta}{(1 - e\cos\theta)^2} = d\cos\theta \qquad d\theta \neq 0$$

or $$l\cos\theta - le\,(\cos^2\theta + \sin^2\theta) = d\cos\theta\,(1 - e\cos\theta)^2$$

$$l\cos\theta - el = d\cos\theta\,(1 - e\cos\theta)^2 \qquad ...(2)$$

Now for the least length of the rod, B coincides with S

or $$AB = 2d = r = \frac{l}{1 - e\cos\theta} \qquad ...(3)$$

Putting the value of d in equation (2), we have,

$$l\cos\theta - el = \frac{l\cos\theta}{2(1 - e\cos\theta)}(1 - e\cos\theta)^2$$

or $$2l\cos\theta - 2el - l\cos\theta\,(1 - e\cos\theta) = 0$$

or $$e\cos^2\theta + \cos\theta - 2e = 0$$

On solving $\cos\theta = \dfrac{-1+\sqrt{1+8e^2}}{2e}$

Putting in (3), $d = \dfrac{l}{\left[2\left[1-e.\left\{\dfrac{-1+\sqrt{1+8e^2}}{2e}\right\}\right]\right]} = \dfrac{l}{3-\sqrt{1+8e^2}}$

or $$2d = \frac{2l}{3-\sqrt{1+8e^2}} = \frac{2l\left[3+\sqrt{1+8e^2}\right]}{8(1-e^2)} \quad ...(4)$$

But semi latus rectum $l = \dfrac{b^2}{a} = \dfrac{a^2(1-e^2)}{a} = a(1-e^2)$

$$\therefore \quad AB = 2d = \frac{2a(1-e^2)}{8(1-e^2)}\,[3+\sqrt{1+8e^2}\,] = \frac{3a}{4} + \frac{1}{4}a\sqrt{1+8e^2}.$$

Example 6. *A heavy elastic string of natural length $2\pi a$ is placed round a smooth cone whose axis is vertical and whose semi vertical angle is α. If W be the weight and λ be the modulus of elasticity, prove that it will be in equilibrium when in the form of a circle its radius is* $a\left[1+\dfrac{W\cot\alpha}{2\pi\lambda}\right]$.

Sol. Let r be the radius of the string placed round the cone with vertex A. Let O′ be the centre of the circle of the band. Measuring distance from A after giving a small displacement, the equation of virtual work is

$$-Td(l) + Wd(AO') = 0 \qquad (l = 2\pi r)$$

$$-Td(2\pi r) + Wd(r\cot\alpha) = 0$$

or $$2\pi T dr = W dr \cot\alpha \qquad (\because dr \neq 0)$$

$$T = W\cot\alpha/2\pi \quad ...(1)$$

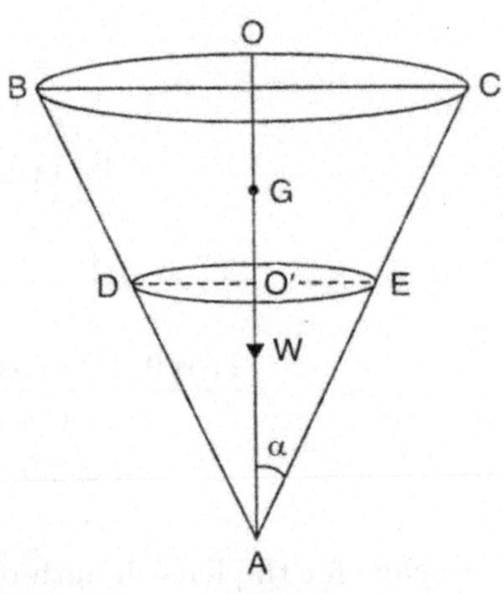

Fig. 19

But by Hooke's Law

$$T = \lambda\frac{(l-a)}{a} = \lambda\left(\frac{2\pi r - 2\pi a}{2\pi a}\right) \quad ...(2)$$

where natural length is $2\pi a$ and extended length is $2\pi r$.

Comparing (1) and (2)

$$\frac{W\cot\alpha}{2\pi} = \lambda\frac{(r-a)}{a}$$

$$r - a = \frac{aW\cot\alpha}{2\pi\lambda} \quad \text{or} \quad r = a\left[1+\frac{W\cot\alpha}{2\pi\lambda}\right].$$

Example 7. *One end of a uniform rod AB, of length 2a and weight W, is attached by a frictionless joint to a smooth vertical wall and the other end B is smoothly joined to an equal rod BC. The middle points of the rods are joined by an elastic string of natural length a and modulus of elasticity 4W. Prove that the system can rest in equilibrium in a vertical plane with C in contact with the wall below A and the angle between the rods is 2 $\sin^{-1}(3/4)$.*

Sol. Let T be the tension in the string DE, joining middle points of rods AB and BC. 2W is the weight of the rods acting at G.

Let $\angle ABL = \theta$

$\because$ A is fixed, therefore the distances are measured from A.

Let the system be given a small virtual displacement such that θ changes to $\theta + \delta\theta$.

$\therefore$ Equation of virtual work is

$$2Wd(AL) - Td(DE) = 0 \qquad ...(1)$$

$$AL = 2a \sin\theta \qquad \because AB = 2a$$

$$DE = \frac{1}{2} AC = AL = 2a \sin\theta \qquad ...(2)$$

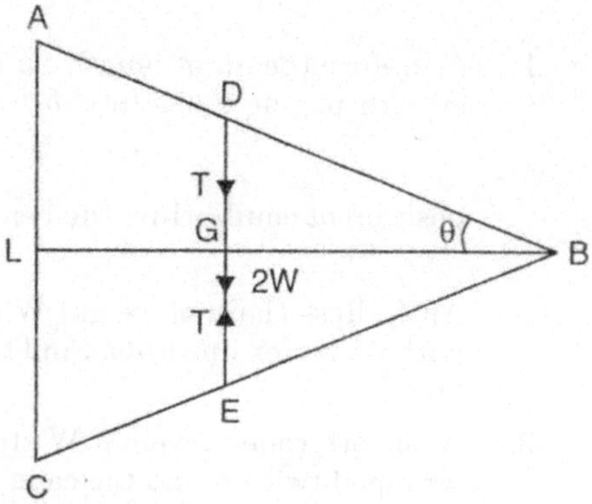

Fig. 20

Putting in (1), we get

$$\therefore \quad 2Wd(2a \sin\theta) - Td(2a \sin\theta) = 0$$

$$\therefore \quad T = 2W \qquad ...(3)$$

But by Hooke's Law

$$T = \lambda \frac{(DE - a)}{a}$$

$$T = 4W \frac{(2a \sin\theta - a)}{a} \qquad ...(4)$$

Comparing (3) and (4), we get

$$2W = 4W\,(2 \sin\theta - 1)$$

$$6W = 8W \sin\theta \quad \text{or} \quad \sin\theta = \frac{3}{4}$$

or

$$\theta = \sin^{-1}\left(\frac{3}{4}\right).$$

Example 8. *A heavy rod AB of length 2l rests upon a fixed smooth peg at C with its end B upon a smooth curve. If it rests in all positions, show that its polar equation is*

$$r = l + a/\sin\theta.$$

Sol. Rod AB rests on peg C and the end B on curve. Taking C as fixed point (pole). Let coordinate of B be (r, θ).

Weight of the rod W acts at middle point G of rod AB.

If rod is slightly displaced, then only W does the work.

The equation of virtual work

$$Wd(GL) = 0$$

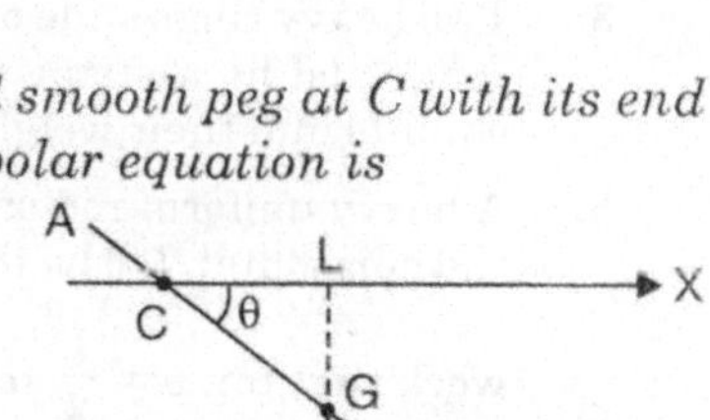

Fig. 21

or $$Wd(CG \sin \theta) = 0$$

or $$Wd\{(BC - BG) \sin \theta\} = 0$$

or $$d\{(r - l) \sin \theta\} = 0$$

On integration, we get $(r - l) \sin \theta = a$ (constant)

or $$r = l + a/\sin \theta.$$

EXERCISE 2

1. A uniform beam of length $2a$ rests in equilibrium against a smooth vertical wall and upon a smooth peg at a distance b from the wall. Show by the principle of virtual work that in the position of equilibrium the beam is inclined to the wall at an angle $\sin^{-1}\left(\frac{b}{a}\right)^{\frac{1}{3}}$.

2. An endless chain of weight W rests in the form of a circular band round a smooth vertical cone with its vertex upwards. Find the tension in the chain due to weight. Vertical angle of the cone be 2α.

3. A smooth cone of weight W stands inverted in a circular hole with its axis vertical. A string is wrapped twice round the cone just above the hole and pulled tight. What must be the tension in the string so that it will just raise the cone.

4. Two uniform rods AB and AC smoothly joined at A are in equilibrium in a vertical plane. B and C rest on a smooth horizontal plane and middle points of AB and AC are connected by a string. Show that the tension in the string is $W/(\tan B + \tan C)$ where W is the total weight of both the rods.

5. One end of a uniform rod rests against a smooth vertical wall and the other end on a smooth curve in a vertical plane perpendicular to the wall. If the beam rests in all position the equation of the curve is an ellipse whose major axis lies along the horizontal line described by C.G. of the beam.

6. Four equal rods each of weight w are joined to form a rhombus, the whole system hangs from a fixed point A and kept in shape by a light rod BD. A weight W hangs from C. Show that the thrust in the rod is $(2w + W) \tan \frac{A}{2}$.

7. Two small smooth rings of equal weight slide on a fixed elliptical wire whose major axis is vertical. They are connected by a string which passes over a smooth peg at the upper focus. Show that the weights will be in equilibrium wherever they are placed.

8. Two heavy rings slide on a smooth parabolic wire with axis horizontal and plane vertical and are connected by a string passing round a smooth peg at the focus. Prove that in the position of equilibrium their weights are proportional to the vertical depths below the axis.

9. A heavy uniform rod of length $2a$, rests on two smooth inclined planes of inclinations α and β to the horizontal. If θ be the inclination of the rod to the horizontal, prove by the principal of virtual work that $\tan \theta = \frac{1}{2} (\cot \alpha - \cot \beta)$. *(K.U. 2005)*

10. Four equal uniform bars, each of weight W are jointed to form a rhombus. If it is suspended from one of the angular point and a sphere of weight P is balanced inside the rhombus so as to save it

from collapsing. Show that if 2θ be the angle at the fixed joint in the position of equilibrium then

$$\frac{\sin^3\theta}{\cos\theta} = \frac{Pr}{4(P+2W)a}$$, where r is the radius of the sphere and a is the length of each bar.

11. The middle points of opposite sides of a quadrilateral are connected by light rods of length l and l'. If T and T′ be the tensions in these rods, prove that $\frac{T}{l} + \frac{T'}{l'} = 0$.

12. Four equal rods, each of length $2a$ and weight W, are freely joined to form a square ABCD, which is kept in shape by a light rod BD and is supported in a vertical plane with BD horizontal, A above C and AB and AD in contact with two pegs at distance $2b$ apart on the same level. Show that stress in BD is $2(a - b\sqrt{2})W/a$.

ANSWERS

2. $\frac{W\cot\alpha}{2\pi}$ 3. $\frac{W\cot\alpha}{4\pi}$.

3

*Catenary

The word catenary is derived from the latin word catena, meaning chain. In this chapter we study the strings, ropes or cables suspended from two fixed points not in the same vertical line. Ropes of suspension bridges, cables are the examples of suspended strings.

Flexible Strings. Strings which offer no resistance on bending at any point are called flexible strings. A chain with quite small and smooth links can be considered as flexible string.

The normal section of the chain is so small that it may be considered as a curve line.

Catenary. The curve in which a heavy uniform strings hangs freely under gravity between two points (points are not in the vertical line) is called a catenary. *(K.U. 2007)*

Common Catenary. When weight per unit length of the string is constant, the catenary is called the common catenary.

1. To discuss Equilibrium of a String in a Plane under the Action of a given force

Let A be the fixed point on the string $\overset{\frown}{AB}$.

Let m denotes mass/length of the string.

P and Q be two points $\overset{\frown}{PQ} = \delta s$.

Let X and Y be the resolved parts of forces per unit mass parallel to X and Y-axis.

$\therefore$ Component of forces on PQ along X and Y axes are $m\ \delta s$ X and $m\ \delta s$ Y respectively. Also tensions at P and Q be T and T + δT respectively.

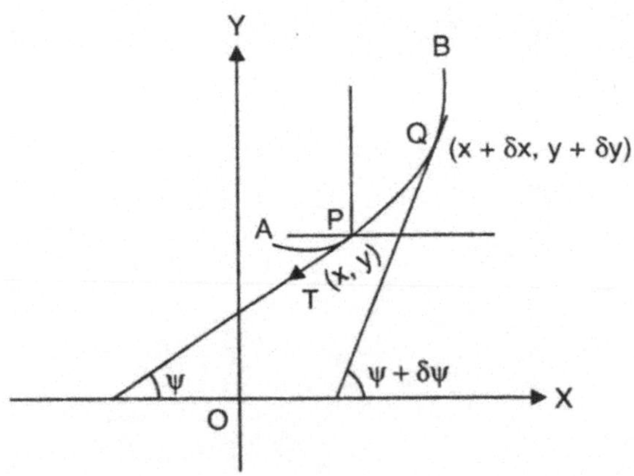

Fig. 1

Resolved parts of T at P are

$$T \cos \psi = T.\ \frac{dx}{ds} = f(s)$$

$$T \sin \psi = T\ .\ \frac{dy}{ds}$$

Resolved part of T + δT along X-axis

$$= f(s + \delta s)$$

$$= f(s) + \delta s\, f'(s) + \frac{\delta s^2}{2!}\, f''(s) + \ldots\ldots \qquad \text{(Taylor's Theorem)}$$

*Not included in M.D.U. 2007–2008 syllabus.

$$= T\frac{dx}{ds} + \delta s\frac{d}{ds}\,T\frac{dx}{ds} \qquad \text{(neglecting higher power of } \delta s\text{)}$$

For the equilibrium of PQ, resolving the forces along x-axis.

Component of T along X-axis = Component of external force along x-axis + Component of T + δT along x-axis.

$$\therefore \quad T\frac{dx}{ds} = m\,\delta s\,X + T\frac{dx}{ds} + \delta s\frac{d}{ds}\left(T\frac{dx}{ds}\right) \qquad \ldots(1)$$

or
$$m\,\delta s\,X + \delta s\frac{d}{ds}\left(T\frac{dx}{ds}\right) = 0$$

or
$$\frac{d}{ds}\left(T\frac{dx}{ds}\right) + m\,X = 0 \qquad \ldots(2)$$

Similarly,
$$\frac{d}{ds}\left(T\frac{dy}{ds}\right) + m\,Y = 0 \qquad \ldots(3)$$

From (2) and (3),

$$T\frac{d^2x}{ds^2} + \frac{dT}{ds}\cdot\frac{dx}{ds} + m\,X = 0 \qquad \ldots(4)$$

$$T\frac{d^2y}{ds^2} + \frac{dT}{ds}\frac{dy}{ds} + m\,Y = 0 \qquad \ldots(5)$$

Multiplying (4) by $\frac{dx}{ds}$ and (5) by $\frac{dy}{ds}$ and adding, we get

$$T\left[\frac{d^2x}{ds^2}\cdot\frac{dx}{ds} + \frac{d^2y}{ds^2}\cdot\frac{dy}{ds}\right] + \frac{dT}{ds}\left[\left(\frac{dx}{ds}\right)^2 + \left(\frac{dy}{ds}\right)^2\right] + m\left(X\frac{dx}{ds} + Y\frac{dy}{ds}\right) = 0 \qquad \ldots(6)$$

$$\cos^2\psi + \sin^2\psi = \left(\frac{dx}{ds}\right)^2 + \left(\frac{dy}{ds}\right)^2 = 1 \qquad \ldots(7)$$

$$\Rightarrow \quad 2\left[\frac{dx}{ds}\frac{d^2x}{ds^2} + \frac{dy}{ds}\frac{d^2y}{ds^2}\right] = 0 \qquad \ldots(8)$$

Putting (7) and (8) in (6), we get

We have,
$$\frac{dT}{ds} + m\left(X\frac{dx}{ds} + Y\frac{dy}{ds}\right) = 0$$

or
$$T = -\,m\int(X\,dx + Y\,dy) + C$$

Hence, knowing m, X and Y we can find T at every point.

2. Intrinsic Equation of Common Catenary

(K.U. 2005, 2007)

Let ACB be the uniform string suspended from A and B . C is lowest point of the string. Let weight per unit length of the string be w. Let P be any point on CA such that CP = s. The weight of string CP is sw acts vertically downward at C.G. of string CP. The portion CP of the string is in equilibrium under the action of following forces.

(*i*) Tension T_0 acting horizontally at C.

(*ii*) Tension T acting along the tangent at P.

(*iii*) Weight sw acting vertically downward at the C.G. of arc CP.

∵ Equilibrium is under three coplaner forces, therefore they are concurrent.

Resolving the forces horizontally and vertically

$$T\cos\psi = T_0 \qquad ...(1)$$

$$T\sin\psi = sw \qquad ...(2)$$

From (1) and (2), we get

$$\tan\psi = \frac{sw}{T_0}$$

Tension T_0 at the lowest point C is equal to weight of length c of the string *i.e.* cw

$$\therefore \quad T_0 = cw \qquad ...(3)$$

$$\therefore \quad \tan\psi = \frac{sw}{cw} = \frac{s}{c}$$

or $$\boxed{\mathbf{s = c\tan\psi}} \qquad ...(4)$$

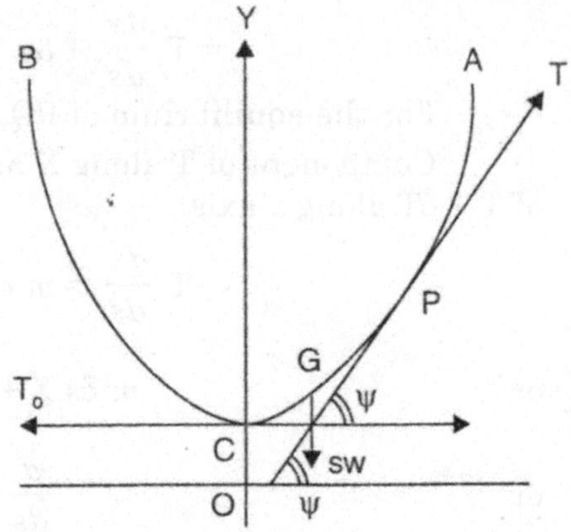

Fig. 2

Equation (4) is a relation between s and ψ. Hence it is the **intrinsic equation** of the common catenary where c is the parameter.

In the above discussion we have seen that

(*i*) Horizontal component of tension $T\cos\psi$ at any point is constant and is equal to cw

(*ii*) Vertical component of tension $T\sin\psi$ at any point is equal to weight of the length of the string from lowest point to the point under consideration *i.e.*, sw.

3. Equation of the Common Catenary in Cartesian Coordinates

A heavy uniform string hangs freely under gravity. Refer to Fig. 2. (K.U. 2004)

Condition of equilibrium

$$\frac{d}{ds}\left(T\frac{dx}{ds}\right) + mX = 0 \qquad ...(1)$$

$$\frac{d}{ds}\left(T\frac{dy}{ds}\right) + mY = 0 \qquad ...(2)$$

$\because$ String hangs under gravity therefore $X = 0, Y = -g$

$$\therefore \quad \frac{d}{ds}\left(T\frac{dx}{ds}\right) = 0 \Rightarrow T\frac{dx}{ds} = \text{constant} = \lambda \text{ (say)} \qquad ...(3)$$

$$\frac{d}{ds}\left(T\frac{dy}{ds}\right) = mg$$

or $$\frac{d}{ds}\left(T\frac{dx}{ds}\cdot\frac{dy}{dx}\right) = mg \qquad \text{Using (3)}$$

$\Rightarrow$ $$\frac{d}{ds}\left(\lambda\frac{dy}{dx}\right) = mg$$

or $$\frac{d^2y}{dx^2}\cdot\frac{dx}{ds} = \frac{mg}{\lambda}$$

or $$\frac{d^2y}{dx^2} = \frac{mg}{\lambda}\cdot\frac{ds}{dx}$$

or $$\frac{d^2y}{dx^2} = \frac{mg}{\lambda}\sqrt{1+\left(\frac{dy}{dx}\right)^2} \qquad \left(\because \frac{ds}{dx} = \sqrt{1+\left(\frac{dy}{dx}\right)^2}\right)$$

From calculus

or $$\frac{dz}{dx} = \frac{mg}{\lambda}\sqrt{1+z^2} \qquad \text{where } z = \frac{dy}{dx}$$

$$\frac{dz}{\sqrt{1+z^2}} = \frac{1}{c}\,dx \qquad \text{Taking } \frac{mg}{\lambda} = \frac{1}{c}$$

Integrating both sides

$$\sinh^{-1} z = \frac{x}{c} + k$$

At C $\quad x = 0, \quad z = \frac{dy}{dx} = 0 \;\Rightarrow\; k = 0$

$\therefore \quad z = \sinh\left(\frac{x}{c}\right) \;\Rightarrow\; \frac{dy}{dx} = \sinh x/c$

Again integrating $\quad y = c \cosh x/c + k'$

At $\quad x = 0, \quad y = c \qquad \therefore \; k' = 0$

$$\therefore \quad \boxed{\mathbf{y = c\cosh\frac{x}{c}}} \qquad \text{...(4)}$$

This is required equation of catenary in cartesian from.

Aliter

Cartesian Equation of Common Catenary

Intrinsic equation of common catenary is

$$s = c\tan\psi \qquad \text{...(1)}$$

$$\tan\psi = \frac{dy}{dx} \quad \text{or} \quad s = c\,\frac{dy}{dx}$$

Differentiating both sides w.r.t. x, we get

$$\frac{ds}{dx} = c\,\frac{d^2y}{dx^2}$$

$$\therefore \quad \sqrt{1+\left(\frac{dy}{dx}\right)^2} = c\,\frac{d^2y}{dx^2} \qquad \frac{ds}{dx} = \sqrt{1+\left(\frac{dy}{dx}\right)^2} \text{ From calculus}$$

Putting $\quad \frac{dy}{dx} = z, \; \frac{d^2y}{dx^2} = \frac{dz}{dx},$

we have, $\quad \sqrt{1+z^2} = c\,\frac{dz}{dx} \;\Rightarrow\; \frac{c\,dz}{\sqrt{1+z^2}} = dx$

Integrating both sides

$$x = c\sinh^{-1} z + \text{A} \qquad \text{...(2)}$$

Where A is the constant of integration. Choose the vertical line through C as Y-axis choose origin O at depth c below the lowest point of catenary. Draw X-axis through O and perpendicular to Y-axis.

$\therefore$ At C $x = 0, z = \dfrac{dy}{dx} = 0$ as the tangent at C is parallel to X-axis.

Putting in equation (2),

$$0 = 0 + A \quad \Rightarrow \quad A = 0$$

$$\therefore \quad x = c \sinh^{-1} z \quad \Rightarrow \quad z = \sinh (x/c)$$

$$\Rightarrow \quad \frac{dy}{dx} = \sinh x/c$$

Integrating both sides again

$$y = c \cosh x/c + A'$$

$\because$ Coordinate of C are $(0, c)$

$$\therefore \quad c = c \cosh + A' \quad \Rightarrow \quad A' = 0$$

$$\therefore \quad \boxed{\mathbf{y = c \cosh x/c}} \quad \text{...(3)}$$

is the required equation of common catenary.

4. Properties of the Common Catenary

Intrinsic equation of common catenary is

$$\mathbf{s = c \tan \psi} \quad \text{...(1)}$$

Cartesian equation of catenary is

$$\mathbf{y = c \cosh x/c} \quad \text{...(2)}$$

Relation between s and x

$$\Rightarrow \quad \frac{dy}{dx} = \sinh x/c$$

But

$$\frac{dy}{dx} = \tan \psi$$

$$\Rightarrow \quad \tan \psi = \sinh x/c.$$

$$\therefore \quad \boxed{\mathbf{s = c \sinh x/c}} \quad \text{...(3)}$$

Relation between s and y

From (2) and (3), $y^2 - s^2 = c^2 \cosh^2 x/c - c^2 \sinh^2 x/c$

$$\therefore \quad y^2 - s^2 = c^2$$

or

$$\boxed{\mathbf{y^2 = s^2 + c^2}} \quad \text{...(4)}$$

Relation between y and ψ

Putting s from equation (1)

$$y^2 = c^2 \tan^2 \psi + c^2 = c^2(1 + \tan^2 \psi) = c^2 \sec^2 \psi$$

$$\therefore \quad \boxed{\mathbf{y = c \sec \psi}} \quad \text{...(5)}$$

Relation between x and ψ

$$s = c \tan \psi$$

Differentiating w.r.t. s

$$1 = c \sec^2 \psi \frac{d\psi}{ds} = c \sec^2 \psi \frac{d\psi}{dx} \cdot \frac{dx}{ds} \qquad \left(\frac{dx}{ds} = \cos \psi\right)$$

$$= c \sec^2 \psi \cos \psi \frac{d\psi}{dx} \qquad \left(\frac{dy}{ds} = \sin \psi\right)$$

$\therefore$ $$dx = c \sec \psi \, d\psi \qquad \text{From calculus}$$

$$\int dx = c \int \sec \psi \, d\psi + A$$

$$x = c \log (\sec \psi + \tan \psi) + A \qquad \text{(At } c \text{ tangent is parallel to X-axis)}$$

At c $\quad x = 0, \psi = 0$

$0 = 0 + A \quad \therefore \quad A = 0$

$\therefore$ $$\boxed{x = c \log (\sec \psi + \tan \psi)} \qquad ...(6)$$

Further $$x = c \log \left(\frac{y}{c} + \frac{s}{c}\right)$$

$\therefore$ $$\frac{x}{c} = \log \left(\frac{y+s}{c}\right)$$

$\Rightarrow$ $$\boxed{y + s = ce^{x/c}} \qquad ...(7)$$

Relation between T and y

$$T \cos \psi = T_0 \qquad \text{where } T_0 = c\omega$$

$\therefore$ $$T = c\omega \sec \psi = c\omega \frac{y}{c}$$

$\therefore$ $$\boxed{T = \omega y} \qquad ...(8)$$

Definitions

Vertex. Lowest point C of the catenary is called the **vertex.**

Axis. The vertical line through the lowest point C of the catenary is called the **axis.**

Span. If points of suspension A and B are in horizontal line then AB is called the **span.**

Sag. The distance between span and vertex is **sag.**

Directrix. The line perpendicular to axis of catenary at the depth c below the vertex is **Directrix.** Where c is the parameter.

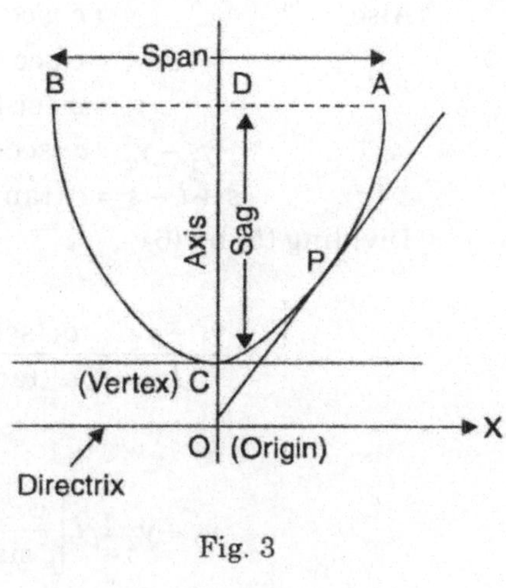

Fig. 3

Example 1. *A cable 20 m long weighs 25 N/m. It hangs between two points A and B at the same level. If the sag of the cable is 5 m find the span. Also find the maximum tension in the cable.*

Sol.

$$AB = 2x$$
$$OC = c$$
$$\overparen{ACB} = 20 \text{ m}$$
$$\therefore \quad \overparen{AC} = 10 \text{ m}$$
$$OD = y = 5 + c$$
$$y^2 = c^2 + s^2$$
$$(c+5)^2 = c^2 + 100 \quad \Rightarrow \quad c = 7.5 \text{ m}$$

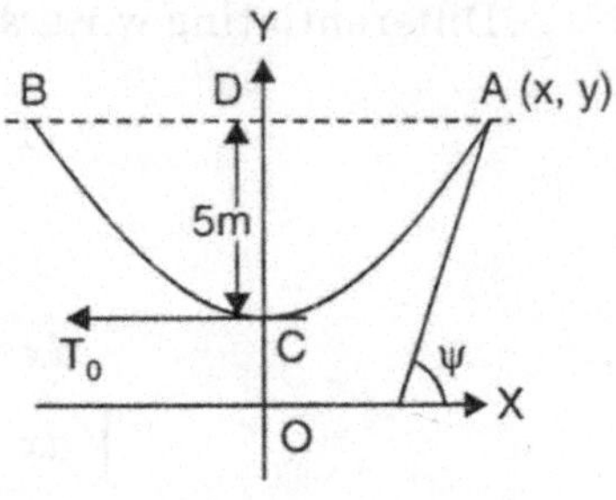

Fig. 4

Horizontal Tension $T_0 = \omega c = 25 \times 7.5$ N.

Vertical Tension at $A = \omega s = 25 \times 10 = 250$ N.

$$\therefore \quad \tan \psi = \frac{s}{c} = \frac{10}{7.5} = \frac{4}{3}, \ \sec \psi = \frac{5}{3}$$

The distance between the supports AB = Span

$$2x = 2c \log(\sec \psi + \tan \psi) = 2 \times 7.5 \log\left(\frac{5}{3} + \frac{4}{3}\right) = 15 \log 3$$

Also tension at A, $T = \omega y = 25\,(5 + 7.5) = 312.5$ N.

Example 2. *If α and β be the inclinations to the horizon of the tangents at the extremities of an arc of common catenary and l be its length show that the height of one end of the arc above the other is* $l \sin \dfrac{(\alpha+\beta)}{2} \Big/ \cos \dfrac{(\alpha-\beta)}{2}$. *The arc is on one side of the vertex.*

Sol. Let $\overparen{AP} = l, \ CP = s$

$$\therefore \quad \overparen{AC} = s + l$$
$$\because \quad s = c \tan \psi$$
$$\therefore \quad s = c \tan \beta \quad \ldots(1)$$
$$s + l = c \tan \alpha \quad \ldots(2)$$

Also,

$$y = c \sec \psi$$
$$y_1 = c \sec \alpha \quad \ldots(3)$$
$$y_2 = c \sec \beta \quad \ldots(4)$$
$$\therefore \quad y_1 - y_2 = c\,(\sec \alpha - \sec \beta) \quad \ldots(5)$$
$$s + l - s = c\,(\tan \alpha - \tan \beta) \quad \ldots(6)$$

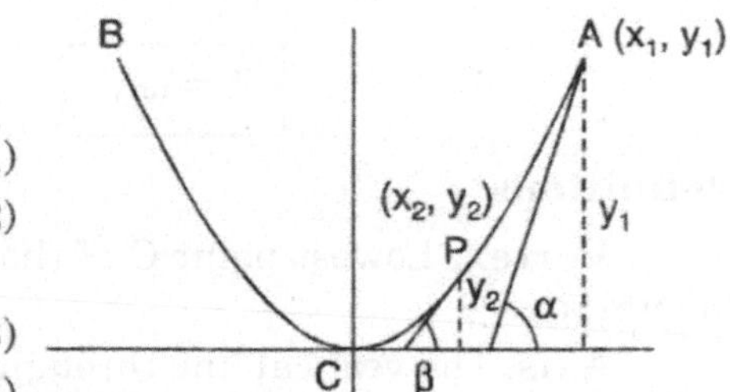

Fig. 5

Dividing (5) by (6)

$$\frac{y_1 - y_2}{l} = \frac{c}{c}\frac{(\sec \alpha - \sec \beta)}{(\tan \alpha - \tan \beta)} = \frac{\dfrac{1}{\cos \alpha} - \dfrac{1}{\cos \beta}}{\dfrac{\sin \alpha}{\cos \alpha} - \dfrac{\sin \beta}{\cos \beta}}$$

$$\therefore \quad y_1 - y_2 = l\left[\frac{\cos \beta - \cos \alpha}{\sin(\alpha - \beta)}\right] = \frac{l\,.\,2 \sin \dfrac{\alpha+\beta}{2}\,.\,\sin \dfrac{\alpha-\beta}{2}}{2 \sin \dfrac{\alpha-\beta}{2} \cos \dfrac{\alpha-\beta}{2}}$$

$$\therefore \qquad y_1 - y_2 = \frac{l \sin \frac{(\alpha+\beta)}{2}}{\cos \frac{(\alpha-\beta)}{2}}.$$

Example 3. *The extremities of a heavy uniform string of length 2l and weight 2lω are attached to two small rings which can slide on a fixed horizontal wire. Each ring is acted on by a horizontal force lω. Show that the distance between the rings is 2l log $(1+\sqrt{2})$.*

Proof. Let weight/length = ω

A and B are the rings where the string is attached.

$$\overset{\frown}{AC} = l.$$

Considering the whole system and resolving vertically

$$R + R = 2l\omega \Rightarrow R = l\omega.$$

$\therefore$ Tension at A is the resultant of R and $l\omega$. If ψ is the angle which the tangent makes with X-axis then

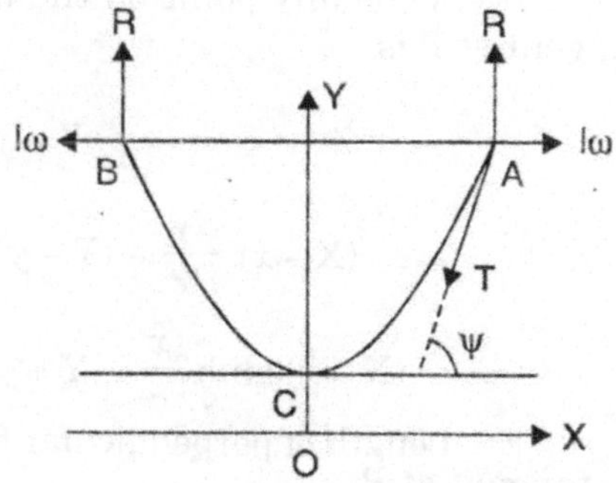

Fig. 6

$$T \cos \psi = l\omega$$
$$T \sin \psi = R$$

$$\Rightarrow \qquad \tan \psi = \frac{R}{l\omega} = \frac{l\omega}{l\omega} = 1$$

$$\therefore \qquad \psi = \frac{\pi}{4}$$

Also, $\qquad s = c \tan \psi$

At A, $\qquad s = l \qquad \therefore \quad l = c \tan \frac{\pi}{4} \quad \Rightarrow \quad c = l$

$\therefore$ Distance between the rings AB = 2x

$$2x = 2c \log (\tan \psi + \sec \psi) = 2l \log \left(\tan \frac{\pi}{4} + \sec \frac{\psi}{4} \right) = 2l \log (1+\sqrt{2}).$$

Example 4. *If tension at point A of a catenary is n times at the vertex then the span of the catenary ACB is* $\frac{2l}{\sqrt{n^2-1}} \log (n + \sqrt{n^2-1})$ *where 2l is the length of the catenary.* (K.U. 2004)

Sol. Let length of the catenary ACB = 2l.

Tension at A = T

and Tension at C = T_0

$$T = n T_0$$
$$T = n T \cos \psi$$

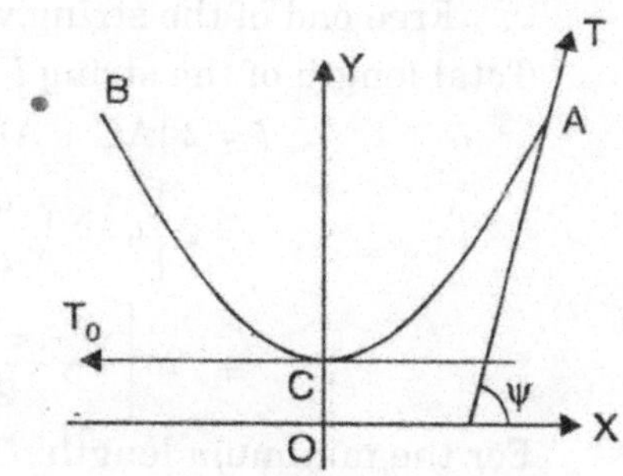

Fig. 7(i)

$$\therefore \qquad \cos \psi = \frac{1}{n}$$

$$\Rightarrow \qquad \tan \psi = \frac{\sqrt{n^2-1}}{1}$$

$$\sec \psi = \frac{n}{1}$$

Also at A, $s = c \tan \psi$

$$l = c\sqrt{n^2 - 1}$$

$$\Rightarrow \qquad c = \frac{l}{\sqrt{n^2 - 1}}$$

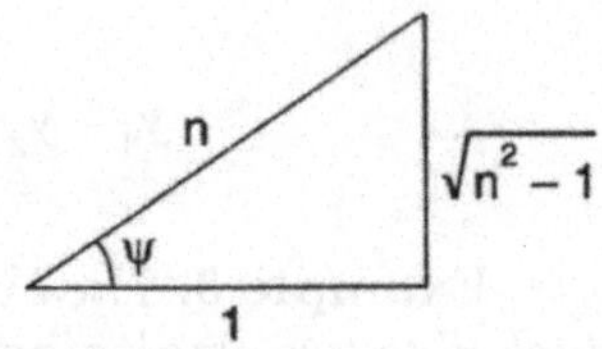

Fig. 7(*ii*)

$\therefore$ Span $\quad 2x = 2c \log(\tan\psi + \sec\psi) = \dfrac{2l}{\sqrt{n^2-1}} \log\left(\sqrt{n^2-1} + n\right).$

Example 5. *Prove that the catenary y = c cosh x / c the length of perpendicular from foot of ordinate is of constant length.*

Proof. Let ACB be the catenary $y = c \cosh x/c$ ...(1)

$P(x, y)$ be any point on the catenary equation of the tangent at P is

$$Y - y = \frac{dy}{dx}(X - x) \qquad ...(2)$$

$$(X - x)\frac{dy}{dx} - (Y - y) = 0$$

or $$(X - x)\sinh\frac{x}{c} - Y + y = 0$$

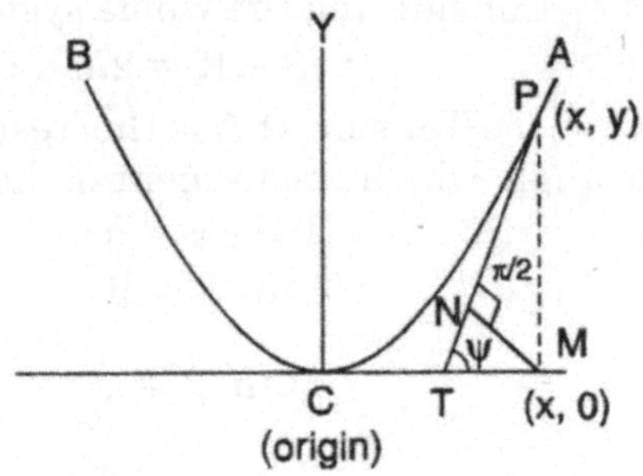

Fig. 8

$\therefore$ Length of perpendicular from foot of ordinate $(x, 0)$ on tangent at P

$$= \frac{(x - x)\sinh\frac{x}{c} - 0 + y}{\sqrt{\sinh^2 x/c + 1}} = \frac{y}{\cosh x/c} = \frac{c\cosh x/c}{\cosh x/c} = c.$$

Example 6. *A heavy uniform string hangs over two small smooth pegs at same level and distance 2a apart. Ends are hanging freely and central portion hangs in a catenary, show that equilibrium is possible only if the length of string is at least 2ae.*

Proof. Let A and B be two smooth pegs $AB = 2a$. Let ω be weight/ length coordinate of A be (a, y). Tension at A due to catenary is $T = \omega y$.

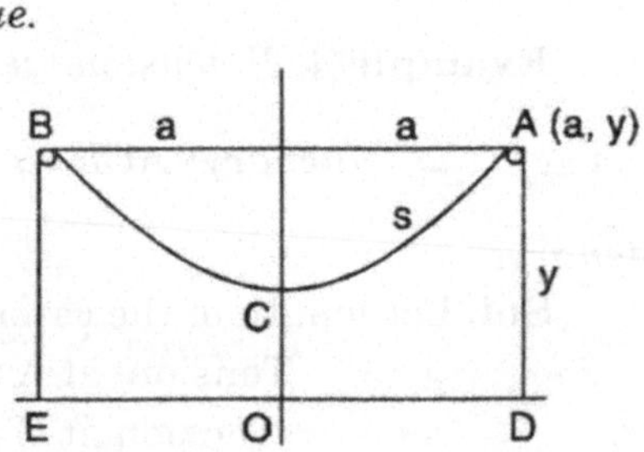

Fig. 9

If length l hanging then $T = \omega l$.

$$\therefore \qquad \omega l = \omega y \;\Rightarrow\; l = y$$

$\therefore$ Free end of the string will be on the directrix.

Total length of the string l

$$l = 2[AC + AD] = 2[s + y]$$

$$= 2\left[c\sinh\frac{a}{c} + c\cosh\frac{a}{c}\right]$$

$$= 2c\left[\frac{e^{a/c} - e^{-a/c}}{2} + \frac{e^{a/c} + e^{-a/c}}{2}\right] = 2ce^{a/c}$$

For the minimum length

$$\frac{dl}{dc} = 2\left[e^{a/c} + c\,.\,e^{a/c}\left(-\frac{a}{c^2}\right)\right] = 0 \quad \Rightarrow \quad 1 - \frac{a}{c} = 0 \;\Rightarrow\; c = a$$

$$\frac{d^2l}{dc^2} = 2\left[e^{a/c}\left(-\frac{a}{c^2}\right) - \left\{e^{a/c}.\left(-\frac{a}{c^2}\right).\frac{a}{c} + e^{a/c}\left(-\frac{a}{c^2}\right)\right\}\right] = 2\,\frac{a^2}{c^2}\,e^{a/c} > 0$$

$\therefore$ Length is minimum when $c = a$

and minimum length is $2a \,.\, e^{a/a} = 2ae$.

Example 7. *The end links of a uniform chain slide along a fixed rough horizontal rod. Prove that the ratio of the maximum span to the length of the chain is* $\mu \log\left[\dfrac{1+\sqrt{1+\mu^2}}{\mu}\right]$, *where μ is the coefficient of friction.*

Sol. Let the chain ACB is of length $2l$ hanging in the form of a catenery with C as the vertex. The end links of the chain slide on the rod AB.

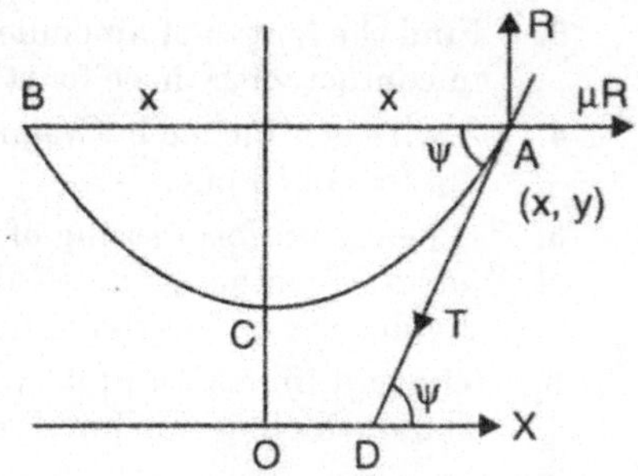

Fig. 10

Forces acting at A are :

(*i*) Tension T acting along the tangent AD.

(*ii*) Reaction R acting perpendicular to AB.

(*iii*) Friction force μR acting along BA.

Resolving the forces horizontally and vertically

$$T\cos\psi = \mu R$$

$$T\sin\psi = R$$

Eliminating T, we get

$$\tan\psi = \frac{1}{\mu} \qquad \therefore \quad \sec\psi = \sqrt{1+\tan^2\psi} = \sqrt{1+\frac{1}{\mu^2}} = \frac{\sqrt{1+\mu^2}}{\mu}$$

Now, $\qquad s = c\tan\psi$

$$s = \widehat{AC} = l$$

$\therefore \qquad l = c\,.\,\dfrac{1}{\mu} \quad \Rightarrow \quad c = l\mu$

Now span $\qquad AB = 2x = 2c\log(\sec\psi + \tan\psi)$

$$= 2l\mu\log\left(\frac{\sqrt{1+\mu^2}}{\mu} + \frac{1}{\mu}\right) = 2\mu l\log\left[\frac{1+\sqrt{1+\mu^2}}{\mu}\right]$$

$$\therefore \quad \frac{\text{Maximum span}}{\text{Length of chain}} = \frac{AB}{2l} = \frac{2\mu l\log\left[\dfrac{1+\sqrt{1+\mu^2}}{\mu}\right]}{2l} = \mu\log\left[\frac{1+\sqrt{1+\mu^2}}{\mu}\right].$$

EXERCISE 1

1. A chain of length $2l$ is suspended between two points in the same horizontal line. If the sag of the catenary formed is b show that horizontal component of tension is $\frac{\omega}{2b}(l^2 - b^2)$ where ω is the weight per unit length of the chain.
2. The ends of the string of length l are attached with two rings which can slide on a smooth horizontal rod. If each ring is acted upon by a horizontal force $\frac{1}{2}\omega l$, show that span of the catenary formed is $l \log(\sqrt{2}+1)$, where ω is weight per unit length of the string.
3. Find the length of an endless chain which will hang over a circular pulley of radius a so as to be in contact with three fourth of the circumference of pulley.
4. A wire is attached between two pegs 50 m. apart Find the necessary length of wire if the central dip (sag) is 5 m.
5. A heavy uniform string of length 90 cm. hangs over two smooth pegs at different heights. The parts which hangs vertically are of length 30 cm and 33 cm. Prove that vertex of the catenary divides the whole string in the ratio 4 : 5. Also find horizontal distance between the pegs.
6. The end links of a uniform chain of length l can slide on two smooth rods in the same vertical plane which are inclined in opposite direction at equal angle ϕ to the vertical. Prove that sag in the middle is $\frac{l}{2}\tan\frac{\phi}{2}$.
7. If normal at any point P of the catenary $y = c\cosh x/c$ meets the directrix at Q, show that $PQ = \rho = c\sec^2\psi$. *(K.U. 2005)*
8. A uniform string hangs freely under gravity. Prove that mass per unit length is proportional to $\sec^2\psi/\rho$, where ρ is the radius of curvature.
9. A uniform chain of length l is suspended from two points in the same horizontal line. If the tension at the highest point is twice that at the lowest point, show that the span is
$$\frac{l}{\sqrt{3}}\log(2+\sqrt{3}).$$
10. If T be the tension at any point P of a catenary and T_0 that at the lowest point C, show that $T^2 - T_0^2 = W^2$, where W is the weight of arc CP of the catenary.
11. A uniform chain of length l rests in a straight line on a rough horizontal table. One end of the chain is raised to height h above the table and the chain is on the point of motion. Show that the length of the straight part on the table is $l + \mu h - [(\mu^2+1)h^2 + 2\mu hl]^{1/2}$, where μ is coefficient of friction.
12. If a chain of length $2l$ is suspended from two points A and B on the same level and sag in the middle is l/n, show that the horizontal span AB is equal to
$$l\left(n - \frac{1}{n}\right)\log\left(\frac{n+1}{n-1}\right).$$

ANSWERS

3. $a\left[\frac{3\pi}{2} + \frac{\sqrt{2}}{\log(1+\sqrt{2})}\right]$.

4. 51.33 m.

CATENARY OF UNIFORM STRENGTH

Definition. The form of the curve in which a chain hangs when its weight at each point is proportional to the tension at that point is called as the catenary of uniform strength.

Def. A chain of uniform material suspended from two fixed points, and such that the area of its cross-section is proportional to the tension is the catenary of uniform strength.

The two definitions are equivalent as the weight of an element is proportional to the cross section. If $\omega\, \delta s$ denotes the weight of an element δs at a place where the tension is T, then $T = \omega c$ where c is a constant.

5. Equation of the Catenary of Uniform Strength

If T_0 and T denote the tensions at the lowest point C and at a point P, where $CP = s$ and tangent of P makes an angle ψ with x-axis the resolving T horizontally and vertically

$$T \cos \psi = T_0 \quad \text{...(1)}$$

$$T \sin \psi = \int_0^s \omega\, ds \quad \text{...(2)}$$

$$\therefore \quad \tan \psi = \frac{1}{T_0} \int_0^s \omega\, ds \qquad \because \; T = \omega c$$

$$= \frac{1}{T_0} \int_0^s \frac{T}{c}\, ds \qquad \text{using (1)}$$

$$= \frac{1}{T_0} \frac{1}{c} \int_0^s \frac{T_0}{\cos \psi}\, ds$$

$$\therefore \quad \tan \psi = \frac{1}{c} \int_0^s \sec \psi\, ds \quad \text{...(3)}$$

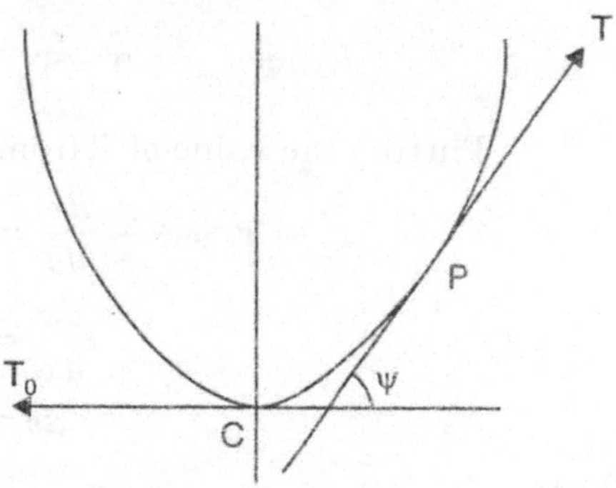

Fig. 11

Differentiating (3) both sides

$$\sec^2 \psi \frac{d\psi}{ds} = \frac{1}{c} \sec \psi$$

or

$$\frac{d\psi}{ds} = \frac{\cos \psi}{c} = \frac{1}{c} \frac{dx}{ds}$$

or

$$dx = c\, d\psi \quad \Rightarrow \quad x = c\psi + k$$

At $x = 0, \psi = 0$ $\quad\because$ Tangent is parallel to x-axis

$\therefore$ $x = 0$

$$\therefore \quad \boxed{\mathbf{x = c\psi}} \quad \text{...(4)}$$

$$\therefore \quad \frac{dy}{dx} = \tan \psi = \tan \frac{x}{c} \qquad \text{using (4)}$$

Integrating both sides, we have

$$\boxed{\mathbf{y = c \log \left(\sec \frac{x}{c} \right)}}$$

$$y \to \infty \quad \Rightarrow \quad x \to \pm \frac{\pi c}{2}$$

$\therefore$ $x = \pm \dfrac{\pi c}{2}$ are two asymptotes parallel to y-axis.

$\therefore$ Span of the catenary can not exceed πc.

Theorem. *In case of catenary of uniform strength the mass per unit length is proportional to cosh* $\left(\frac{s}{c}\right)$.

Proof. We know that in catenary of uniform strength

$$T \cos \psi = T_0 \quad \text{...(1)}$$

$$y = c \log \left(\sec \frac{x}{c}\right) \quad \text{...(2)}$$

$$x = c\psi \quad \text{...(3)}$$

From (1), $T \frac{dx}{ds} = T_0$ ...(4)

$$T = T_0 \frac{ds}{dx} = T_0 \sec \psi \qquad \because \frac{dx}{ds} = \cos \psi$$

$$T = T_0 \sec \left(\frac{x}{c}\right) \quad \text{using (3)} \quad \text{...(5)}$$

Putting the value of T from (5) into (4)

$$T_0 \sec \frac{x}{c} \frac{dx}{ds} = T_0$$

$$\frac{ds}{dx} = \sec \left(\frac{x}{c}\right)$$

$$ds = \sec x/c \, dx$$

Integrating, $s = c \log \left(\sec \frac{x}{c} + \tan \frac{x}{c}\right) + k$

If arc length is measured from lowest point C,

we have, $x = 0 \; s = 0$ at C $\Rightarrow$ $k = 0$

$$\therefore \quad \boxed{\mathbf{s = c \log \left(\sec \frac{x}{c} + \tan \frac{x}{c}\right)}} \quad \text{...(6)}$$

or $\sec \frac{x}{c} + \tan \frac{x}{c} = e^{s/c}$ ...(7)

$$\because \sec^2 \theta = \tan^2 \theta = 1$$

$$\sec \frac{x}{c} - \tan \frac{x}{c} = e^{-s/c} \quad \text{...(8)}$$

Adding (7) and (8)

$$2 \sec \frac{x}{c} = e^{s/c} + e^{-s/c}$$

or $\sec \frac{x}{c} = \frac{e^{s/c} + e^{-s/c}}{2}$ or $\sec x/c = \cosh s/c$ ...(9)

From (1), $T \cos \psi = T_0$

$$T = T_0 \sec \psi = T_0 \sec s/c$$

$$T = T_0 \cosh s/c$$

or $T \propto \cosh s/c \Rightarrow m \propto \cosh s/c$

$\because$ **Mass is proportional to tension in a catenary of uniform strength.**

6. Approximation of the Common Catenary

The equation of the catenary is

$$y = c \cosh x/c \qquad \because \quad \cosh x = 1 + \frac{x^2}{2!} + \frac{x^4}{4!} + \ldots\ldots.$$

$$= c\left[1 + \frac{x^2}{2!\,c^2} + \frac{x^4}{4!\,c^4} + \ldots\ldots.\right]$$

If $\frac{x}{c}$ is small, we can neglect higher powers

$$\therefore \qquad y \simeq c\left[1 + \frac{x^2}{2\,c^2}\right]$$

$$\therefore \qquad y \simeq c + \frac{x^2}{2c}$$

This shows that the curve coincides with a parabola of latus rectum $2c = \frac{2\,T_o}{\omega}$

$\because \; T_o = \omega c$

If x is very large $e^{-x/c}$ approaches to zero

In that case $y = c\left[\frac{e^{x/c} + e^{-x/c}}{2}\right]$ becomes

$$y = \frac{c}{2}\, e^{x/c}$$

Thus for large values of x curve behaves as an **exponential curve.**

7. Sag of Tightly Stretched Wires

Now we consider tightly stretched strings which appear nearly to be a straight line such as telegraph wires, electric wires.

Let the wire ABC be attached between two points A and B. Let l be the length of the wire and AB = a, CD = k.

Let T_0 be the tension at C. Now Arc AB is in equilibrium under the following forces.

(*i*) Tension T at A along the tangent at A.

(*ii*) Weight $\frac{W}{2}$ at middle point G of AC.

AG $\simeq$ $l/4$.

(*iii*) Tension T_0 at lowest point C horizontally.

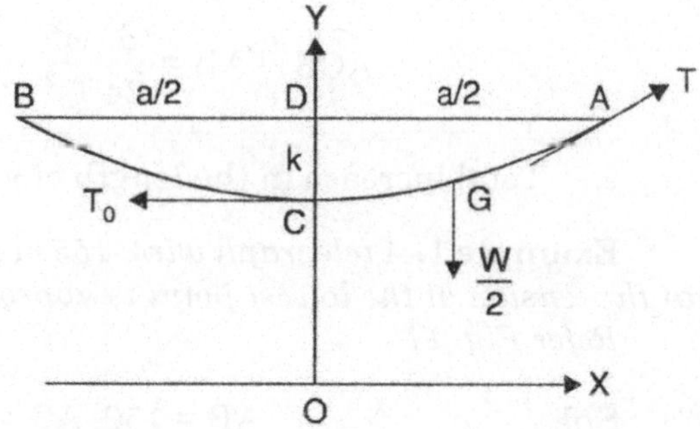

Fig. 12

Taking moment of the forces about A

$$\frac{W}{2} \cdot \frac{l}{4} - T_0 \,.\, k = 0$$

$$\therefore \qquad T_0 = \frac{Wl}{8k} \qquad \ldots(1)$$

The intrinsic equation of catenary is

$$s = c \tan \psi$$

$$\frac{ds}{d\psi} = c \sec^2 \psi$$

If radius of curvature is ρ then $\rho = \frac{ds}{d\psi}$

$$\Rightarrow \qquad \rho = c \sec^2 \psi \qquad ...(2)$$

or

$$c = \rho \cos^2 \psi$$

c will be very large if ρ is large.

∴ Curve is flat near the vertex

Also, $$s = c \sinh x/c = c\left[\frac{x}{c} + \frac{\left(\frac{x}{c}\right)^3}{3!} + \frac{\left(\frac{x}{c}\right)^5}{5!} + \ldots\ldots\right]$$

Neglecting higher power of $\frac{x}{c}$

$$s = x + \frac{x^3}{6c^2} \qquad ...(3)$$

$$s - x = \frac{x^3}{6c^2} = \frac{x^3\,\omega^2}{6\,.\,T_0^{\,2}} \qquad (\because \; T_0 = c\omega) \qquad ...(4)$$

Putting $x = \frac{a}{2}$ in (3)

$$s - \frac{a}{2} = \frac{a^3\,\omega^2}{48\,T_0^{\,2}}$$

$$2s - a = \frac{a^3\,\omega^2}{24\,T_0^{\,2}}$$

$$\overset{\frown}{ACB} - AB = \frac{a^3\,\omega^2}{24\,T_0^{\,2}}$$

∴ Total increase in the length of wire due to sagging is $\frac{a^3\,\omega^2}{24\,T_0^{\,2}}$.

Example 1. *A telegraph wire, 155 m length is suspended between two poles 150 m. Show that the tension at the lowest point is approx 1.08 times the weight of the string.*

Refer Fig. 11

Sol. $AB = 150,\; AC = \frac{155}{2}$ m.

$$s = c \sinh \frac{x}{c} = c\left[\frac{x}{c} + \frac{x^3}{3!\,c^3} + \ldots\ldots\right]$$

$$\therefore \qquad s = x + \frac{x^3}{6c^2} \text{ approximately}$$

$$\frac{155}{2} = 75 + \frac{(75)^3}{6c^2}$$

$$\frac{5}{2} = \frac{(75)^3}{6c^2} \quad \Rightarrow \quad c = 75\sqrt{5} \text{ m.}$$

Let ω be weight per meter of wire, weight of wire = W = 155 ω

Tension at tangent point $T_0 = c\omega = 75\sqrt{5}\ \dfrac{W}{155}$.

Example 2. *A telegraph wire stretched between two poles a meter apart sags n metre in the middle. Prove that tension at the ends is* $\omega\left[\dfrac{a^2}{8n} + \dfrac{7n}{6}\right]$ *where* ω *is weight per metre.*

(M.D.U. 2005)

Sol. We know that sag $n = y - c$

$$\therefore \quad n = c\left[\cosh\frac{x}{c} - 1\right] = c\left[1 + \frac{x^2}{2!c^2} + \frac{x^4}{4!c^4} + \ldots\ldots - 1\right] = \frac{x^2}{2c} + \frac{x^4}{24c^3}$$

But $\quad x = \dfrac{a}{2}$ at the ends

$$\therefore \quad n = \frac{a^2}{8c} + \frac{a^4}{24 \times 16c^3} \qquad \ldots(1)$$

$\because$ c is very large hence taking first order approx.

$$n = \frac{a^2}{8c} \quad \text{or} \quad c = \frac{a^2}{8n} \qquad \ldots(2)$$

Putting c in second term of (1)

$$n = \frac{a^2}{8c} + \frac{a^4}{24 \times 16 \times (a^2/8n)^3} = \frac{a^2}{8c} + \frac{(8n)^3}{24 \times 16 \times a^2}$$

or
$$\frac{a^2}{8c} = n - \frac{4n^3}{3a^2} = n\left(1 - \frac{4n^2}{3a^2}\right)$$

or
$$c = \frac{a^2}{8n}\left[1 - \frac{4n^2}{3a^2}\right]^{-1}$$

$$c = \frac{a^2}{8n}\left[1 + \frac{4n^2}{3a^2}\right] \qquad \ldots(3)$$

$\therefore$ Tension at A = $\omega\ y_1$, where y_1 is the ordinate of A = $\omega(n + c)$

$$= \omega\left[n + \frac{a^2}{8n}\left\{1 + \frac{4n^2}{3a^2}\right\}\right] = \omega\left[n + \frac{a^2}{8n} + \frac{n}{6}\right] = \omega\left[\frac{a^2}{8n} + \frac{7n}{6}\right].$$

Example 3. *A uniform string of weight W is suspended from two points at the same level and a weight W′ is attached to the lowest point. If* α *and* β *are now the inclinations to the horizontal of the tangents at the highest and lowest points, prove that*

$$\frac{\tan\alpha}{\tan\beta} = \frac{W + W'}{W'}.$$

Sol. Let the string of length l and weight W hangs freely in the form of catenary ADB. When weight W′ is attached D descends to P and $\widehat{AP}$ and $\widehat{BP}$ are two equal parts of another catenary.

Let vertex of the catenary $\widehat{AP}$ be at C.

Let T and T′ be the tensions at A and P.

Considering the equilibrium of the whole system and resolving the forces vertically

$$2T \sin \alpha = W + W' \qquad ...(1)$$

For the equilibrium of P

$$2T' \sin \beta = W' \qquad ...(2)$$

Resolving horizontally

$$T \cos \alpha = T' \cos \beta = \omega c \qquad ...(3)$$

where ω is weight/length.

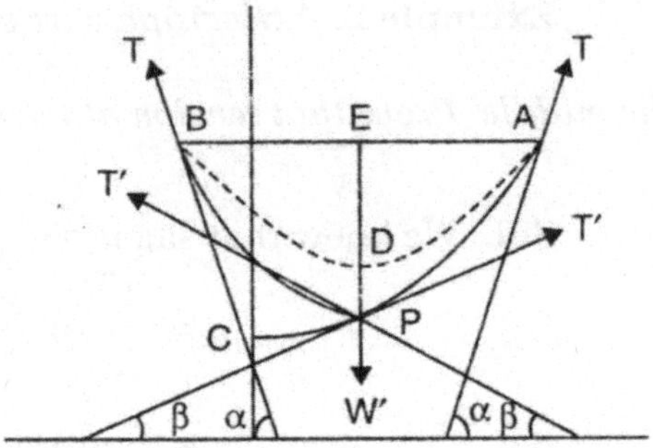

Fig. 13

From (1) and (3) $\quad 2 \tan \alpha = \dfrac{W + W'}{\omega c} \qquad ...(4)$

From (2) and (3) $\quad 2 \tan \beta = \dfrac{W'}{\omega c} \qquad ...(5)$

Dividing (4) by (5) $\quad \dfrac{\tan \alpha}{\tan \beta} = \dfrac{W + W'}{W'}$.

Example 4. *The distance between the points of suspension of uniform strength is a and length is l. Show that the parameter c of the catenary is given by tanh (l/2c) = tan (a/4c).*

Sol. $\quad AB = a$ and $\widehat{ACB} = l \Rightarrow AD = a/2,\ \widehat{AC} = l/2$

$$\therefore \quad \frac{l}{2} = c \log (\sec \psi + \tan \psi) \qquad \because \ s = c \log (\sec \psi + \tan \psi)$$

$$\therefore \quad \frac{l}{2c} = \log \left[\frac{1 + \sin \psi}{\cos \psi}\right] = \log \left[\frac{1 - \cos (\pi/2 + \psi)}{\sin (\pi/2 + \psi)}\right]$$

$$= \log \left[\frac{2 \sin^2 (\pi/4 + \psi/2)}{2 \sin (\pi/4 + \psi/2) \cos (\pi/4 + \psi/2)}\right]$$

$$\frac{l}{2c} = \log \tan \left(\frac{\pi}{4} + \frac{\psi}{2}\right) \qquad \text{or} \qquad e^{l/2c} = \tan \left(\frac{\pi}{4} + \frac{\psi}{2}\right)$$

$$\frac{e^{l/4c}}{e^{-l/4c}} = \frac{1 + \tan \dfrac{\psi}{2}}{1 - \tan \dfrac{\psi}{2}} \qquad \because \ x = c\psi,\ \frac{a}{2} = c\psi$$

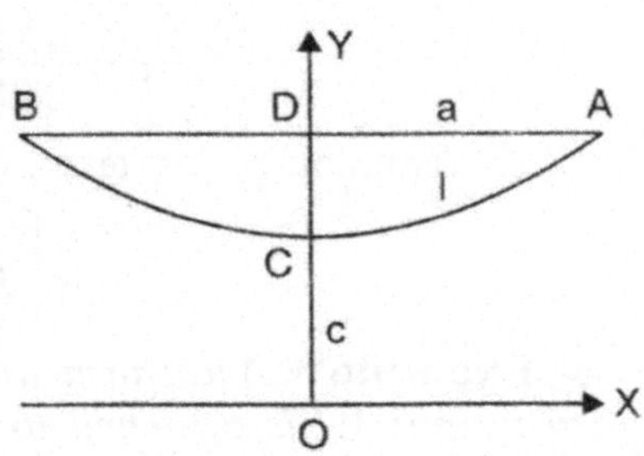

Fig. 14

Applying dividendo and componendo

$$\frac{e^{l/4c} - e^{-l/4c}}{e^{l/4c} + e^{-l/4c}} = \frac{1 + \tan \dfrac{\psi}{2} - 1 + \tan \dfrac{\psi}{2}}{1 + \tan \dfrac{\psi}{2} + 1 - \tan \dfrac{\psi}{2}}$$

$$\tanh(l/4c) = \frac{2\tan\psi/2}{2}$$

or $$\tanh(l/4c) = \tan\left(\frac{a}{2c}\right).$$

Example 5. *A string of length l is attached to a fixed point A and the other end B is pulled with a force ωa inclined at an angle α to the horizon, ω being weight per unit length. Show that vertical distance of A above B is* $\sqrt{l^2 + a^2 + 2la\sin\alpha} - a$.

Sol. AB is the string of length l and the other end B is pulled with a force ωa inclined at an angle α with horizontal.

Let C be the vertex of the catenary.

Let $\overset{\frown}{CB} = s$

At B, $T = \omega y_2 = \omega a$

$\therefore \quad y_2 = a \quad$ ...(1)

Also $y_2 = c\sec\alpha \quad (\because \; y = c\sec\psi)$

or $c = a\cos\alpha \quad$ ...(2)

Also $s = c\tan\alpha \quad (\because \; s = c\tan\psi)$

$$= a\cos\alpha\tan\alpha = a\cos\alpha \cdot \frac{\sin\alpha}{\cos\alpha}$$

$\therefore \quad s = a\sin\alpha \quad$ ...(3)

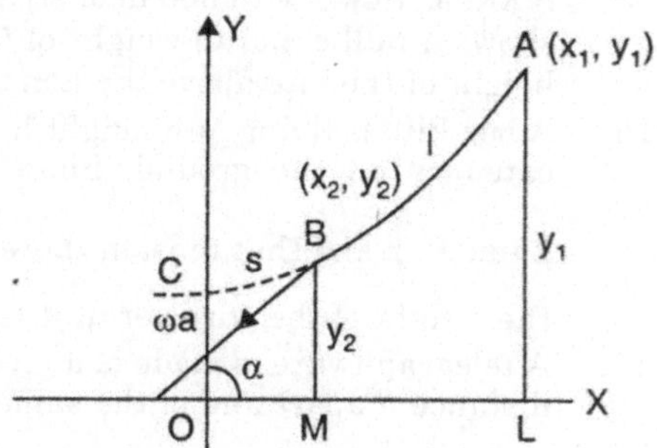

Fig. 15

At A, $y^2 = c^2 + s^2$

i.e., $$y_1^2 = c^2 + (s + l)^2$$
$$= a^2\cos^2\alpha + (a\sin\alpha + l)^2$$

or $$y_1 = \sqrt{a^2 + l^2 + 2al\sin\alpha}$$

Vertical distance between A and B

$$= y_1 - y_2 = \sqrt{a^2 + l^2 + 2al\sin\alpha} - a.$$

EXERCISE 2

1. A telegraph wire is attached between the points 100 m. apart and sags 1m. in the middle. Show that the tension at the lowest point is approx. 187.5 kg wt. (weight of wire per metre is .15 kg).
2. A chain is suspended from two points 462 m. apart. If tangent at the either terminal is $\tan^{-1}\frac{3}{4}$ to the horizontal find the length of the chain ($\log_e 2 = 0.693$).
3. A telephone wire has a span 40 m, find the sag in the middle if tension is not to exceed 400 kg. wt. (weight of wire is 1 kg/m.)
4. Prove that in the catenary of uniform strength, the mass per unit length is proportional to $\cosh\frac{s}{c}$.
5. A uniform chain of length l is tightly stretched over a river, the middle point just touching the surface of water, while the ends of the chain have elevation k above the surface. Show that difference between the length of the chain and width of the river is approx. $8k^2/3l$.

6. A uniform chain of length l and weight W hangs between two fixed points at the same level, and weight W′ is attached at the middle point. If k be the sag in the middle, prove that pull on either point of support is

$$\frac{k}{2l}W + \frac{l}{4k}W' + \frac{l}{8k}W.$$

7. A uniform string hangs freely under gravity. Show that mass per unit length is proportional to $\frac{\sec^2 \psi}{\rho}$, where ρ is radius of curvature.

8. A boat is towed by means of a rope attached to ship and lower end the rope makes an angle 30° with the horizontal. If the length of the rope is 36 m and height difference between upper and lower end is 20 m. Find the resistance of the water to the motion of the boat if weight of rope is 500 gm per metre.

9. A kite is flown with 300 m of string from the hand to kite and a spring balance held in the hand shows a pull equal to weight of 50 m of string inclined at 30° to the horizon. Find the vertical height of the kite above the hand. *(K.U. 2007)*

10. A box kite is flying at a height h with a length l of the wire paid out, and with the vertex of the catenary on the ground. Show that at the kite, the inclination of the wire to the ground is $2\tan^{-1}\frac{h}{l}$ and that tension there and at the ground are $\frac{\omega(l^2+h^2)}{2h}$ and $\frac{\omega(l^2-h^2)}{2h}$, where ω is the weight of the wire per unit length.

11. A telegraph wire is made of a given material and such a length l is stretched between two posts, distance d apart and of the same height as will produce the least possible tension at the posts. Show that $l = \frac{d}{\lambda}\sinh\lambda$, where λ is given by the equation $\lambda \tanh \lambda = 1$. *(M.D.U. 2005)*

ANSWERS

2. 500 m. **9.** 256.6 m approx.

SECTION II

4

Forces in Three Dimensions

Rigid body. A rigid body is composed of many particles, such that the distance between the two particles remains constant, no matter what forces act on it.

No perfectly rigid body exists in nature. Bodies change their shapes and sizes under the stress of great forces.

1. Forces in Three Dimensions

When a rigid body is being acted upon by forces in different directions in space, we say that body is under the action of forces in the three dimensions.

If X, Y, Z be the components of a force R along x, y and z-axis respectively then

$$R^2 = X^2 + Y^2 + Z^2$$

Direction cosine of line of action of R are

$$\frac{X}{R}, \frac{Y}{R}, \frac{Z}{R}$$

Conversely if direction cosines of line of action of force R are l, m, n, then the components along the axes are lR, mR, nR respectively.

Theorem 1. *Any given system of forces acting at different points of a rigid body can be reduced to a single force through any arbitrary chosen point and a couple whose axis passes through that point.* (M.D.U. 2005)

Proof. Let S be a rigid body. Let the forces $\vec{F_1}, \vec{F_2}, \ldots\ldots, \vec{F_n}$ be acting on it at $A_1, A_2, \ldots\ldots, A_n$ respectively.

Consider an arbitrary point O on S and choose it as origin. Let $\vec{a_1}, \vec{a_2}, \ldots\ldots, \vec{a_n}$ be the position vectors of the points $A_1, A_2, \ldots\ldots A_n$ respectively. Consider a force $\vec{F_r}$ acting at A_r. Apply two equal and opposite forces $\vec{F_r}$ and $-\vec{F_r}$ at O. The forces $\vec{F_r}$ at A_r and $-\vec{F_r}$ at O form a couple of moment $\vec{a_r} \times \vec{F_r}$ and we are left with a single force $\vec{F_r}$ at O. Similarly the forces $\vec{F_1}, \vec{F_2}, \ldots\ldots, \vec{F_n}$ acting at $A_1, A_2, \ldots\ldots, A_n$ reduces to couples of moment $\vec{a_1} \times \vec{F_1}$, $\vec{a_2} \times \vec{F_2}$ ...... and forces $\vec{F_1}, \vec{F_2}, \ldots\ldots, \vec{F_n}$ at O.

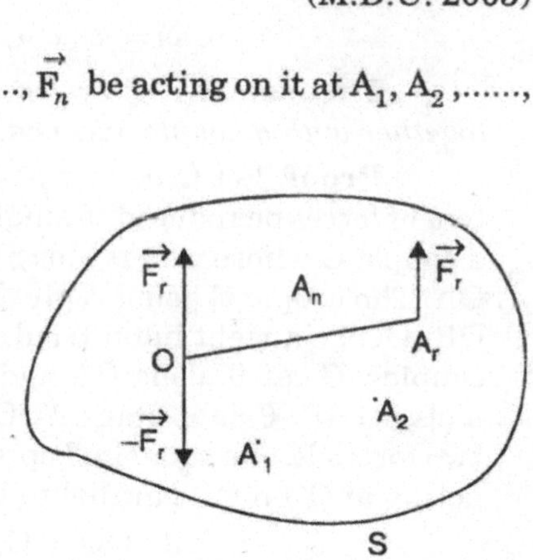

Fig. 1

Thus the given system of forces is equivalent to $\vec{R} = \vec{F_1} + \vec{F_2} + ... + \vec{F_n}$ together with a couple of moment $\vec{G} = \vec{a_1} \times \vec{F_1} + \vec{a_2} \times \vec{F_2} + ... + \vec{a_n} \times \vec{F_n}$ whose axis passes through G.

2. Dyname. The combination of a force $\vec{R}$ and a couple $\vec{G}$ is called a dyname denoted by $(\vec{R}, \vec{G})$.

If $\vec{F_r} = X_r \hat{i} + Y_r \hat{j} + Z_r \hat{k}$ be the forces acting at

$$\vec{a_r} = x_r \hat{i} + y_r \hat{j} + z_r \hat{k}, \quad r = 1, 2, ... n$$

then resultant force R is given by

$$\vec{R} = (\Sigma X_r) \hat{i} + (\Sigma Y_r) \hat{j} + (\Sigma Z_r) \hat{k} = X\hat{i} + Y\hat{j} + Z\hat{k}$$

and resultant couple G is given by

$$\vec{G} = \Sigma \vec{a_r} \times \vec{F_r} = \Sigma \begin{vmatrix} i & j & k \\ x_r & y_r & z_r \\ X_r & Y_r & Z_r \end{vmatrix}$$

$$= \Sigma \{\hat{i} (y_r Z_r - z_r Y_r) + \hat{j} (z_r X_r - x_r Z_r) + \hat{k} (x_r Y_r - y_r X_r)\}$$

$$\vec{G} = L\hat{i} + M\hat{j} + Z\hat{k}$$

X, Y, Z and L, M, N are components of R and G along the axes.

where $X = \Sigma X_r$, $Y = \Sigma Y_r$, $Z = \Sigma Z_r$

$$L = \Sigma (y_r Z_r - z_r Y_r), \quad M = \Sigma (z_r X_r - x_r Z_r)$$

$$N = \Sigma (x_r Y_r - y_r X_r).$$

Theorem 2. *Any system of forces acting on a rigid body can be reduced to a single force together with a couple axis is along the direction of the forces.*

Proof. Let O be any point of the body. Let the system of forces be reduced to single force R at O along OA and a couple G whose axis is along OL making an angle θ with OA. The couple G being represented by vector OL. Let OA, OB, OC be a right hand triad. G can be resolved into two couples, G cos θ along OA and G sin θ about OB as their axes. G sin θ is ⊥ plane AOC. G sin θ can be replaced by two forces R, one equal and opposite to R at O and the other acting at O′ on OC parallel to R such that

$$R \cdot OO' = G \sin\theta$$

$$OO' = \frac{G \sin\theta}{R}$$

The forces R and R at O balance each other and we are left with a force R at O′ parallel to OA and couple G cos θ about OA. Since the axis of couple can be shifted

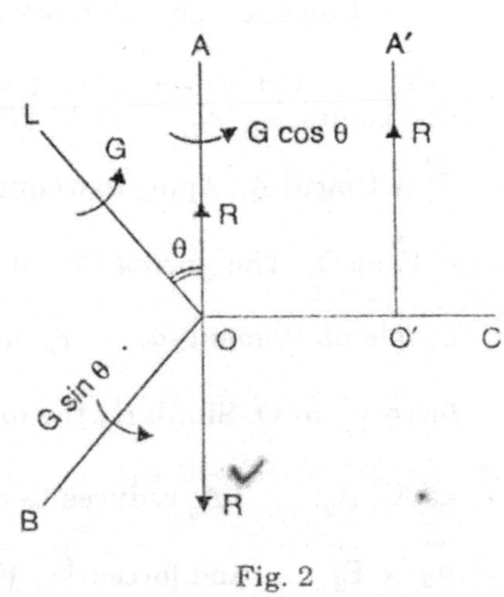

Fig. 2

anywhere parallel to itself, we can take it along O′A′ in place of OA. Thus the system of force is reduced to a force R along O′A′ and couple G cos θ whose axis is along O′A′.

Wrench. A single force R, together with a couple G whose axis coincides with the line of action of R is said to form a **wrench**.

Poinsot's central axis. The axis of single couple and the line of action of R to which the system of force is reduced, is called the **Poinsot's central axis** or simply **central axis**.

Central axis for a system of forces is unique. When a system (R, G) at O is reduced to a force R to O′ together with a couple G cos θ whose axis is along the force R at O′ Fig. (2). If we write K = G cos θ. It is said to be reduced to wrench (R, K) at O′. The ratio p = K/R is called **Pitch** of the system.

The single force R is called **Intensity** of the wrench.

Screw. The straight line along which the single force acts when considered together with the pitch is called **screw**.

3. General Condition of Equilibrium

It is necessary for equilibrium that there shall be neither translation nor rotation in the body. *i.e.,* there should be neither resultant force R nor couple G.

$$\therefore \quad R = 0 \quad \text{and} \quad G = 0$$

$$R = 0 \Rightarrow X^2 + Y^2 + Z^2 = 0$$

$$\Rightarrow \quad X = 0, Y = 0, Z = 0$$

$$G = 0 \Rightarrow L^2 + M^2 + N^2 = 0$$

$$\Rightarrow \quad L = 0, M = 0, N = 0$$

Conversely if $X = Y = Z = 0 \Rightarrow R^2 = 0 \Rightarrow R = 0$

$$L = M = N = 0 \Rightarrow G^2 = 0 \Rightarrow G = 0.$$

Hence, we can say that necessary and sufficient condition for the equilibrium of a rigid body under the action of a system of forces are that sum of the resolved parts of the forces along the three mutually perpendicular axes must vanish and also the sum of their moments about the three axes must vanish separately.

4. Condition that a General System of Force Reduces to a Single Force

We know that any system of force acting on a rigid body can be reduced to a single force R acting at an arbitrary chosen point O along the line OA together with a couple moment G about an axis OL Fig. (2). Let X, Y, Z be the components of R and L, M, N be the components of G along the coordinate axes. The direction cosines of OA are $\frac{X}{R}, \frac{Y}{R}, \frac{Z}{R}$ and the direction cosines of OL are $\frac{L}{G}, \frac{M}{G}, \frac{N}{G}$.

$\therefore$ Angle θ between OA and OL is given by

$$\cos\theta = \frac{X}{R}\cdot\frac{L}{G} + \frac{Y}{R}\cdot\frac{M}{G} + \frac{Z}{R}\cdot\frac{N}{G} \qquad \boxed{\cos\theta = ll' + mm' + nn'}$$

$$= \frac{LX + MY + NZ}{RG}$$

$$\therefore \quad LX + MY + NZ = RG\cos\theta \qquad \text{...(1)}$$

We know that this system can be further reduced to a single force R along O′A′ and couple of moment G cos θ about O′A′ as axis.

In order that system reduces to a single force,

$$G\cos\theta = 0 \Rightarrow \theta = \pi/2$$

From (1) $\boxed{LX + MY + NZ = 0}$ is the required condition.

5. Equation of the Central Axis of any given System of Forces

Let (R, G) be the reduction of any given system of forces $F_1, F_2, \ldots\ldots, F_n$ acting at $A_1, A_2, \ldots\ldots A_n$.

$$R^2 = X^2 + Y^2 + Z^2,$$
$$G^2 = L^2 + M^2 + N^2$$

where $A_1(x_1, y_1, z_1), A_2(x_2, y_2, z_2) \ldots\ldots, A_n(x_n, y_n, z_n)$

and

$$X = \Sigma X_1,\ Y = \Sigma Y_1,\ Z = \Sigma Z_1$$
$$L = \Sigma(y_1 Z_1 - z_1 Y_1)$$
$$M = \Sigma(z_1 X_1 - x_1 Z_1)$$
$$N = \Sigma(x_1 Y_1 - y_1 X_1)$$

[From article (2)]

Let this system be reduced to a wrench with O′A′ be the central axis where $O' \equiv (f, g, h)$. R remains the same along O′A′. Let component of new couple be L′, M′, N′. To find the values of L′, M′, N′ we have to replace x_1, y_1, z_1 by $x_1 - f, y_1 - g, z_1 - h$ respectively in the values of L, M, N.

$$\left.\begin{aligned} L' &= \Sigma[(y_1 - g)Z_1 - (z_1 - h)Y_1] \\ &= \Sigma[y_1Z_1 - z_1Y_1] - g\Sigma Z_1 + h\Sigma Y_1 = L - gZ + hY. \\ M' &= \Sigma[(z_1 - h)X_1 - (x_1 - f)Z_1] \\ &= \Sigma(z_1X_1 - x_1Z_1) - h\Sigma X_1 + f\Sigma Z_1 = M - hX + fZ \\ N' &= \Sigma(x_1Y_1 - y_1X_1) = \Sigma[(x_1 - f)Y_1 - (y_1 - g)X_1] \\ &= \Sigma(x_1Y_1 - y_1X_1) - f\Sigma Y_1 + g\Sigma X_1 = N - fY + gX \end{aligned}\right\} \quad \ldots(1)$$

∴ Axis of couple and line of action of R coincide,

$$\therefore \quad \frac{L'}{X} = \frac{M'}{Y} = \frac{N'}{Z}$$

Putting the values of L′, M′, N′

$$L'X + M'Y + N'Z = (L - gZ + hY)X + (M - hX + fZ)Y + (N - fY + gX)Z$$
$$= LX + MY + NZ. \quad \ldots(2)$$

$$\therefore \quad L'X + M'Y + N'Z = LX + MY + NZ = RG\cos\theta$$ [From equation article 4)

$$\therefore \quad \frac{L'}{X} = \frac{M'}{Y} = \frac{N'}{Z} = \frac{L'X + M'Y + N'Z}{X^2 + Y^2 + Z^2} = \frac{RG\cos\theta}{R^2} = \frac{k}{R} = p. \quad \ldots(3)$$

Putting L′, M′, N′ from (1) into (3), we get

$$\frac{L - gZ + hY}{X} = \frac{M - hX + fZ}{Y} = \frac{N - fY + gX}{N} = \frac{k}{R} = p$$

Replacing (f, g, h) by (x, y, z) for generalisation, equation of central axis is

$$\mathbf{\frac{L - yZ + zY}{X} = \frac{M - zX + xZ}{Y} = \frac{N - xY + yX}{N} = \frac{k}{R} = p.} \quad \ldots(4)$$

Method for solving problems on wrenches and cental axis.

(*i*) Write the equation of the line of action of the forces $\vec{F}_r$ in the form

$$\frac{x - x_r}{l_r} = \frac{y - y_r}{m_r} = \frac{z - z_r}{n_r}$$

where l, m, n are direction cosine of the line.

(*ii*) Write the component of the forces along the axes as

$$X_r = F_r l_r,\ Y_r = F_r m_r,\ Z_r = F_r n_r$$

(*iii*) Write the resultant force

$$\vec{R} = (\Sigma X_r)\hat{i} + (\Sigma Y_r)\hat{j} + (\Sigma Z_r)\hat{k} = Xi + Yj + Z\hat{k}$$

(*iv*) Write the compoment of the couple as

$$L_r i + M_r \hat{j} + N_r \hat{k} = \begin{vmatrix} i & j & k \\ x_r & y_r & z_r \\ X_r & Y_r & Z_r \end{vmatrix}$$

$$\therefore \quad L\hat{i} + M\hat{j} + N\hat{k} = (\Sigma L_r)\hat{i} + (\Sigma M_r)\hat{j} + (\Sigma N_r)\hat{k}$$

(*v*) Resultant force $\quad R = \sqrt{X^2 + Y^2 + Z^2}$

Resultant couple $\quad G = \sqrt{L^2 + M^2 + N^2}$

Equation of central axis is

$$\frac{L - yZ + zY}{X} = \frac{M - zX + xZ}{Y} = \frac{N - xY + yX}{N}$$

Corrolary. *To show that (LX + MY + NZ) and ($X^2 + Y^2 + Z^2$) are invariants for any given system of forces, whatever axes and origin may be chosen.* (K.U. 2005)

We know that central axis of the system is unique. Hence the force R and the couple G cos θ, both are invariant, because R and G cos θ, are the force and couple about central axis of the system.

$$\because \quad R = \sqrt{X^2 + Y^2 + Z^2}$$

$\therefore$ R is invariable $\Rightarrow$ $\sqrt{X^2 + Y^2 + Z^2}$ is invariable

$\Rightarrow$ $X^2 + Y^2 + Z^2$ is invariant.

$\because$ G cos θ is invariable

Hence $G\left[\frac{L'X + M'Y + N'Z}{GR}\right]$ is invariant. $\qquad \left(\because \cos\theta = \frac{L'X + M'Y + N'Z}{GR}\right)$

(From eq. (3) article 5. Page 68)

$$= \frac{(L - gZ + hY)X + (M - hX + fZ)Y + (N - fY + gX)Z}{R}$$

$$= \frac{LX + MY + NZ}{R} \text{ is invariant.}$$

R is invariant $\Rightarrow$ LX + MY + NZ is invariant.

Hence proved.

Example 1. *Three equal forces act on a rigid body at (a, 0, 0), (0, b, 0), (0, 0, c) parallel to Y, Z and X-axis respectively. Find the resultant wrench and central axis.*

Sol. Let the forces each equal to P act at A, B, C.

The equations of their lines of action are

$$\frac{x-a}{0}=\frac{y}{1}=\frac{z}{0}$$

$$\frac{x}{0}=\frac{y-b}{0}=\frac{z}{1}$$

$$\frac{x}{1}=\frac{y}{0}=\frac{z-c}{0}$$

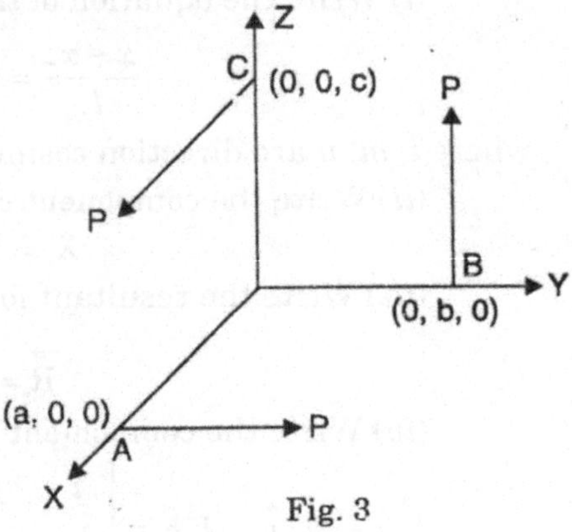

Fig. 3

Components of forces parallel to axes are

$$X_1=0 \qquad Y_1=P \qquad Z_1=0$$
$$X_2=0 \qquad Y_2=0 \qquad Z_2=P$$
$$X_3=P \qquad Y_3=0 \qquad Z_3=0$$
$$X=\Sigma Xi=P \qquad Y=\Sigma Yi=P \qquad Z=\Sigma Zi=P \qquad \ldots(1)$$

$$L_1\hat{i}+M_1\hat{j}+N_1\hat{k}=\begin{vmatrix} i & j & k \\ x_1 & y_1 & z_1 \\ X_1 & Y_1 & Z_1\end{vmatrix}=\begin{vmatrix} i & j & k \\ a & 0 & 0 \\ 0 & P & 0\end{vmatrix}=aP\,\hat{k}.$$

$$L_2\hat{i}+M_2\hat{j}+N_2\hat{k}=\begin{vmatrix} i & j & k \\ 0 & b & 0 \\ 0 & 0 & P\end{vmatrix}=bP\,\hat{i}$$

$$L_3\hat{i}+M_3\hat{j}+N_3\hat{k}=\begin{vmatrix} i & j & k \\ 0 & 0 & c \\ P & 0 & 0\end{vmatrix}=cP\,\hat{j}$$

$$\vec{L}=(\Sigma L_1)\hat{i}+(\Sigma M_1)\hat{j}+(\Sigma N_1)\hat{k}$$

$$=bP\,\hat{i}+cP\,\hat{j}+aP\,\hat{k}=L\hat{i}+M\hat{j}+N\hat{k} \qquad \ldots(2)$$

$$R^2=X^2+Y^2+Z^2=P^2+P^2+P^2$$

$$R=\sqrt{3}\,P.$$

The resultant wrench is given by

$$\frac{k}{R}=\frac{LX+MY+NZ}{R^2}$$

$$k=\frac{LX+MY+NZ}{R}=\frac{(a+b+c)P^2}{P\sqrt{3}}=\frac{(a+b+c)P}{\sqrt{3}}$$

Equation of central axis is

$$\frac{L-(yZ-zY)}{X}=\frac{M-(zX-xZ)}{Y}=\frac{N-(xY-yX)}{Z}=\frac{k}{R}$$

Putting the values of X, Y, Z and L, M, N from (1) and (2).

$$\frac{bP-yP+Pz}{P}=\frac{cP-Pz+Px}{P}=\frac{aP-Px+Py}{P}=\frac{(a+b+c)P}{\sqrt{3}\,P\sqrt{3}}$$

$$\frac{b-y+z}{1}=\frac{c-z+x}{1}=\frac{a-x+y}{1}=\frac{a+b+c}{3}$$

or $$x+\frac{a+2b+3c}{3}=y+\frac{b+2c+3a}{3}=z+\frac{c+2a+3b}{3}.$$

Example 2. *Equal forces act along the coordinate axes and the line* $\frac{x-\alpha}{l}=\frac{y-\beta}{m}=\frac{z-\gamma}{n}$. *Find the resultant wrench and the equation of the central axis.* (K.U. 2004)

Sol. Let the force each equal to P act along $\frac{x}{1}=\frac{y}{0}=\frac{z}{0}$, $\frac{x}{0}=\frac{y}{1}=\frac{z}{0}$, $\frac{x}{0}=\frac{y}{0}=\frac{z}{1}$ and $\frac{x-\alpha}{l}=\frac{y-\beta}{m}=\frac{z-\gamma}{n}$, l, m, n are d.c. of the line.

Component of force P along $\frac{x}{1}=\frac{y}{0}=\frac{z}{0}$ are

$$X_1 = P.1, \quad Y_1 = P.0, \quad Z_1 = P.0$$

i.e., $$X_1 = P, \quad Y_1 = 0, \quad Z_1 = 0$$

Similarly component of P along other three lines are

$$X_2 = 0, \quad Y_2 = P, \quad Z_2 = 0$$
$$X_3 = 0, \quad Y_3 = 0, \quad Z_2 = P$$
$$X_4 = Pl, \quad Y_4 = Pm, \quad Z_4 = Pn$$

$\therefore$ Resultant force $\vec{R} = P(1+l)\hat{i} + P(1+m)\hat{j} + P(1+n)\hat{k} = X\vec{i} + Y\vec{j} + Z\vec{k}$ ...(1)

Component of couple

$$L_1\hat{i} + M_1\hat{j} + N_1\hat{k} = \begin{vmatrix} i & j & k \\ 0 & 0 & 0 \\ P & 0 & 0 \end{vmatrix} = 0\vec{i} + 0\vec{j} + 0\vec{k}$$

Similarly, $L_2\hat{i} + M_2\hat{j} + N_2\hat{k} = 0\hat{i} + 0\hat{j} + 0\hat{k}$

$$L_3\hat{i} + M_3\hat{j} + N_3\hat{k} = 0\hat{i} + 0\hat{j} + 0\hat{k}$$

$$L_4\hat{i} + M_4\hat{j} + N_4\hat{k} = \begin{vmatrix} i & j & k \\ \alpha & \beta & \gamma \\ lP & mP & nP \end{vmatrix}$$

Resultant couple

$$= (\beta n - \gamma m)P\hat{i} + (\gamma l - \alpha n)P\hat{j} + (\alpha m - \beta l).P\hat{k}$$

$$\therefore \quad L\hat{i} + M\hat{j} + N\hat{k} = (\Sigma L_1)\hat{i} + (\Sigma M_1)\hat{j} + (\Sigma N_1)\hat{k}$$

$$= (\beta n - \gamma m)P\hat{i} + (\gamma l - \alpha n)P\hat{j} + (\alpha m - \beta l)P\hat{k} \quad ...(2)$$

Putting the values of X, Y, Z, L, M, N from (1) and (2)

$$\therefore \quad k = \frac{LX + MY + NZ}{R}$$

$$k = \frac{(\beta n - \gamma m)\,\mathrm{P}\,.\,\mathrm{P}\,(1+l) + (\gamma l - \alpha n)\,\mathrm{P}\,.\,\mathrm{P}\,(1+m) + (\alpha m - \beta l)\,\mathrm{P}\,.\,\mathrm{P}\,(1+n)}{\mathrm{P}\sqrt{(1+l)^2 + (1+m)^2 + (1+n)^2}}$$

Equation of central axis is

$$\frac{\mathrm{L} - y\mathrm{Z} + z\mathrm{Y}}{\mathrm{X}} = \frac{\mathrm{M} - z\mathrm{X} + x\mathrm{Z}}{\mathrm{Y}} = \frac{\mathrm{N} - x\mathrm{Y} + y\mathrm{X}}{\mathrm{Z}} = \frac{k}{\mathrm{R}}$$

Putting the values of L, M, N and X, Y, Z we get

$$\frac{(\beta n - \gamma m)\,\mathrm{P} - \mathrm{P}(1+n)\,y + \mathrm{P}(1+m)\,z}{\mathrm{P}(1+l)} = \frac{(\gamma l - \alpha n)\,\mathrm{P} - \mathrm{P}(1+l)\,z + \mathrm{P}(1+n)\,x}{\mathrm{P}(1+m)}$$

$$= \frac{(\alpha m - \beta l)\,\mathrm{P} - \mathrm{P}(1+m)\,x + \mathrm{P}(1+l)\,y}{\mathrm{P}(1+n)}$$

i.e.,

$$\frac{(\beta n - \gamma m) - (1+n)\,y + (1+m)\,z}{1+l} = \frac{(\gamma l - \alpha n) - (1+l)\,z + (1+n)\,x}{1+m}$$

$$= \frac{(\alpha m - \beta l) - (1+m)\,x + (1+l)\,y}{1+n}.$$

Example 3. *Forces P, Q, R act along three non intersecting edges of a cube. Find the equation of the central axis.*

Sol. Let a be length of the edge of the cube. Let OA, OB, OC be the axis. Let the force P, Q, R act along OA, CA′ and C′O′ respectively.

The lines of action are

$$\frac{x}{1} = \frac{y}{0} = \frac{z}{0} \quad \text{for P}$$

$$\frac{x}{0} = \frac{y}{1} = \frac{z-a}{0} \quad \text{for Q}$$

$$\frac{x-a}{0} = \frac{y-a}{0} = \frac{z}{1} \quad \text{for R.}$$

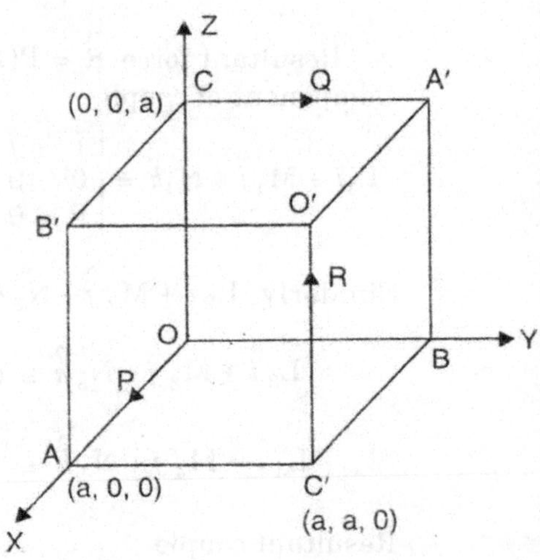

Fig. 4

The forces are $\mathrm{P}\hat{i}$, $\mathrm{Q}\hat{j}$ and $\mathrm{R}\hat{k}$

Resultant force

$$= \mathrm{P}\hat{i} + \mathrm{Q}\hat{j} + \mathrm{R}\hat{k} = \mathrm{X}\hat{i} + \mathrm{Y}\hat{j} + \mathrm{Z}\hat{k}$$

$$\mathrm{L}_1\hat{i} + \mathrm{M}_1\hat{j} + \mathrm{N}_1\hat{k} = \begin{vmatrix} i & j & k \\ 0 & 0 & 0 \\ \mathrm{P} & 0 & 0 \end{vmatrix}$$

$$= 0\hat{i} + 0\hat{j} + 0\hat{k}$$

$$\mathrm{L}_2\hat{i} + \mathrm{M}_2\hat{j} + \mathrm{N}_2\hat{k} = \begin{vmatrix} i & j & k \\ 0 & 0 & a \\ 0 & \mathrm{Q} & 0 \end{vmatrix} = -a\mathrm{Q}\,\hat{i}$$

$$\mathrm{L}_3\hat{i} + \mathrm{M}_3\hat{j} + \mathrm{N}_3\hat{k} = \begin{vmatrix} i & j & k \\ a & a & 0 \\ 0 & 0 & \mathrm{R} \end{vmatrix} = a\mathrm{R}\,\hat{i} - a\mathrm{R}\,\hat{j}$$

$\therefore$ Resultant couple

$$L\hat{i}+M\hat{j}+N\hat{k} = (aR-aQ)\hat{i}-aR\hat{j}+0\hat{k}$$

$\therefore$ Equation of central axis is

$$\frac{L-yZ+zY}{X} = \frac{M-zX+xZ}{Y} = \frac{N-xY+yX}{Z}$$

$$\frac{a(R-Q)-yR+zQ}{P} = \frac{-aR-Pz+Rx}{Q} = \frac{0-xQ+yP}{R}.$$

Example 4. *Two forces P and Q act along the straight lines $y = x \tan \alpha$, $z = c$ and $y = -x \tan \alpha$, $z = -c$ respectively. Show that the central axis lies along the straight line*

$$y = x\left(\frac{P-Q}{P+Q}\right)\tan\alpha,\ \frac{z}{c} = \frac{(P^2-Q^2)}{P^2+2PQ\cos 2\alpha+Q^2}$$

If P and Q vary this line generates the surface
$(x^2+y^2)\, z \sin 2\alpha = 2\, cxy$.

Sol. Line of action of P and Q can be written as

$$\frac{x}{\cos\alpha} = \frac{y}{\sin\alpha} = \frac{z-c}{0} \quad \text{and} \quad \frac{x}{\cos\alpha} = \frac{y}{-\sin\alpha} = \frac{z+c}{0} \text{ respectively.}$$

Components of P and Q are (P cos α, P sin α, 0) and (Q cos α, – Q sin α, 0).

The resultant force is

$$\vec{R} = (P+Q)\cos\alpha\,\hat{i} + (P-Q)\sin\,\hat{j} + 0\,\hat{k}$$

$$L_1\hat{i}+M_1\hat{j}+N_1\hat{k} = \begin{vmatrix} i & j & k \\ 0 & 0 & c \\ P\cos\alpha & P\sin\alpha & 0 \end{vmatrix} = -Pc\sin\alpha\,\hat{i}+Pc\cos\alpha\,\hat{j}+0\,\hat{k}$$

$$L_2\hat{i}+M_2\hat{j}+N_2\hat{k} = \begin{vmatrix} i & j & k \\ 0 & 0 & -c \\ Q\cos\alpha & -Q\sin\alpha & 0 \end{vmatrix} = -Qc\sin\alpha\,\hat{i}-Qc\cos\alpha\,\hat{j}+0\,\hat{k}$$

$$L\hat{i}+M\hat{j}+N\hat{k} = -(P+Q)c\sin\alpha\,\hat{i}+(P-Q)c\cos\alpha\,\hat{j}+0\,\hat{k}$$

$\therefore$ Equation of central axis is

$$\frac{L-yZ+zY}{X} = \frac{M-zX+xZ}{Y} = \frac{N-xY+yX}{N}$$

$$= \frac{(P+Q)c\sin\alpha-0+(P-Q)\sin\alpha\, z}{(P+Q)\cos\alpha}$$

$$= \frac{(P-Q)c\cos\alpha-(P+Q)\cos\alpha\, z}{(P-Q)\sin\alpha}$$

$$= 0-\frac{(P-Q)\sin\alpha\, x+(P+Q)\cos\alpha\, y}{0} = p$$

$$\Rightarrow \quad -(P-Q)\,x\sin\alpha+(P+Q)\,y\cos\alpha = 0$$

$$\Rightarrow \quad y = x\,\frac{(P-Q)}{(P+Q)}\tan\alpha \qquad \text{...(1)}$$

Also, $$\frac{-(P+Q)c\sin\alpha+(P-Q)z\sin\alpha}{(P+Q)\cos\alpha}=\frac{(P-Q)c\cos\alpha-(P+Q)z\cos\alpha}{(P-Q)\sin\alpha}$$

$$\Rightarrow \quad -(P^2-Q^2)c\sin^2\alpha+(P-Q)^2 z\sin^2\alpha=(P^2-Q^2)c\cos^2\alpha-(P+Q)^2 z\cos^2\alpha$$

$$\Rightarrow \quad z[(P-Q)^2\sin^2\alpha+(P+Q)^2\cos^2\alpha]=c[(P^2-Q^2)\cos^2\alpha+(P^2-Q^2)\sin^2\alpha \qquad \ldots(2)$$

$$\Rightarrow \quad z[P^2+Q^2+2PQ(\cos^2\alpha-\sin^2\alpha)]=c(P^2-Q^2)$$

$$\Rightarrow \quad \frac{z}{c}=\frac{P^2-Q^2}{P^2+Q^2+2PQ\cos 2\alpha} \qquad \ldots(3)$$

To find out the surface generated we have to eliminate P and Q from the equation of central axis

From equation (1)

$$\frac{P-Q}{P+Q}=\frac{y\cos\alpha}{x\sin\alpha} \qquad \ldots(4)$$

From (2) $$\frac{z}{c}=\frac{P^2-Q^2}{[(P-Q)^2\sin^2\alpha+(P+Q)^2\cos^2\alpha]}$$

$$=\frac{(P-Q)(P+Q)}{(P+Q)^2\left[\left(\frac{P-Q}{P+Q}\right)^2\sin^2\alpha+\cos^2\alpha\right]}$$

$$=\frac{\frac{P-Q}{P+Q}}{\left(\frac{P-Q}{P+Q}\right)^2\sin^2\alpha+\cos^2\alpha}$$

$$=\frac{\frac{y\cos\alpha}{x\sin\alpha}}{\frac{y^2\cos^2\alpha}{x^2\sin^2\alpha}\sin^2\alpha+\cos^2\alpha}=\frac{\frac{y\cos\alpha}{x\sin\alpha}}{\frac{y^2\cos^2\alpha+x^2\cos^2\alpha}{x^2}}$$

$$=\frac{xy}{\cos\alpha\sin\alpha(x^2+y^2)}=\frac{2xy}{(x^2+y^2)\sin 2\alpha}$$

$$\therefore \quad z(x^2+y^2)\sin 2\alpha=2cxy$$

which is the required surface.

Example 5. *A single force is equivalent to component forces X, Y and Z along the coordinate axes and to couples L, M, N about the coordinate axes. Prove that the magnitude of the single force is* $\sqrt{X^2+Y^2+Z^2}$ *and its line of action is*

$$\frac{yZ-zY}{L}=\frac{zX-xZ}{M}=\frac{xY-yX}{N}=1.$$

Sol. The condition that system reduces to a single force is LX + MY + NZ = 0

Equation of central axis is

$$\frac{L - yZ + zY}{X} = \frac{M - zX + xZ}{Y} = \frac{N - xY + yX}{Z} = \frac{LX + MY + NZ}{X^2 + Y^2 + Z^2} = 0$$

$$\therefore \quad L - yZ + zY = 0 \quad \Rightarrow \quad \frac{yZ - zY}{L} = 1$$

$$M - zX + xZ = 0 \quad \Rightarrow \quad \frac{zX - xZ}{M} = 1$$

$$N - xY + yX = 0 \quad \Rightarrow \quad \frac{xY - yX}{N} = 1$$

$$\therefore \quad \frac{yZ - zY}{L} = \frac{zX - xZ}{M} = \frac{xY - yX}{N} = 1$$

$\because$ Component of force are X, Y, Z

$\therefore$ Resultant force = $\sqrt{X^2 + Y^2 + Z^2}$.

Example 6. *A force* $\vec{R} = X\hat{i} + Y\hat{j} + Z\hat{k}$ *acting along the generator of hyperbolic paraboloid* $\frac{x^2}{a^2} - \frac{y^2}{b^2} = \frac{2z}{c}$ *is equivalent to an equal force* $\vec{R}$ *at origin together with a couple* $\vec{G} = L\hat{i} + M\hat{j} + N\hat{k}$.

Show that $aL \pm bM = 0,\ bX \pm aY = 0,\ cN \pm abZ = 0$.

Sol. The generators of $\frac{x^2}{a^2} - \frac{y^2}{b^2} = \frac{2z}{c}$ are

$$\frac{x}{a} + \frac{y}{b} = \frac{2z}{c\mu},\ \frac{x}{a} - \frac{y}{b} = \mu \qquad \text{...(1)}$$

$$\frac{x}{a} - \frac{y}{b} = \frac{2z}{c\lambda},\ \frac{x}{a} + \frac{y}{b} = \lambda \qquad \text{...(2)}$$

Let us consider generator (1) and reduce it to symmetrical form.

Let d.c. of (1) be l, m, n:

$$\frac{l}{a} + \frac{m}{b} - \frac{2n}{c\mu} = 0 \qquad \text{...(3)}$$

$$\frac{l}{a} - \frac{m}{b} = 0 \quad \Rightarrow \quad \frac{l}{a} = \frac{m}{b} = k \qquad \text{(say)}$$

$$\Rightarrow \quad l = ak,\ m = bk$$

Putting in (3), we have $n = \mu c k$

To find a point on (1) take Z = 0

We have, $\frac{x}{a} + \frac{y}{b} = 0,\ \frac{x}{a} - \frac{y}{b} = \mu$

Solving we have, $x = \frac{a\mu}{2}\ y = -\frac{b\mu}{2}$

$\therefore$ Equation (1) can be written as

$$\frac{x-\frac{a\mu}{2}}{ak}=\frac{y+\frac{b\mu}{2}}{b\mu}=\frac{z-0}{c\mu\, k}$$

$\therefore$ Components of R are

$$X = ak\,R,\ Y = bk\,R,\ Z = c\mu k\,R$$

Component of couple

$$L\hat{i}+M\hat{j}+N\hat{k}=\begin{vmatrix} i & j & k \\ \frac{a\mu}{2} & -b\frac{\mu}{2} & 0 \\ ak\,R & bk\,R & c\mu k\,R \end{vmatrix}$$

$$=\frac{-bc\mu^2\,k\,R}{2}\hat{i}-\frac{ac\mu^2\,k\,R}{2}\hat{j}+ab\mu k R\,\hat{k}$$

$$\therefore \quad aL-bM=a\left(\frac{-bc\mu^2\,k\,R}{2}\right)-b\left(\frac{-ac\mu^2\,k\,R}{2}\right)$$

$$=\frac{-abc\mu^2\,k\,R}{2}\frac{abc\mu^2\,k\,R}{2}=0$$

$$bX-aY=bak\,R-abk\,R=0$$

$$cN-abZ=c\,ab\,\mu\,k\,R-abc\,\mu\,k\,R=0$$

$$\therefore \quad aL-bM=bX-aY=cN-abZ=0$$

Similarly, if we consider the second generator we will have

$$aL+bM=0 \quad bX+aY=0 \quad cN+abZ=0.$$

Example 7. *Any number of wrenches of the same pitch act along generators of the same system of the hyperboloid*

$$\frac{x^2}{a^2}+\frac{y^2}{b^2}-\frac{z^2}{c^2}=1$$

Show that they will reduce to a single resultant provided their central axis is parallel to a generator of the cone

$$\left(p+\frac{bc}{a}\right)x^2+\left(p+\frac{ca}{b}\right)y^2+\left(p-\frac{ab}{c}\right)z^2=0.$$

Sol. Let the generators of the hyperboloid be

$$\frac{x-a\cos\theta_r}{a\sin\theta_r}=\frac{y-b\sin\theta_r}{-b\cos\theta_r}=\frac{z-0}{c} \quad \text{...(1)}$$

where $r = 1, 2, 3, \ldots\ldots$.

Let the d.c. of the generators be $\frac{a\sin\theta_r}{\mu_r}, \frac{-b\cos\theta_r}{\mu_r}, \frac{c}{\mu_r}$

where $\mu_r=\sqrt{a^2\sin^2\theta_r+b^2\cos^2\theta_r+c^2}$ $\quad (l^2+m^2+n^2=1)$

Along these generators, suppose the wrenches

$$(R_1, k_1), (R_2, k_2), (R_3, k_3), \ldots\ldots \text{ act.}$$

pitch $\quad p=\frac{k_1}{R_1}=\frac{k_2}{R_2}=\frac{k_3}{R_3}\ \ldots\ldots=\frac{k_r}{R_r}$

$\therefore \quad k_r = pR_r$

Let the system reduces to (X, Y, Z, L, M, N)

where
$$X = \sum R_r \frac{a \sin \theta_r}{\mu_r}$$
$$Y = \sum R_r \left(\frac{-b \cos \theta_r}{\mu_r}\right)$$
$$Z = \sum R_r \frac{c}{\mu_r}$$

L = sum of moment of forces R_r along x-axis + sum of component of couple about x-axis

$$= \sum (y_r Z_r - z_r Y_r) + \sum k_r \frac{a \sin \theta_r}{\mu_r}$$
$$= \sum \left[b \sin \theta_r \,.\, \frac{cR_r}{\mu_r} - 0 \,.\, R_r \left(\frac{-b \cos \theta_r}{\mu}\right)\right] + \sum pR_r \frac{a \sin \theta_r}{\mu_r}$$
$$= bc \sum R_r \frac{\sin \theta_r}{\mu_r} + p \sum R_r \frac{a \sin \theta_r}{\mu_r}$$
$$bc \frac{X}{a} + pX = \left(p + \frac{bc}{a}\right) X$$
$$M = \sum (z_r X_r - x_r Z_r) + \sum k_r \left(\frac{-b \cos \theta_r}{\mu_r}\right)$$
$$= \sum \left(0 - a \cos \theta_r R_r \frac{c}{\mu_r}\right) - p \sum R_r \frac{b \cos \theta_r}{\mu_r}$$
$$= \frac{ac}{b} Y + pY = \left(p + \frac{ac}{b}\right) Y$$
$$N = \sum (x_r Y_r - y_r X_r) + \sum kr \frac{c}{\mu_r}$$
$$= \sum \left\{a \cos \theta_r \left(-R_r \frac{b \cos \theta_r}{\mu_r}\right) - b \sin \theta_r \,.\, R_r \frac{a \sin \theta_r}{\mu_r}\right\} + p \sum R_r \frac{c}{\mu_r}$$
$$= \sum -\frac{ab}{c}\left[\frac{cR_r}{\mu_r}(\cos^2 \theta_r + \sin^2 \theta_r)\right] + pz$$
$$= -\frac{ab}{c} z + pz = \left(p - \frac{ab}{c}\right) z$$

$\because$ System reduces to a single resultant

$$\therefore \quad LX + MY + NZ = 0$$

$$\therefore \quad \left(p + \frac{bc}{a}\right) X^2 + \left(p + \frac{ca}{b}\right) Y^2 + \left(p - \frac{ab}{c}\right) Z^2 = 0 \qquad ...(2)$$

∵ Direction cosine of central axis are proportional to X, Y, Z.

∴ Equation of a straight line parallel to central axis and passing through (0, 0, 0) is

$$\frac{x-0}{X}=\frac{y-0}{Y}=\frac{z-0}{Z} \quad \text{...(3)}$$

Substituting the values of X, Y, Z from (3) in (2), we have

$$\left(p+\frac{bc}{a}\right)x^2+\left(p+\frac{ca}{b}\right)y^2+\left(p-\frac{ab}{c}\right)z^2=0 \quad \text{...(4)}$$

which is the equation of a cone with (3) as the generator.

Thus, the wrenches reduces to a single force if their central axis is parallel to the generator of the cone given by (4).

Example 8. *Show that the following system of forces reduces to a single force*

Force $5\sqrt{2}$ *acting at* $\left(0,0,\frac{1}{5}\right)$ *in the direction of* $3i+4\hat{j}-5\hat{k}$,

Force $3\sqrt{11}$ *acting at* $\left(\frac{1}{3},0,0\right)$ *in the direction of* $5i+5\hat{j}-7\hat{k}$,

and Force $\sqrt{11}$ *acting at* $\left(0,\frac{1}{4},0\right)$ *in the direction of* $-3i+\hat{j}+\hat{k}$.

Sol. The lines of action of the forces $5\sqrt{2}, 3\sqrt{11}$ and $\sqrt{11}$ are

$$\frac{x-0}{3/5\sqrt{2}}=\frac{y-0}{4/5\sqrt{2}}=\frac{z-\frac{1}{5}}{-5/5\sqrt{2}}$$

$$\frac{x-\frac{1}{3}}{\frac{5}{3\sqrt{11}}}=\frac{y-0}{\frac{5}{3\sqrt{11}}}=\frac{z-0}{-7/3\sqrt{11}}$$

$$\frac{x-0}{-3/\sqrt{11}}=\frac{y-\frac{1}{4}}{1/\sqrt{11}}=\frac{z-0}{1/\sqrt{11}}$$

Components of the forces along the axis are

$$X_1=5\sqrt{2}\,\frac{3}{5\sqrt{2}}=3, \qquad Y_1=4, \qquad Z_1=-5$$

$$X_2=3\sqrt{11}\,\frac{5}{3\sqrt{11}}=5, \qquad Y_2=5, \qquad Z_2=-7$$

$$X_3=\sqrt{11}\left(\frac{-3}{\sqrt{11}}\right)=-3, \qquad Y_3=1, \qquad Z_3=1$$

∴ Resultant force is $\vec{R}=X\hat{i}+Y\hat{j}+Z\hat{k}=(3+5-3)\,\hat{i}+(4+5+1)\,\hat{j}+(-5-7+1)\hat{k}$

$$=5\hat{i}+10\hat{j}-11\hat{k}$$

$$L_1\hat{i} + M_1\hat{j} + N_1\hat{k} = \begin{vmatrix} i & j & k \\ 0 & 0 & \frac{1}{5} \\ 3 & 4 & -5 \end{vmatrix} = \frac{-4}{5}\hat{i} + \frac{3}{5}\hat{j}$$

$$L_2\hat{i} + M_2\hat{j} + N_2\hat{k} = \begin{vmatrix} i & j & k \\ \frac{1}{3} & 0 & 0 \\ 5 & 5 & -7 \end{vmatrix} = \frac{7}{3}\hat{j} + \frac{5}{3}\hat{k}$$

$$L_3\hat{i} + M_3\hat{j} + N_3\hat{k} = \begin{vmatrix} i & j & k \\ 0 & \frac{1}{4} & 0 \\ -3 & 1 & 1 \end{vmatrix} = \frac{1}{4}\hat{i} + \frac{3}{4}\hat{k}$$

$\therefore$ Adding, we get $L\hat{i} + M\hat{j} + N\hat{k} = \left(-\frac{4}{5} + \frac{1}{4}\right)\hat{i} + \left(\frac{3}{5} + \frac{7}{3}\right)\hat{j} + \left(\frac{5}{3} + \frac{3}{4}\right)\hat{k}$

$$= \frac{-11}{20}\hat{i} + \frac{44}{15}\hat{j} + \frac{29}{12}\hat{k}$$

Now, $\quad LX + MY + NZ = 5 \times \frac{-11}{20} + 10 \times \frac{44}{15} - 11 \times \frac{29}{12} = 0$

$\therefore$ System reduces to a single force.

EXERCISE 1

1. Forces P, Q, R act along three non intersecting edges of a cube, find the central axis.
2. Equal forces act along the axes and along AB, where A is (1, 2, 3) and B (3, 5, 9). Find the equation of central axis.
3. Three forces each of magnitude P and acting in the positive direction of the axes having their lines of action $-y = z = a$, $-z = x = a$, $-x = y = a$.

   Prove that they are equivalent to a force $P\sqrt{3}$ at the origin and a couple.
4. Three equal forces act along the straight lines $x = 0, y - z = 5$; $y = 0, z - x = 5$; $z = 0, x - y = 5$. Show that they can not reduce to a couple. Prove also that if the system reduces to a single force, its line of action must lie on the surface $x^2 + y^2 + z^2 - 2yz - 2zx - 2xy = 25$.
5. A force P acts along the axis of x and another force 2P along a generator of cylinder $x^2 + y^2 = 4$. Show that the surface on which central axis lies $4(2x - z)^2 + 25y^2 = 64$.
6. If a force (X, Y, Z) acts along a generator of the hyperbolic paraboloid $\frac{x^2}{a^2} - \frac{y^2}{b^2} = \frac{2z}{c}$ and be equivalent to an equal force (X, Y, Z) at the origin together with a couple (L, M, N). Show that $aL \pm bM = 0$, $bX \pm aY = 0$, $cN \pm abZ = 0$.
7. Two equal forces act along each of the straight line

$$\frac{x \pm a\cos\theta}{a\sin\theta} = \frac{x - b\sin\theta}{\pm b\cos\theta} = \frac{z}{c}$$

   Show that their central axis must lie on the surface $y\left(\frac{x}{z} + \frac{z}{x}\right) = b\left(\frac{a}{c} + \frac{c}{a}\right)$ for all values of θ.

8. Forces X, Y, Z act along the three lines given as $y = 0, z = c ; z = 0, x = a ; x = 0, y = b$ respectively.

Prove that pitch of the wrench is $\dfrac{aYZ + bZX + cXY}{X^2 + Y^2 + Z^2}$.

9. Equal forces act along two perpendicular diagonals of opposite faces of a cube of side a. Show that they are equivalent to a single force R acting along the line through the centre of the cube and a couple $\dfrac{aR}{2}$ with the same line for axis. *(K.U. 2007)*

10. Two forces act along the lines $y = 0, z = 0$ and $x = 0, z = c$. As the forces vary, their central axis generates the surface $(x^2 + y^2) z = cy^2$.

11. Forces X, Y and Z act along the straight lines $y = b, z = -c ; x = -a, z = c$ and $x = a, y = -b$ respectively. Show that the system reduces a single resultant if $\dfrac{a}{X} + \dfrac{b}{Y} + \dfrac{c}{Z} = 0$ and the line of action is given by any two of the three

$$\frac{y}{Y} - \frac{z}{Z} - \frac{a}{X} = 0, \ \frac{z}{Z} - \frac{b}{Y} - \frac{x}{X} = 0, \ \frac{x}{X} - \frac{y}{Y} - \frac{c}{Z} = 0.$$

12. OA, OB, OC are three coterminus edges of a cube and AA′, BB′, CC′, OO′ are diagonals. Forces X, Y, Z and R act along BC, CA′, AB′ and OO′ respectively. Show that they are equivalent to a single resultant if $(XY + YZ + ZX)\sqrt{3} + R(X + Y + Z) = 0$

13. A force P act along X-axis and another force nP act along a generator of the cylinder $x^2 + y^2 = a^2$. Show that the central axis lies on the surface $n^2 (nx - z)^2 + (1 + n^2)^2 y^2 = n^4 a^2$.

14. Three forces act along the straight lines $x = 0, y - z = a ; y = 0, z - x = a ; z = 0, x - y = a$. Show that they can not reduce to a couple. Prove also that if the system reduces to a single force, its line of action must lie on $x^2 + y^2 + z^2 - 2xy - 2yz - 2zx = a^2$.

15. Two forces P and Q act along the straight lines $y = x, z = c ; y = -x, z = -c$ respectively. Show that equation of central axis is given by $y = \left(\dfrac{P-Q}{P+Q}\right) x, \ \dfrac{z}{c} = \dfrac{P^2 - Q^2}{P^2 + Q^2}$ and if P and Q vary surface generated by central axis is $(x^2 + y^2) z = 2c\, xy$.

ANSWERS

1. $\dfrac{a(R - Q) - yR + zQ}{P} = \dfrac{-aR - Pz + Rx}{Q} = \dfrac{0 - xQ + yP}{R}$, a being the edge of cube.

3. $\dfrac{3 - 13y + 10z}{9} = \dfrac{-6 - 9z + 13x}{10} = \dfrac{-1 - 10x + 9y}{13}$.

5

Screws and Wrenches

1. Screw and Pitch

If a body rotates about a straight line and at the same times moves forward along this line the body is said to be **serewed** about the line. The axis of the screw is often called a **screw.**

The word is taken from ordinary serew in every day use. As the screw is driven in, it moves forward a length x and turns through an angle θ the ratio x/θ is called the **pitch** of the screw.

Theorem 1. *A given system of force can be replaced by two forces equivalent to the given system, in an infinite number of ways and that the tetrahedron formed by the two forces is of constant volume.* (K.U. 2004)

Proof. Let the given system of forces be reduced to the wrench (R, k) with central axis along OZ. R is the resultant force along OZ. k is the resultant couple about OZ.

Draw $AB \perp OZ$

Let $OA = a,\ OC = b$

Let the given system be equivalent to a force P at A along AC in a plane $\perp$ OA and a force Q at B $\perp$ OB. Let these forces make angle θ and ϕ with Z respectively.

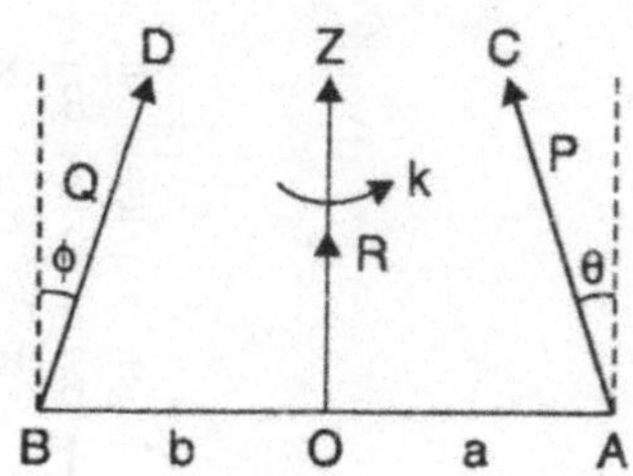

Fig. 1

Resolving forces along and perpendicular to OZ

$$P\cos\theta + Q\cos\phi = R \qquad ...(1)$$

$$P\sin\theta - Q\sin\phi = O \qquad ...(2)$$

Taking moment of the forces about OZ and perpendicular to it.

$$aP\sin\theta + b\,Q\sin\phi = k \qquad ...(3)$$

$$aP\cos\theta - b\,Q\cos\phi = O \qquad ...(4)$$

$$\therefore \quad P\cos\theta\,.\,a = Q\cos\phi\,.\,b$$

$$\frac{P\cos\theta}{b} = \frac{Q\cos\phi}{a} = \frac{P\cos\theta + Q\cos\phi}{a+b} = \frac{R}{a+b} \qquad ...(5)$$

Also, $\quad P\sin\theta = Q\sin\phi$

$$\therefore \quad \frac{aP\sin\theta}{a} = \frac{bQ\sin\phi}{b} = \frac{aP\sin\theta + bQ\sin\phi}{a+b} = \frac{k}{a+b} \qquad ...(6)$$

From (5) and (6), we get

$$P\cos\theta = \frac{bR}{a+b},\quad P\sin\theta = \frac{ak}{a\,(a+b)} \qquad ...(7)$$

Squaring and adding

$$P^2 = \frac{b^2 R^2 + k^2}{(a+b)^2} \quad ...(8)$$

Again from (5) and (6)

$$Q \cos \phi = \frac{aR}{a+b}, \quad Q \sin \phi = \frac{k}{a+b} \quad ...(9)$$

$$\Rightarrow \quad Q^2 = \frac{a^2R^2 + k^2}{(a+b)^2} \quad ...(10)$$

From (7) we have $\tan \theta = \dfrac{k}{bR}$ ...(11)

From (9) we have $\tan \phi = \dfrac{k}{aR}$ ...(12)

Thus, we have P, Q , θ and φ in terms of *a*, *b*, R and *k*.

∵ *a* and *b* are arbitrary we have infinite solutions for P, Q, θ and φ.

∴ The system can be replaced by two forces in infinite number of ways.

Further volume of tetrahedron

$$= \frac{1}{3} \text{ Area of } \Delta ABC \times \text{perpendicular from D upon ABC}$$

$$= \frac{1}{3} \times \frac{1}{2} (AB) \times (AC) . BD \sin (\theta + \phi)$$

$$= \frac{1}{6} . (a + b) P . Q \sin (\theta + \phi)$$

$$= \frac{1}{6} (a + b) P . Q (\sin \theta \cos \phi + \cos \theta \sin \phi)$$

$$= \frac{1}{6} (a + b) [P \sin \theta\, Q \cos \phi + P \cos \theta . Q \sin \phi]$$

$$= \frac{1}{6} (a + b) \left[\frac{k}{a+b} \frac{aR}{a+b} + \frac{bR}{a+b} . \frac{k}{a+b}\right]$$

$$= \frac{1}{6} kR \frac{(a+b)(a+b)}{(a+b)^2} = \frac{1}{6} kR$$

= A constant quantity.

2. Resultant Wrench of two given Wrenches

Suppose there are two wrenches (R_1, k_1) and (R_2, k_2) with their central axes along AB and CD. Let θ be the angle between AB and CD. Let AC = *a*. Let resultant wrench (R, *k*) be along OZ ⊥ AC. Let OA = *x*, OC = *a* – *x*. If AB makes an angle θ with OZ then CD makes θ – α with OZ. Resolving along and perpendicular to OZ.

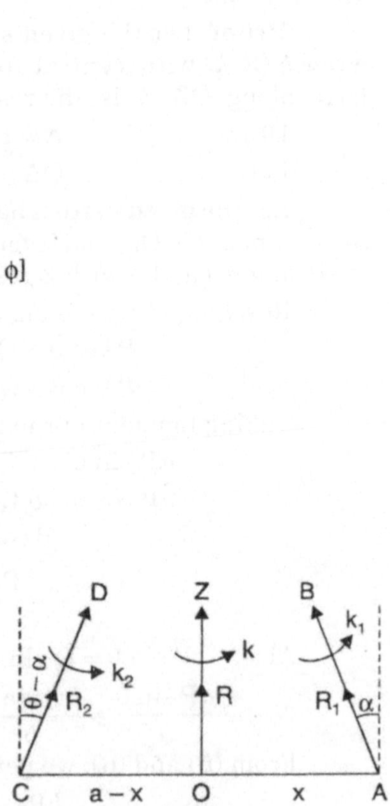

Fig. 2

$$R_1 \cos \alpha + R_2 \cos (\theta - \alpha) = R \quad ...(1)$$

$$R_1 \sin \alpha - R_2 \sin (\theta - \alpha) = 0 \quad ...(2)$$

Taking moment along and perpendicular to OZ

$$k_1 \cos\alpha + k_2 \cos(\theta-\alpha) + R_1 \sin\alpha \,.\, x + R_2 \sin(\theta-\alpha)(a-x) = k \qquad ...(3)$$

$$k_1 \sin\alpha - k_2 \sin(\theta-\alpha) - R_1 \cos\alpha \,.\, x + R_2 \cos(\theta-\alpha)\,.\,(a-x) = 0 \qquad ...(4)$$

Squaring and adding (1) and (2), we have

$$R_1^2 + R_2^2 + 2R_1 R_2 \{\cos\alpha \cos(\theta-\alpha) - \sin\alpha \sin(\theta-\alpha)\} = R^2$$

$$\mathbf{R^2 = R_1^2 + R_2^2 + 2R_1 R_2 \cos\theta} \qquad ...(5)$$

Using (2), equation (3) becomes

$$k_1 \cos\alpha + k_2 \cos(\theta-\alpha) + xR_1 \sin\alpha + (a-x) R_1 \sin\alpha = k \qquad ...(6)$$

Using (1), equation (4) becomes

$$k_1 \sin\alpha - k_2 \sin(\theta-\alpha) - x\{R - R_2\cos(\theta-\alpha)\} + (a-x) R_2 \cos(\theta-\alpha) = 0$$

$$k_1 \sin\alpha - k_2 \sin(\theta-\alpha) - xR + aR_2 \cos(\theta-\alpha) = 0 \qquad ...(7)$$

From (2), $R_1 \sin\alpha = R_2 [\sin\theta \cos\alpha - \cos\theta \sin\alpha]$

$\sin\alpha (R_1 + R_2 \cos\theta) = R_2 \sin\theta \cos\alpha$

$$\frac{\sin\alpha}{R_2 \sin\theta} = \frac{\cos\alpha}{R_1 + R_2\cos\theta} = \frac{\sqrt{\sin^2\alpha + \cos^2\alpha}}{\sqrt{R_2^2 \sin^2\theta + R_1^2 + R_2^2 \cos^2\theta + 2R_1 R_2 \cos\theta}}$$

$$\therefore \quad \frac{\sin\alpha}{R_2 \sin\theta} = \frac{\cos\alpha}{R_1 + R_2\cos\theta} = \frac{1}{\sqrt{R_1^2 + R_2^2 + 2R_1 R_2\cos\theta}} = \frac{1}{R} \qquad ...(8)$$

Multiplying (3) by R and using equation (2)

$$Rk = Rk_1 \cos\alpha + Rk_2 (\cos\theta\cos\alpha + \sin\theta\sin\alpha) + a\, R_1 R \sin\alpha.$$

Putting the values of $R\sin\alpha$ and $R\cos\alpha$ from (8)

$$Rk = k_1 (R_2 + R_2\cos\theta) + k_2 \cos\theta (R_1 + R_2\cos\theta) + (k_2 \text{Sin}\,\theta + aR_1 R_2 \sin\theta)$$

$$= (k_1 + k_2 \cos\theta)(R_1 + R_2\cos\theta) + (k_2 \sin\theta + aR_1) R_2 \sin\theta$$

$$\mathbf{Rk = R_1 k_1 + R_2 k_2 + (R_1 k_2 + R_2 k_1) + \cos\theta + a R_1 R_2 \sin\theta} \qquad ...(9)$$

$\because$ θ is given and R is obtained from (5) we can determine the value of k.

To find x. Using (7)

$$xR = k_1 \sin\alpha - k_2 \sin(\theta-\alpha) + aR_2 \cos(\theta-\alpha)$$

$$xR^2 = k_1 R\sin\alpha - k_2 R[\sin\theta\cos\alpha - \cos\theta\sin\alpha] + aR_2 R[\cos\theta\cos\alpha + \sin\theta\sin\alpha]$$

Putting $R\sin\alpha$ and $R\cos\alpha$ from (8)

$$xR^2 = (k_1 + k_2\cos\theta + a R_2 \sin\theta) R_2 \sin\theta + (a R_2 \cos\theta - k_2 \sin\theta)(R_1 + R_2\cos\theta)$$

$$= (R_2 k_1 - R_1 k_2)\sin\theta + aR_2^2 (\sin^2\theta + \cos^2\theta) + aR_1 R_2 \cos\theta$$

$$= (R_2 k_1 - R_1 k_2)\sin\theta + aR_2^2 + aR_1 R_2 \cos\theta$$

$$\mathbf{xR^2 = (R_2 k_1 - R_1 k_2)\sin\theta + aR_2(R_2 + R_1\cos\theta)} \qquad ...(10)$$

From (10) we can determine the position of central axis of resultant wrench.

Corollary. Find resultant wrench of two forces R_1 and R_2 inclined at an angle θ.

Sol. Putting $k_1 = 0$, $k_2 = 0$ in (5), (9) and (10), we have

$$\mathbf{R^2 = R_1^2 + R_2^2 + 2R_1 R_2 \cos\theta} \qquad ...(11)$$

$$Rk = R_1 \,.\, 0 + R_2 \,.0 + (R_1 \,.\, 0 + R_2 \,.0)\cos\theta + aR_1 R_2 \sin\theta$$

$$\therefore \quad \mathbf{Rk = aR_1 R_2 \sin\theta} \qquad ...(12)$$

$$xR^2 = 0 - 0 + aR_2(R_2 + R_1\cos\theta)$$

$$\mathbf{xR^2 = aR_2(R_2 + R_1\cos\theta)} \qquad ...(13)$$

Equations (11), (12) and (13) gives the intensity R of wrench (R, k) and position of the central axis of the wrench.

Theorem 2. *The axes of two given wrenches intersect at right angles. Their intensities are R_1 and R_2 and pitches p_1 and p_2. Find the locus of central axis.*

Sol. Let the axes of two wrenches be OX and OY which are two axes of coordinates.

Let R be the resultant of R_1 and R_2 along OA inclined at angle θ with OX,

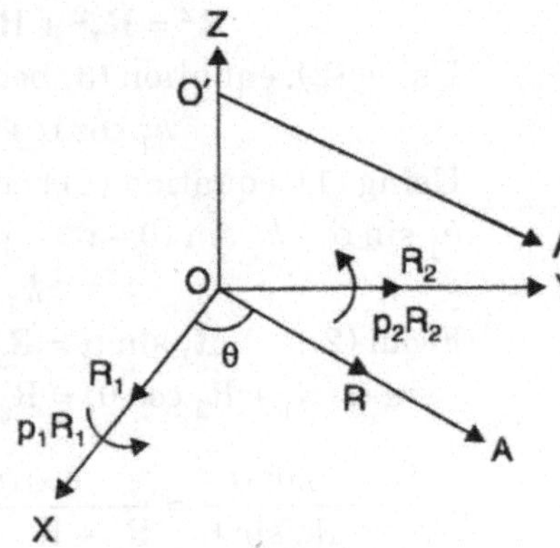

Fig. 3

Resolving the forces

$$R\cos\theta = R_1 \quad \text{...(1)}$$

$$R\sin\theta = R_2$$

$$\therefore \quad R^2 = R_1^{\,2} + R_2^{\,2}$$

$$\tan\theta = \frac{R_2}{R_1}$$

Let the resultant couple G makes an angle ϕ with OA

$$\therefore \quad G\cos\phi = \text{couple about OA}$$

$$= p_1 R_1 \cos\theta + p_2 R_2 \sin\theta$$

$$= (p_1\cos^2\theta + p_2\sin^2\theta)\,R \quad \text{...(2)}$$

$$G\sin\phi = \text{Couple about a line perpendicular to OA}$$

$$= -p_1 R_1 \sin\theta + p_2 R_2 \cos\theta = (p_2 - p_1)\,R\sin\theta\cos\theta \quad \text{...(3)}$$

Hence, we have a single force R at O along OA and a couple of moment G whose axis passes through A. Further the couple G at O and force R along OA are equivalent to a wrench whose axis is a line O′ A′ parallel to OA.

$$OO' = \frac{G\sin\phi}{R} = \frac{(p_2 - p_1)\,R\sin\theta\cos\theta}{R} = (p_2 - p_1)\sin\theta\cos\theta$$

This gives the position of central axis O′ A′ of the wrench whose force is R and a couple G cos ϕ about OA as axis

$$\text{Pitch} \quad p = \frac{k}{R} = \frac{G\cos\phi}{R} = \frac{(p_1\cos^2\theta + p_2\sin^2\theta)\,R}{R}$$

$$\therefore \quad p = p_1\cos^2\theta + p_2\sin^2\theta\,. \quad \text{...(4)}$$

Let A′ (x, y, z) be on O′ A′, $\tan\theta = \frac{y}{x}$

$$\Rightarrow \quad \sin\theta = \frac{y}{\sqrt{x^2+y^2}} \quad \cos\theta = \frac{x}{\sqrt{x^2+y^2}}$$

$$\therefore \quad z = O\,O' = (p_2 - p_1)\sin\theta\cos\theta = (p_2 - p_1)\frac{xy}{x^2+y^2} \quad \text{...(5)}$$

$\therefore$ Central axis generates the surface

$$(x^2 + y^2)\,z = (p_2 - p_1)xy$$

Resolution of a wrench.

Theorem 3. *A wrench may be resolved into two wrenches whose axes intersect at right angle in an infinite number of ways.*

Proof. Refer to Fig. (3).

Let the wrench be (R, pR) about O′ A′ as the axis. Draw a line O′O ⊥ O′ A′. Take O as origin and O O′ as Z-axis. Let OO′ = Z. Draw X and Y respectively taking O as origin.

Let there be two wrenches (R_1, p_1R_1), (R_2, p_2R_2) about OX and OY respectively such that they are equivalent to (R, pR) about O′ A′. Draw a line OA parallel to O′ A′ in XY plane. Let OA makes an angle θ with X-axis.

From (1), (4) and (5) of Theorem 2

$$R_1 = R\cos\theta,\ R_2 = R\sin\theta \qquad \text{...(1)}$$

$$p = p_1\cos^2\theta + p_2\sin^2\theta \qquad \text{...(2)}$$

$$\frac{z}{\sin\theta\cos\theta} = p_2 - p_1 \qquad \text{...(3)}$$

Solving (2) and (3), we have

$$p_1 = p - z\tan\theta$$

$$p_2 = p + z\cot\theta$$

Thus, we have the value of R_1, R_2, p_1R_1 and p_2R_2 in terms of z and θ.

∵ z and θ can take any value hence (R, pR) can be resolved in infinite number of ways.

∴ (R_1, p_1R_1) and (R_2, p_2R_2) with their axes at right angle.

3. Work done by a Wrench about a given Screw

Let O′ A be the axis of the wrench (R, pR) and OX be the axis of the screw whose pitch is p_1. Let OO′ = h be shortest distance between O′ A and OX. Draw O′ M and O′ N parallel to OX and OY respectively such that O′ N ⊥ O′ M and O′O component of R along O′ A are R cos θ along O′M and R sin θ along O′N.

Further R cos θ and R sin θ are equivalent to R cos θ and R sin θ along OX and OY respectively, together with couple hR cos θ along OY and – hr sin θ about OX.

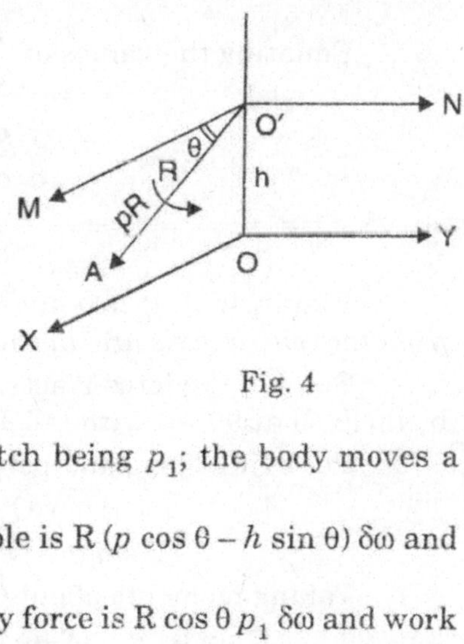

Fig. 4

Couple pR about O′A is equivalent to pR cos θ about O′M and pR sin θ about O′N.

Since the couple can be shifted to any where parallel to its axis, these couples are equivalent to pR cos θ about OX and pR sin θ about OY.

∴ The wrenches (R, pR) is equivalent to pR cos θ – hR sin θ about OX and pR sin θ + hR cos θ about OY.

If a small twist of angle δω is given about OX, the pitch being p_1; the body moves a distance $p_1 \cdot \delta\omega$ along OX.

∴ Due to angular displacement work done by couple is $R(p\cos\theta - h\sin\theta)\,\delta\omega$ and work done by force is zero.

Due to linear displacement $p_1\,\delta\omega$ along OX work done by force is $R\cos\theta\, p_1\,\delta\omega$ and work done by couple is zero.

∴ Total work done $= R(p\cos\theta - h\sin\theta)\,\delta\omega + R\cos\theta\, p_1\,\delta\omega$

$$= R\,\delta\omega\,\{(p + p_1)\cos\theta - h\sin\theta\}.$$

Def. Reciprocal Screw. When a wrench acting on one screw does no work as the body is given a small twist about another screw the two screws are said to be reciprocal.

$\because$ Work done by wrench (R, pR) is

$R\,\delta\omega\,\{(p + p_1)\cos\theta - h\sin\theta\}$

Work done is zero if $(p + p_1)\cos\theta = h\sin\theta$

This is necessary condition for screws of pitch p and p_1 to be reciprocal.

If axis of two reciprocal screws intersect shortest distance, $h = 0$

$\therefore \quad (p + p_1)\cos\theta = 0 \quad \Rightarrow \quad p = -p_1 \quad \text{or} \quad \theta = 90°$

$\Rightarrow$ Either axes are at right angle or pitches are equal and opposite.

Example 1. *If the position of the central axis of forces P and Q is given and line of action of P is inclined at α with the axis. Show that locus of Q is a conicoid.*

Sol. Let the lines of action of P and Q make angles α, β with Z-axis, where α is fixed and β vary. Consider a point $(-x, -y, z)$ on BQ. Let OA = a.

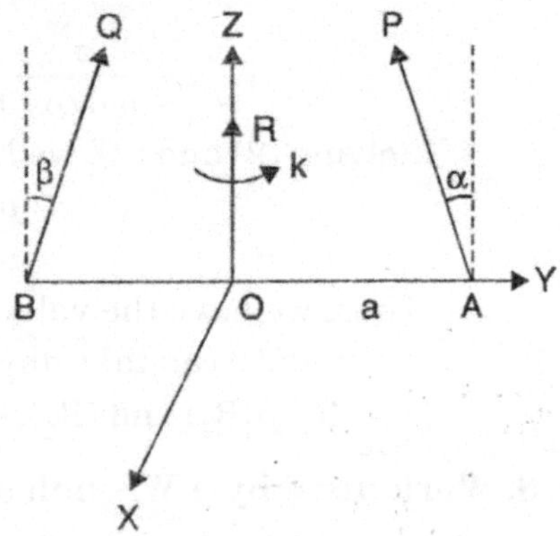

Fig. 5

Resolving the forces along OZ and OX

$$P\cos\alpha + Q\cos\beta = R \quad \text{...(1)}$$

$$P\sin\alpha - Q\sin\beta = 0 \quad \text{...(2)}$$

Taking moment about OZ and OX

$$P\sin\alpha \,.\, a + Q\sin\beta \,.\, y = k \quad \text{...(3)}$$

$$-P\cos\alpha \,.\, a + Q\cos\beta \,.\, y = 0 \quad \text{...(4)}$$

From (4), $\dfrac{P}{Q} = \dfrac{y}{a}\,\dfrac{\cos\beta}{\cos\alpha}$

From (2), $\dfrac{P}{Q} = \dfrac{\sin\beta}{\sin\alpha}$

Equating the values of $\dfrac{P}{Q}$

$$\frac{y}{a}\,\frac{\cos\beta}{\cos\alpha} = \frac{\sin\beta}{\sin\alpha} \quad \Rightarrow \quad a\tan\beta = y\tan\alpha$$

$$\Rightarrow \quad \frac{ax}{z} = y\tan\alpha \quad \Rightarrow \quad ax = yz\tan\alpha.$$

Example 2. *If two non-intersecting forces P and Q are perpendicular, their distances from the central axis are in the ratio as $Q^2 : P^2$.*

Sol. Let the force P acts at A inclined at α with OZ and Q be inclined at $90 - \alpha$ with OZ. Let OA = a, OB = b.

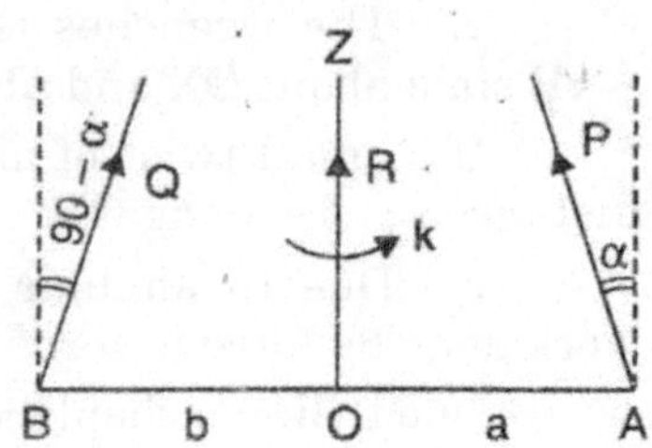

Fig. 6

Resolving along and perpendicular to OZ

$$P\cos\alpha + Q\cos(90 - \alpha) = R \quad \text{...(1)}$$

$$P\sin\alpha - Q\sin(90 - \alpha) = 0 \quad \text{...(2)}$$

Taking moments about OZ and $\perp$ to it

$$P\sin\alpha \,.\, a + Q\sin(90 - \alpha) \,.\, b = k \quad \text{...(3)}$$

$$P\cos\alpha \,.\, a - Q\cos(90 - \alpha) \,.\, b = 0 \quad \text{...(4)}$$

From (2), $$\frac{P}{Q} = \frac{\cos\alpha}{\sin\alpha} \quad ...(5)$$

From (4), $$\frac{P}{Q} = \frac{b\sin\alpha}{a\cos\alpha} \quad ...(6)$$

From (5) and (6), $$\frac{P^2}{Q^2} = \frac{\cos\alpha}{\sin\alpha}\cdot\frac{b\sin\alpha}{a\cos\alpha}$$

$$\frac{P^2}{Q^2} = \frac{b}{a}$$

$\therefore$ $a : b = Q^2 : P^2$.

Example 3. *Show that the distance between two forces P and Q equivalent to wrench (R, k) inclined at $\pi/2$ is* $\frac{kR}{PQ}$.

Sol. Let the forces P and Q act along AC and BD. They are equivalent to wrench (R, k) with OZ as central axis.

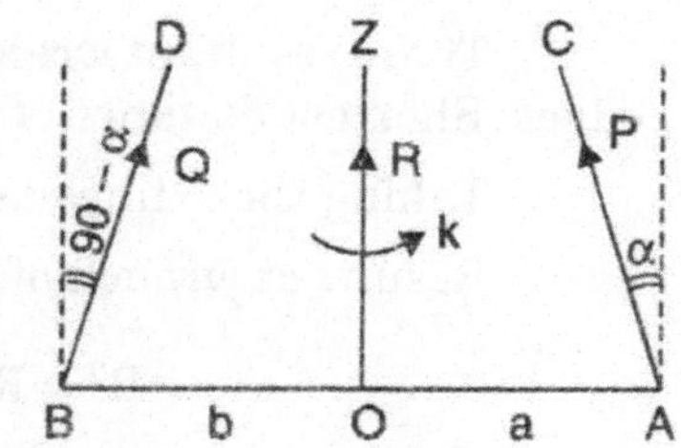

Fig. 7

Resolving along and perpendicular to OZ.

$$P\cos\alpha + Q\cos(90-\alpha) = R \quad ...(1)$$
$$P\sin\alpha - Q\sin(90-\alpha) = 0 \quad ...(2)$$

Taking moment about OZ and $\perp$ to it

$$P\sin\alpha \,.\, a + Q\sin(90-\alpha)\,.\, b = k \quad ...(3)$$
$$-P\cos\alpha\,.\, a + Q\cos(90-\alpha)\,.\, b = 0 \quad ...(4)$$

Multiplying (1) and (3)

$$R\,.\,k = (P\cos\alpha + Q\sin\alpha)(a\,P\sin\alpha + b\,Q\cos\alpha)$$
$$= aP^2\,.\cos\alpha\,.\sin\alpha + b\,P\,Q\cos^2\alpha + a\,PQ\sin^2\alpha + b\,Q^2\sin\alpha\cos\alpha$$

From (2), $P\sin\alpha = Q\cos\alpha$

$$\therefore \quad R.k = a\,P\,.\cos\alpha\,.\,Q\cos\alpha + b\,PQ\cos^2\alpha + a\,PQ\sin^2\alpha + b\,PQ\sin^2\alpha$$
$$= PQ(a+b)\cos^2\alpha + PQ(a+b)\sin^2\alpha$$
$$= (a+b)PQ(\cos^2\alpha + \sin^2\alpha) = (a+b)PQ$$

$$\therefore \quad a + b = \frac{kR}{PQ}.$$

Example 4. *Forces each equal to P act along the edges BC, DA, CA, DB, AB and DC of a regular tetrahedron. Find the resultant wrench.*

Sol. Let the forces P each act along the edges of a regular tetrahedron as shown in the Fig. 8. We know that resultant wrench (R, k) of two forces R_1 and R_2 inclined at α is given by

$$R^2 = R_1^2 + R_2^2 + 2R_1R_2\cos\alpha$$
$$Rk = a\,R_1R_2\sin\alpha \text{ where } a \text{ is shortest distance}$$

between R_1 and R_2

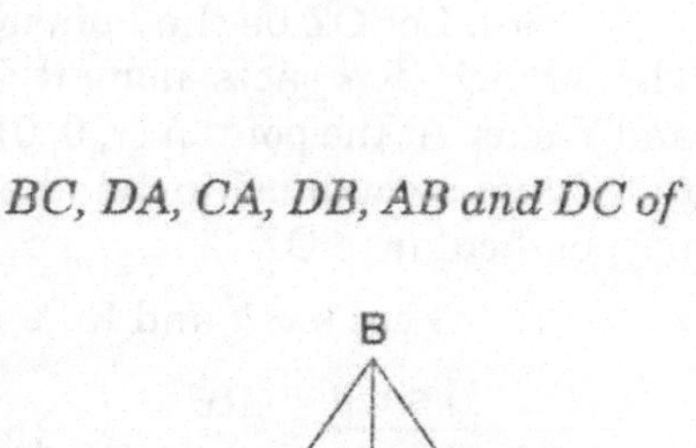

Fig. 8

We know that shortest distance between two opposite edges of regular tetrahedron of side a is $\frac{a}{\sqrt{2}}$ where a is the length of each edge. Also opposite edges are perpendicular to each other.

$\therefore$ Resultant of P along AB and DC is

$$R_1^2 = P^2 + P^2 + 2P.\,P\cos 90^\circ \quad\Rightarrow\quad R_1 = P\sqrt{2}$$

$$R_1\,k_1 = \frac{a}{\sqrt{2}}\,P.P.\sin 90^\circ \quad\Rightarrow\quad R_1\,k_1 = \frac{aP^2}{\sqrt{2}}$$

$$\Rightarrow \qquad k_1 = \frac{aP}{2}$$

Similarly, Resultants of each along DA and BC is R_2

$$= P\sqrt{2}, \qquad k_2 = \frac{aP}{2}$$

and Resultant of P each along CA and DB

$$R_3 = P\sqrt{2}, \qquad k_3 = \frac{aP}{2}$$

These resultant wrenches act along the lines ⊥ to the lines of shortest distance of opposite edges. Shortest distance of opposite edges are mutually perpendicular and concurrent.

Taking these lines as coordinate axes.

Resultant wrench of three wrenches is

$$R^2 = R_1^2 + R_2^2 + R_3^2 = 2P^2 + 2P^2 + 2P^2 \quad\Rightarrow\quad R = P\sqrt{6}$$

$$k^2 = k_1^2 + k_2^2 + k_3^2 = \frac{a^2.P^2}{4} + \frac{a^2.P^2}{4} + \frac{a^2.P^2}{4}$$

$$\Rightarrow \qquad k = \frac{aP\sqrt{3}}{2}.$$

Example 5. *Prove that surface which is traced out by the axis of principal moment at points lying on a straight line which intersects at right angles to the Poinsot's axis of a given system of forces, is a hyperbolic paraboloid.*

Sol. Let OZ be the Poinsot's axis (central axis). The wrench (R, k) acts along it. Take OX and OY on X and Y-axis. At the point A (x, 0, 0), the force R is parallel to OZ and a couple G inclined at an angle θ to R and perpendicular to OX.

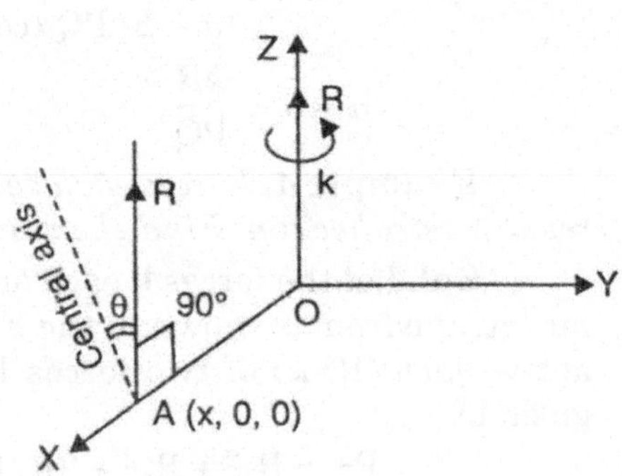

Fig. 9

$\therefore$ $G\cos\theta = k$ and $R.x = G\sin\theta$

$$\therefore \quad \frac{G\sin\theta}{G\cos\theta} = \frac{Rx}{k} \quad\Rightarrow\quad \tan\theta = \frac{Rx}{k} \qquad ...(1)$$

If (x, y, z) be any point on the axis of G, then

$$\tan\theta = \frac{y}{z}$$

Comparing (1) and (2), $\frac{Rx}{k} = \frac{y}{z} \Rightarrow Rxz = yk$

which is a hyperbolic paraboloid.

Example 6. *Two wrenches of pitches p and p′ whose axes are at a distance 2a from each other, have a resultant wrench of pitch ω, whose axis intersects the shortest distance between the axes of the given wrenches at a distance ξ from its middle point. Prove that the angle between the given wrenches is given by* $\tan\theta = \frac{\xi(p-p') - a(2\omega - p - p')}{\xi^2 - a^2 + (\omega - p)(\omega - p')}$.

Sol. Let the two wrenches be (R_1, pR_1), $(R_2, p'R_2)$ and their resultant wrench be $(R, \omega R)$

$$p = \frac{k}{R} \Rightarrow k = pR$$

Let axis of wrench (R_1, pR_1) be taken parallel to X-axis at A and axis of $(R_2, p'R_2)$ makes an angle θ with X-axis at B.

$$AB = 2a \quad \text{(given)}$$

Let C be the middle point of AB.

Let the resultant wrench act at O along the line which has direction cosines (l, m, n). Choose O as origin. $OC = \xi$ (given)

∴ S.D. between R and R_1 and R and R_2 are $OA = a - \xi$

$OB = a + \xi$ respectively.

Resolving forces along X and Y-axis

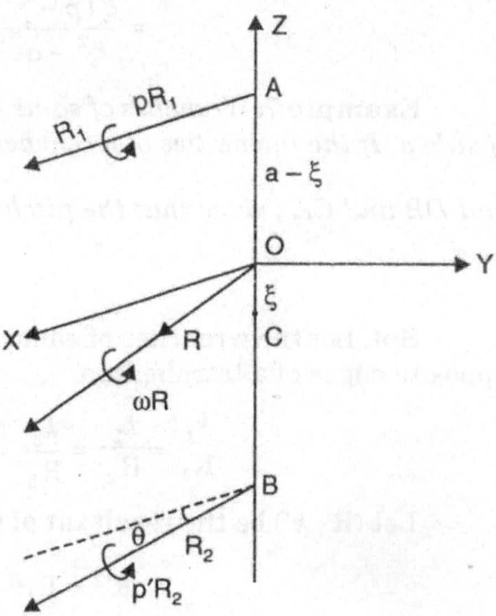

Fig. 10

$$R \,.\, l = R_1 + R_2 \cos\theta \quad ...(1)$$

$$R \,.\, m = R_2 \sin\theta \quad ...(2)$$

Taking moment about X-axis and Y-axis

$$R \,.\, \omega \,.\, l = pR_1 + p'R_2 \cos\theta + R_2 \sin\theta\,(a + \xi) \quad ...(3)$$

$$R \,.\, \omega \,.\, m = R_1 (a - \xi) + p'R_2 \sin\theta - R_2 \cos\theta\,(a + \xi) \quad ...(4)$$

Putting the values of $R \,.\, l$ from (1) into (3)

$$\omega(R_1 + R_2\cos\theta) = pR_1 + p'R_2\cos\theta + R_2\sin\theta\,(a + \xi)$$

or $$(\omega - p)R_1 = R_2\,[p'\cos\theta + (a + \xi)\sin\theta - \omega\cos\theta]$$

or $$(\omega - p)R_1 = R_2\,[(p' - \omega)\cos\theta + (a + \xi)\sin\theta] \quad ...(5)$$

Putting the value of Rm from (2) into (4)

$$\omega R_2 \sin\theta = R_1(a - \xi) + p'R_2\sin\theta - R_2\cos\theta\,(a + \xi)$$

or $$(\xi - a)\,R_1 = R_2\,[(p' - \omega)\sin\theta - (a + \xi)\cos\theta] \qquad ...(6)$$

Dividing (5) by (6)

$$\frac{\omega - p}{\xi - a} = \frac{(p' - \omega)\cos\theta + (a + \xi)\sin\theta}{(p' - \omega)\sin\theta - (a + \xi)\cos\theta}$$

or $$(\omega - p)(p' - \omega)\sin\theta - (\omega - p)(a + \xi)\cos\theta = (\xi - a)(p' - \omega)\cos\theta + (a + \xi)(\xi - a)\sin\theta$$

or $$\{(\omega - p)(p' - \omega) + (a - \xi)(a + x)\}\sin\theta = \{(\xi - a)(p' - \omega) + (\omega - p)(a + \xi)\}\cos\theta$$

$$\tan\theta = \frac{\xi(p' - \omega + \omega - p) - a(p' - \omega + p - \omega)}{(\omega - p)(p' - \omega) + (a^2 - \xi^2)}$$

$$= \frac{\xi(p - p') - a(2\omega - p - p')}{\xi^2 - a^2 + (\omega - p)(\omega - p')}.$$

Example 7. *Wrenches of same pitch p act along the edges of a regular tetrahedron ABCD of side a. If the intensities of wrenches along AB and DC are same and also along BC and DA and DB and CA ; show that the pitch of equivalent wrench is* $p + \dfrac{a}{2\sqrt{2}}$.

(K.U. 2005 ; M.D.U. 2005)

Sol. Let the wrenches of same pitch be (R_1, k_1), (R_2, k_2) and (R_3, k_3) act along the pair of opposite edges of a tetrahedron.

$$\therefore \quad \frac{k_1}{R_1} = \frac{k_2}{R_2} = \frac{k_3}{R_3} = p \qquad ...(1)$$

Let (R', k') be the resultant of (R_1, k_1) along AB and (R_1, k_1) along DC

$$R'^2 = R_1^2 + R_1^2 + 2R_1 . R_1 \cos 90 \;\Rightarrow\; R' = R_1\sqrt{2} \qquad ...(2)$$

$$R'k' = R_1k_1 + R_1k_1 + (R_1k_1 + R_1k_1)\cos 90 + \frac{a}{\sqrt{2}} . R_1R_1 \sin 90$$

$$= 2R_1k_1 + \frac{aR_1^2}{\sqrt{2}} \qquad \text{using (1)}$$

$$= 2R_1 . pR_1 + \frac{aR_1^2}{\sqrt{2}}$$

$$k' = \frac{R_1^2\left(2p + \frac{a}{\sqrt{2}}\right)}{R'} = \frac{R_1^2\left(2p + \frac{a}{\sqrt{2}}\right)}{R_1\sqrt{2}}$$

$$= \left(p\sqrt{2} + \frac{a}{2}\right)R_1 \qquad ...(3)$$

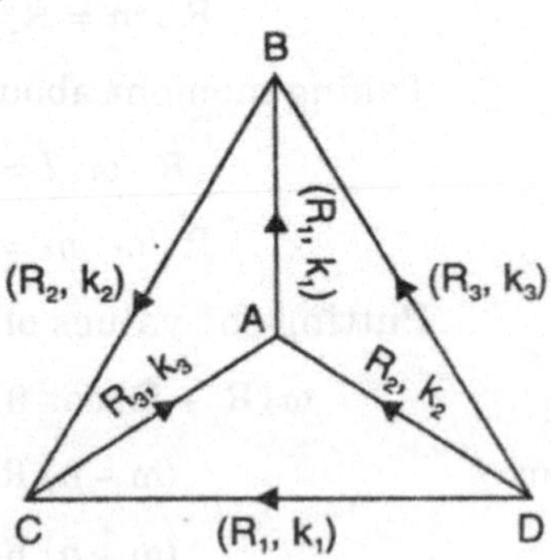

Fig. 12

Similarly, let resultant of (R_2, k_2) and (R_2, k_2) along BC and DA be (R'', x'')

where $R'' = R_2\sqrt{2}$ and $k'' = \left(p\sqrt{2} + \frac{a}{2}\right)R_2$ ...(4)

and similarly, for (R_3, k_3) and (R_3, k_3) resultant wrench (R''', k''') is given as

$$R''' = R_3\sqrt{2} \text{ and } k''' = \left(p\sqrt{2} + \frac{a}{2}\right)R_3 \qquad \text{...(5)}$$

The three resultant wrenches act along the three lines of shortest distance of opposite edges of tetrahedron are mutually perpendicular and concurrent.

Let these line be taken as coordinates axes.

Let (R, k) be the resultant of these wrenches.

$$R^2 = R'^2 + R''^2 + R'''^2 = 2(R_1^2 + R_2^2 + R_3^2)$$

$$k^2 = k'^2 + k''^2 + k'''^2 = \left(p\sqrt{2} + \frac{a}{2}\right)^2 (R_1^2 + R_2^2 + R_3^2)$$

$$\therefore \quad \frac{k^2}{R^2} = \frac{\left(p\sqrt{2} + \frac{a}{2}\right)^2}{\sqrt{2}}$$

$$\text{Pitch} = \frac{k}{R} = \frac{p\sqrt{2} + \frac{a}{2}}{\sqrt{2}} = p + \frac{a}{2\sqrt{2}}.$$

EXERCISE 1

1. If the forces P and 2P act along two non-intersecting perpendicular lines show that the distances of the lines of action of the forces from the central axis are in the ratio of 4 : 1.
2. Show that the minimum distance between two forces which are equivalent to a given system (R, k) and which are inclined at a given angle 2α, is $\frac{2k}{R}\cot\alpha$ and that the forces are then each equal to $\frac{R}{2}\sec\alpha$. *(K.U. 2007)*
3. Forces act along the edges of a regular tetrahedron *viz.* P along BC and DA, Q along CA and DB, and R along AB and DC. Show that the pitch of the equivalent wrench is $\frac{a}{2\sqrt{2}}$ where a is the length of the edge of the tetrahedron.
4. Show that a wrench of which the force is R and the pitch is ϖc may be replaced by forces inclined at an angle 2θ to each other, the shortest distance between them being $2c$ and their magnitudes are $\frac{R}{2}[\sqrt{1+\varpi\tan\theta} \pm \sqrt{1-\varpi\cot\theta}]$.

5. Two forces P and Q are such that their central axis is given in position and the line of action of P is given. Show that the locus of the line of action of Q is a conicoid.

6. Two wrenches of pitches p and q have axes at a distance $2a$ from one another. If resultant wrench is of pitch ω and its axis is equidistant from the axis of component wrenches, show that the angle between them is given by

$$\tan \theta = \frac{a\,(2\omega - p - q)}{a^2 - (\omega - p)\,(\omega - q)}.$$

6

*Null Lines and Null Planes

Definition. Null Line. If the moment of a system of force about a line vanishes, the line is called a null line.

Let the system of forces acting on a rigid body be reduced to single force R to O acting along OB and a couple of moment G whose axis is OA. Draw OC ⊥ OA.

∵ Line of action of R cuts OC, hence moment of R is zero about OC.

∵ OC is ⊥ line of action of G, therefore the component of G about OC is zero.

∴ Sum of moments of (R, G) is zero about OC.

∴ OC is the null line at O and is in a plane normal to OA.

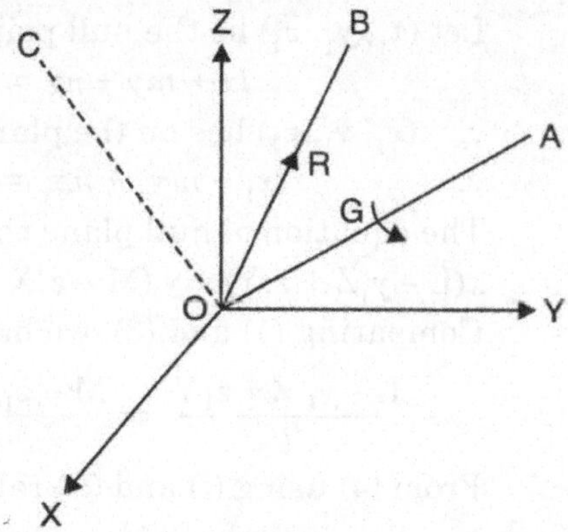

Fig. 1

Def. Null Plane. The plane containing all the null lines is called null plane of the point O.

Def. Null Point. The point O is called the null point.

1. Equation of Null Plane

Let the given system of forces (R, G) have the component X, Y, Z and L, M, N w.r.t. OX, OY, OZ. Let the axes be shifted parallel to itself at O′ (*a*, *b*, *c*). Let moment of couple w.r.t. new axis be L′, M′, N′ where

$$L' = L - bZ + cY$$

$$M' = M - cX + aZ$$

(Refer article 5, *Forces in three dimensions*)

$$N' = N - aY + bX$$

L′, M′, N′ are direction ratio of the axis of the resultant couple at (*a*, *b*, *c*).

Let (*x*, *y*, *z*) be the coordinates of any point in the null plane of (*a*, *b*, *c*).

∴ D.R. of line joining (*x*, *y*, *z*) to (*a*, *b*, *c*) are

$$x - a\ ,\ y - b,\ \ z - c$$

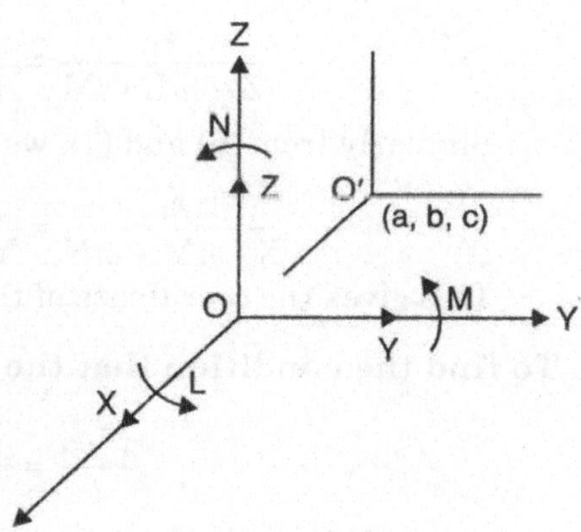

Fig. 2

*Not included in M.D.U. 2007–2008 syllabus.

$\because$ Null line is $\perp$ axis of the couple

$\therefore$ $(x-a)\,L' + (y-b)\,M' + (z-c)\,N' = 0 \quad (l_1 l_2 + m_1 m_2 + n_1 n_2 = 0)$

or $(x-a)(L - bZ + cY) + (y-b)(M - cX + aZ) + (z-c)(N - aY + bX) = 0$

or $x(L - bZ + cY) + y(M - cX + aZ) + Z(N - aY + bX) = aL + bM + cN$ ...(1)

$\therefore$ Equation (1) is the null plane of (a, b, c). It can also be written as

$$\begin{vmatrix} x & y & z \\ a & b & c \\ X & Y & Z \end{vmatrix} = L(x-a) + M(y-b) + N(z-c)$$

2. To find the Null Point of a given Plane (*K.U.* 2007)

Let (x_1, y_1, z_1) be the null point of the plane

$$lx + my + nz = 1 \qquad ...(1)$$

$\therefore$ (x_1, y_1, z_1) lies on the plane

$\therefore$ $$lx_1 + my_1 + nz_1 = 1 \qquad ...(2)$$

The equation of null plane through (x_1, y_1, z_1) is

$x(L - y_1 Z + z_1 Y) + y(M - z_1 X - x_1 Z) + z(N - x_1 Y + y_1 X) = Lx_1 + My_1 + Nz_1$ ...(3)

Comparing (1) and (3), we have Using (2)

$$\frac{L - y_1 Z + z_1 Y}{l} = \frac{M - z_1 X + x_1 Z}{m} = \frac{N - x_1 Y + y_1 X}{n} = \frac{Lx_1 + My_1 + Nz_1}{1} \qquad ...(4)$$

From (4) using (*i*) and (*ii*) ratios

$$(mL - lM) - (my_1 + lx_1)Z + m\,.\,z_1 Y + lz_1 X = 0 \qquad ...(5)$$

Using (*ii*) and (*iii*) and (*iii*) and (*i*) ratios of equation (4) we have

$$(nM - mN) - (nZ_1 + my_1)X + nx_1 Z + mx_1 Y = 0 \qquad ...(6)$$

$$(lN - nL) - (lx_1 + nZ_1)Y + ly_1 X + ny_1 Z = 0 \qquad ...(7)$$

Putting the value from (2) in (5), we have

$$(mL - lM) - (1 - nZ_1)Z + mZ_1 Y + lZ_1 X = 0$$

$\Rightarrow$ $$z_1(lX + mY + nZ) = Z - mL + lM$$

$\therefore$ $$\frac{z_1}{Z - mL + lM} = \frac{1}{lX + mY + nZ} \qquad ...(8)$$

Similarly from (6) and (7), we have

$$\frac{x_1}{X - nM + mN} = \frac{y_1}{Y - lN + nL} = \frac{z_1}{Z - mL + lM} = \frac{1}{lX + mY + nZ}$$

This gives the coordinate of the null point of the given plane.

3. To find the condition that the Straight Line (M.D.U. 2005)

$$\frac{x-a}{l} = \frac{y-b}{m} = \frac{z-c}{n}$$

may be a null line for a given system of force.

Sol. Let the system reduces to a dyname (X, Y, Z, L, M, N) at any point O. Shifting the origin at O′ (a, b, c) component of the couple are

$$L' = L - bZ + cY$$
$$M' = M - cX + aZ$$
$$N' = N - aY + bX$$

(Refer article 5)

Hence the direction ratios of couple at O′ are L′, M′, N′. Direction ratios of line are l, m, n.

$\because$ Line is a null line

$\therefore$ $l L' + m M' + n N' = 0$

$\therefore$ $l(L - bZ + cY) + m(M - cX + aZ) + n(N - aY + bX) = 0$

$\therefore$ $X(mc - nb) + Y(na - lc) + Z(lb - ma) = lL + mM + nN$

This can be written as

$$\begin{vmatrix} X & Y & Z \\ l & m & n \\ a & b & c \end{vmatrix} = lL + mM + nN.$$

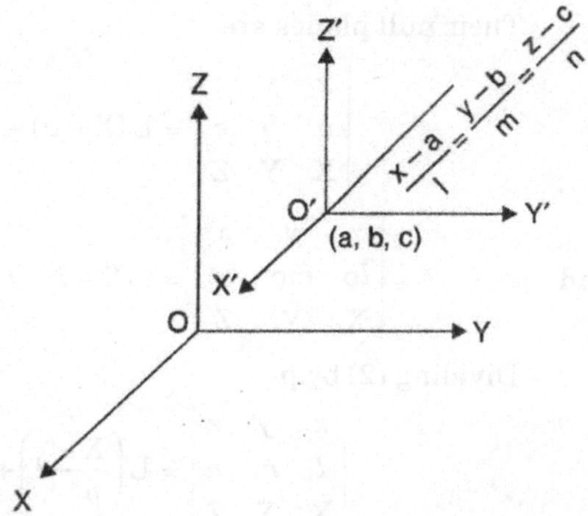

Fig. 3

Which is the required condition, that the given line is a null line

Theorem 1. *A given system of forces may be replaced by two forces, one of which acts along a given line.*

Proof. Let the system of forces be equivalent to dyname (R, G) at a point O.

The plane of couple G is BOD and axis of couple G is OZ which is perpendicular to plane BOD.

Draw a plane through R and given line OA. Let the plane BOD and plane through R and OA intersect in line OC.

Resolving R into two forces P and Q one acting along OA and other along OC.

Resultant of Q along OC and couple G is a force Q parallel to OC in plane BOD.

But the plane BOD is a null plane at O.

Hence the system of forces reduces to P along OA and the force Q in the null plane of O and parallel to OC.

Force P and Q are called conjugate forces and OA and OC are conjugate lines.

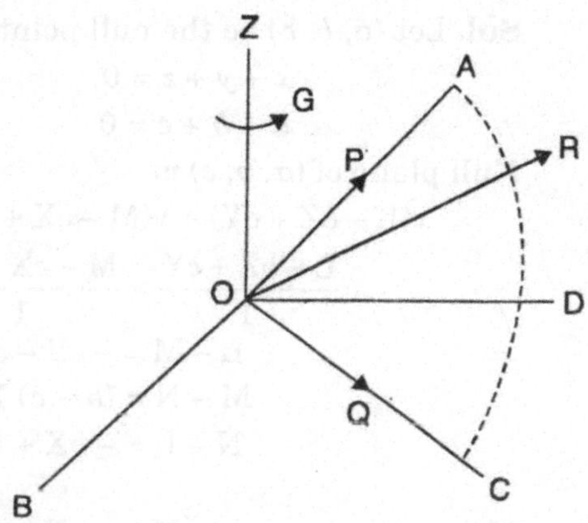

Fig. 4

Method for Finding Conjugate Lines

$\because$ Null plane of OA will always pass through the conjugate line. Consider two points on OA and find their null planes. Intersection of these two planes is the conjugate line of OA.

4. To find the Conjugate Line of $\frac{x-a}{l} = \frac{y-b}{m} = \frac{z-c}{n}$

Sol. Let the system reduces to dyname (X, Y, Z, L, M, N) referred to O as origin. Let (a, b, c) and $(l\rho, m\rho, n\rho)$, $\rho \to \infty$ be two points on the given line.

Their null planes are

$$\begin{vmatrix} x & y & z \\ a & b & c \\ X & Y & Z \end{vmatrix} = L(X-a)+M(Y-b)+N(Z-c) \qquad ...(1)$$

and

$$\begin{vmatrix} x & y & z \\ l\rho & m\rho & n\rho \\ X & Y & Z \end{vmatrix} = L(X-l\rho)+M(Y-m\rho)+N(Z-n\rho) \qquad ...(2)$$

Dividing (2) by ρ

$$\begin{vmatrix} x & y & z \\ l & m & n \\ X & Y & Z \end{vmatrix} = L\left(\frac{X}{\rho}-l\right)+M\left(\frac{Y}{\rho}-m\right)+N\left(\frac{Z}{\rho}-n\right)$$

Taking limit as $\rho \to \infty$

We have,

$$\begin{vmatrix} x & y & z \\ l & m & n \\ X & Y & Z \end{vmatrix} = -(lL+mM+nN) \qquad ...(3)$$

Intersection of planes (1) and (3) is the required conjugate line.

Example 1. *Find the null point of the plane $x + y + z = 0$ for the force system (X, Y, Z ; L, M, N).* (K.U. 2005)

Sol. Let (a, b, c) be the null point of the plane

$$x+y+z=0 \qquad ...(1)$$

$$\therefore \quad a+b+c=0 \qquad ...(2)$$

Null plane of (a, b, c) is

$$x(L-bZ+cY)+y(M-cX+aZ)+z(N-aY+bX)=La+Mb+Nc$$

$$\frac{L-bZ+cY}{1}=\frac{M-cX+aZ}{1}=\frac{N-aY+bX}{1}=\frac{La+Mb+Nc}{0} \qquad ...(3)$$

$$\Rightarrow \quad L-M=-cY-cX+(a+b)Z$$

$$M-N=(b+c)X-aY-aZ$$

$$N-L=-bX+(a+c)Y-bZ.$$

Using (2) $\quad L-M=-cY-cX-cZ \quad \Rightarrow \quad c=\dfrac{M-L}{X+Y+Z}$

$$M-N=-aX-aY-aZ \quad \Rightarrow \quad a=\frac{N-M}{X+Y+Z}$$

$$N-L=-bX-bY-bZ \quad \Rightarrow \quad b=\frac{L-N}{X+Y+Z}$$

∴ Null point is

$$\left(\frac{N-M}{X+Y+Z}, \frac{L-N}{X+Y+Z}, \frac{M-L}{X+Y+Z}\right).$$

Example 2. *A system of forces given by (X, Y, Z ; L, M, N) is replaced by two forces, one of which act along X-axis. Find the magnitude of the forces and line of action of the other force.*

Sol. Let the force acting along X-axis be P

$\because$ Equation of X-axis is $\dfrac{x-0}{1}=\dfrac{y-0}{0}=\dfrac{z-0}{0}$ ...(1)

$\therefore$ Component of force are (P, 0, 0)

and component of couple are

$$L_1 i + M_1 j + N_1 k = \begin{vmatrix} i & j & k \\ 0 & 0 & 0 \\ P & 0 & 0 \end{vmatrix} = 0$$

Component of other forces are X – P, Y, Z.

Let this force act at $(a, b, 0)$

$$\therefore \quad L_2\hat{i} + M_2\hat{j} + N_2\hat{k} = \begin{vmatrix} i & j & k \\ a & b & 0 \\ X-P & Y & Z \end{vmatrix} = bZ\hat{i} - aZ\hat{j} + (aY - bX + bP)\hat{k}$$

$$L\hat{i} + M\hat{j} + N\hat{k} = (L_1 + L_2)\hat{i} + (M_1 + M_2)\hat{j} + (N_1 + N_2)\hat{k}$$

$$\therefore \quad L = bZ, \quad M = -aZ, \quad N = aY - bX + bP$$

$$\Rightarrow \quad N = -\frac{M}{Z}Y - \frac{L}{Z}(X - P)$$

$$\Rightarrow \quad NZ = -MY - LX + LP$$

$$\Rightarrow \quad P = \frac{L + MY + NZ}{L} \qquad ...(2)$$

Component of other force are

$$(X - P, Y, Z) = \left(X - \frac{L + MY + NZ}{L}, Y, Z\right) = \left(\frac{-(MY + NZ)}{L}, Y, Z\right)$$

$\therefore$ Magnitude of 2nd force

$$= \sqrt{\left(\frac{MY + NZ}{L}\right)^2 + Y^2 + Z^2} = \frac{[(MY + NZ)^2 + L^2(Y^2 + Z^2)]^{\frac{1}{2}}}{L} \qquad ...(3)$$

To find line of action of second force null plane at (0, 0, 0)

$$= \begin{vmatrix} x & y & z \\ 0 & 0 & 0 \\ X-P & Y & Z \end{vmatrix} = L(x - 0) + M(y - 0) + N(z - 0)$$

$$= Lx + My + Nz = 0 \qquad ...(4)$$

Point at infinity on line (1) are $(\rho, 0, 0)$ $\rho \to \infty$

Null plane at $(\rho, 0, 0)$

$$= \begin{vmatrix} x & y & z \\ \rho & 0 & 0 \\ X-P & Y & Z \end{vmatrix} = L(x - \rho) + M(y - 0) + N(z - 0)$$

$$\Rightarrow \quad -\rho Zy + \rho Yz = L(x - \rho) + My + Nz$$

$$\Rightarrow \qquad -yZ + Yz = L\left(\frac{x}{\rho} - 1\right) + \frac{M}{\rho}y + \frac{N}{\rho}z$$

Taking $\rho \to \infty$ $\quad -yZ + YZ = -L$

$$yZ - zY = L \qquad ...(5)$$

$\therefore$ (4) and (5) gives line of action of 2nd force.

EXERCISE 1

1. A straight line is given by the equations $Ax + By + Cz = D$, $A'x + B'y + C'z = D'$. Show that conjugate line is given by equation to zero any two of the determinants

$$\begin{vmatrix} L' & M' & N' & Lx + My + Nz \\ A & B & C' & D \\ A' & B' & C' & D' \end{vmatrix} = 0$$

where L, M, N are component of couple at origin and L′, M′, N′ are those at (x', y', z').

2. Show that the Dyname (X, Y, Z ; L, M, N) is equivalent to two forces, one act along the line $x = y = z$ and other along the line given by

$$Lx + My + Nz = 0,$$
$$x(Y - Z) + y(Z - X) + z(X - Y) = L + M + N$$

Find the magnitude of the forces.

3. Show that among the null lines of any system of forces, four are generators of any hyperboloid, two belonging to one system of generators and two to other system.
4. A sphere is described with its centre on the central axis of a system of force. Show that null points of tangent planes to the sphere lie on a hyperboloid of revolution.

7

Stable and Unstable Equilibrium

Let a heavy sphere be at rest on a horizontal surface. Let O be its centre and G be the position of centre of gravity. Let A be the point of contact of the sphere with the plane. If sphere is uniform the C.G. of sphere G coincides with geometric centre O. If sphere is not uniform its centre of gravity is either below O or above O when the sphere is in equilibrium. When the sphere is rotated vertical diameter AC turns the angle. Sphere has new point B in contact with the plane. Forces on the sphere are reaction W at B and weight W at G. Three cases arise.

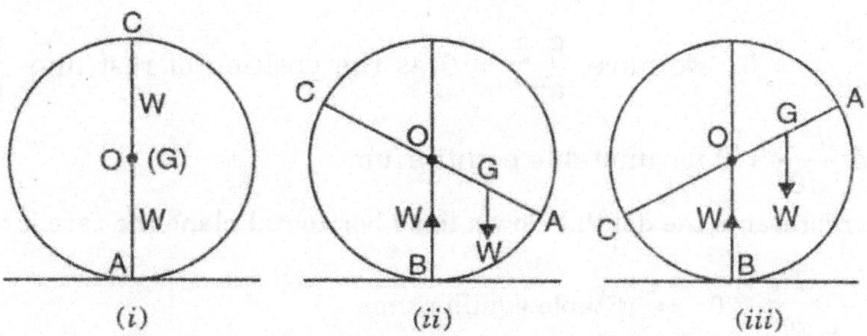

Fig. 1

Case (*i*) When O coincides with G (Fig. 1(*i*)), reaction and weight both pass through O and balance each other and sphere is in equilibrium. The equilibrium is neutral.

Case (*ii*) When G lies below O (Fig. 1(*ii*)), the reaction W at B and weight W at G form a couple which has the tendency to bring the sphere back to its initial position of equilibrium. In this case sphere is said to be in stable equilibrium.

Case (*iii*) When G lies above O (Fig. 1(*iii*)), the couple formed has the tendency to move away from the initial position of equilibrium. In this case equilibrium is unstable.

1. Energy Test for Stability

Principle of conservation of energy states that under a conservative system of forces sum of kinetic and potential energy is constant during any motion.

It follows from this principle that whenever a body begins to move from equilibrium its K.E. increases and since sum of K.E. and P.E. is constant it must loose the same amount of P.E.

If a system is in the position of minimum P.E. and is slightly displaced and set free it moves and looses P.E. Since the P.E. can not be below minimum it will move back to the position of minimum P.E. Thus the equilibrium is stable.

If the system is in the position of maximum P.E. and is displaced it gains K.E. and looses P.E. Due to K.E. it moves farther away from the position of maximum P.E. and the equilibrium is unstable.

Thus the energy test of stability states that positions of maximum P.E. are positions of unstable equilibrium and positions of minimum P.E. are positions of stable equilibrium.

If the centre of gravity is at the height z its P.E. is mgz. Therefore, stable equilibrium is when P.E. is at minimum level and unstable when P.E. is at maximum level. To determine the nature of stability, express the height of C.G. z in terms of a varible θ and apply the condition of maxima and minima. *i.e.,*

(*i*) $\dfrac{dz}{d\theta} = 0$ gives the position of rest $\theta = \alpha, \beta, \gamma$ ....

(*ii*) The equilibrium is stable for $\theta = \alpha$ if $\dfrac{d^2z}{d\theta^2} > 0$ (+) ve

(*iii*) The equilibrium is unstable for $\theta = \alpha$ if $\dfrac{d^2z}{d\theta^2} < 0$ (–) ve

Similar test apply for other solutions $\theta = \beta, \gamma$, ...

In case $\dfrac{d^2z}{d\theta^2} = 0$ we have $\dfrac{d^3z}{d\theta^3} = 0$ as the position of rest and $\dfrac{d^4z}{d\theta^4} > 0$ for stable equilibrium and $\dfrac{d^4z}{d\theta} < 0$ for unstable equilibrium .

Note. If z represents the depth below a fixed horizontal plane the case is reversed *i.e.,*

$$\frac{d^2z}{d\theta^2} < 0 \Rightarrow \text{Stable equilibrium.}$$

$$\frac{d^2z}{d\theta^2} > 0 \Rightarrow \text{unstable equilibrium.}$$

2. To Determine the condition of stable and unstable equilibrium if a heavy body rests on a fixed body

We assume that cross section of the two bodies by a plane are circles of radii r (upper) and R (lower) respectively. The bodies moves along this section. The upper body moves on the lower body without sliding.

Let O and O_1 be the centres of the lower and upper bodies. G_1 be the centre of gravity of the upper body. Let C_1 be the point of contact. Let a slight displacement be given to the upper body. Let C_2 be the new point of contact. Let C_1 rises to new position C'. Let $\angle C_1OC_2 = \theta$, $\angle C'O_2O = \phi$, $C_1G_1 = h = C'G_2$.

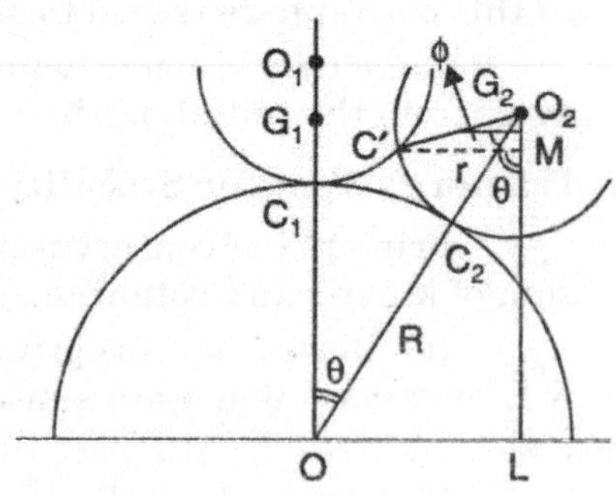

Fig. 2

Where G_2 be the new position of C.G. of upper body.

$\therefore$ Upper Body rolls on lower without sliding

$\therefore \qquad \overset{\frown}{C_1C_2} = \overset{\frown}{C_2C'} \quad \Rightarrow \quad R\theta = r\phi$ ...(1)

In order to discuss the nature of equilibrium.

We have to find out the height z of C.G. in new position above O

$$z = ML = O_2L - O_2M = OO_2 \cos\theta - O_2G_2 \cos(\theta + \phi)$$

$$= (R + r)\cos\theta - (O_2C' - G_2C')\cos(\theta + \phi)$$

$$= (R + r)\cos\theta - (r - h)\cos(\theta + \phi)$$

$$= (R + r)\cos\theta - (r - h)\cos\left(\theta + \frac{R\theta}{r}\right) \qquad \text{Using (1)}$$

$$\therefore \qquad z = (R + r)\cos\theta - (r - h)\cos\left\{\frac{\theta(R + r)}{r}\right\} \qquad ...(2)$$

For equilibrium $\dfrac{dz}{d\theta} = 0$

$\therefore$ Differentiating w.r.t. θ

$$\frac{dz}{d\theta} = -(R + r)\sin\theta + (r - h)\frac{(R + r)}{r}\sin\left\{\frac{\theta(R + r)}{r}\right\}$$

$$= (R + r)\left[-\sin\theta + \frac{(r - h)}{r}\sin\left\{\frac{\theta(R + r)}{r}\right\}\right] = 0$$

$\theta = 0$ is the solution for the equation.

Differentiating again w.r.t. θ

$$\frac{d^2z}{d\theta^2} = (R + r)\left[-\cos\theta + \frac{(r - h)}{r} \cdot \frac{(R + r)}{r}\cos\left\{\frac{\theta(R + r)}{r}\right\}\right] \qquad ...(3)$$

Putting $\theta = 0$ in (3)

$$\frac{d^2z}{d\theta^2} = (R + r)\left[-1 + \frac{(r - h)(R + r)}{r^2}\right] = \frac{(R + r)^2}{r^2}\left[-\frac{r^2}{R + r} + r - h\right]$$

$$= \left(\frac{R + r}{r}\right)^2\left[\frac{-r^2 + rR + r^2 - h(R + r)}{R + r}\right]$$

$$\frac{d^2z}{d\theta^2} = \left(\frac{R + r}{r}\right)^2\left[\frac{rR}{R + r} - h\right] \qquad ....(4)$$

$$\frac{d^2z}{d\theta^2} > 0 \qquad \text{if} \qquad \frac{rR}{R + r} > h$$

or

$$\frac{r + R}{rR} < \frac{1}{h} \quad \Rightarrow \quad \frac{1}{R} + \frac{1}{r} < \frac{1}{h}$$

$\therefore$ Equilibrium is stable if $\dfrac{1}{h} > \dfrac{1}{R} + \dfrac{1}{r}$ ...(5)

Similarly equilibrium is unstable if $\dfrac{1}{h} < \dfrac{1}{R} + \dfrac{1}{r}$ ...(6)

If $$\frac{d^2z}{d\theta^2} = 0 \Rightarrow h = \frac{rR}{r+R}$$

We find $$\frac{d^3z}{d\theta^3} = (R+r)\left[\sin\theta - \left(\frac{r-h}{r}\right)\left(\frac{R+r}{r}\right)^2 \sin\left\{\theta\left(\frac{R+r}{r}\right)\right\}\right]$$

$$\frac{d^3z}{d\theta^3} = 0 \quad \text{at } \theta = 0$$

$$\frac{d^4z}{d\theta^4} = (R+r)\left[\cos\theta - \left(\frac{r-h}{r}\right)\left(\frac{R+r}{r}\right)^3 \cos\left\{\theta\left(\frac{R+r}{r}\right)\right\}\right]$$

$$= (R+r)\left[1 - \frac{\left(r - \frac{rR}{r+R}\right)}{r}\left(\frac{R+r}{r}\right)^3\right] \quad \text{at } \theta = 0$$

$$= (R+r)\left[1 - \left(\frac{R+r}{r}\right)^2\right] = (R+r)\left[1 - \left(1 + \frac{R}{r}\right)^2\right] < 0$$

∴ Equilibrium is unstable.

Therefore, we can conclude that

(*i*) $\frac{1}{h} > \frac{1}{r} + \frac{1}{R}$, **Equilibrium is stable.**

(*ii*) $\frac{1}{h} \le \frac{1}{r} + \frac{1}{R}$, **Equilibrium is unstable.**

Particular cases

(*i*) When upper body has a plane face in contact with the lower body

$$r \to \infty \quad \frac{1}{r} = 0$$

Equilibrium is stable if

$$\frac{1}{h} > \frac{1}{R}$$

or $$h < R.$$

Unstable if $h \ge R$.

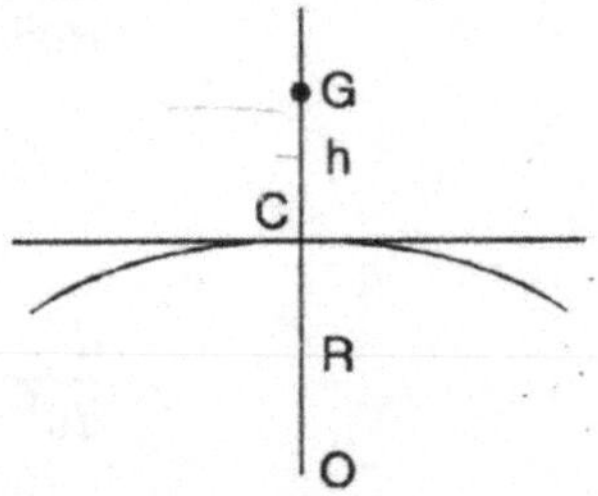

Fig. 3

(*ii*) If lower body is having the plane surface in contact

$$R \to \infty \quad \text{or} \quad \frac{1}{R} = 0$$

Equilibrium is stable if $h < r$

Unstable if $h \ge r$.

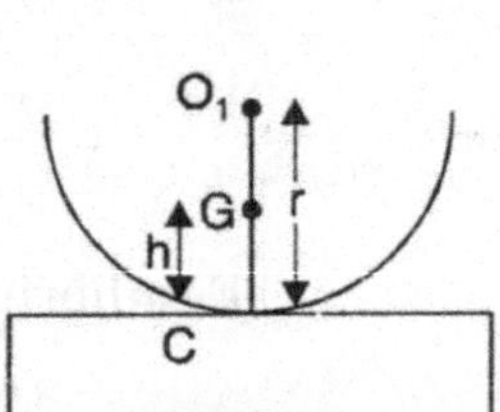

Fig. 4

3. To discuss the condition of stable and unstable equilibrium when a body rests inside other fixed concave body

We assume that cross-sections of the two bodies by a plane are circles.

Let O be the centre of the fixed spherical surface of lower body. The body resting inside it has centre O_1. The figure is a section of the bodies by a vertical plane through G_1. G_1 is the C.G. of the body resting inside. Let $OC_1 = R$, $O_1C_1 = r_1 = O_2C'$ where O_2 is the centre of inside body after slight displacement. C_2 is new point of contact. Let C_1 be displaced to position C'. Let $\angle C_1OC_2 = \theta$

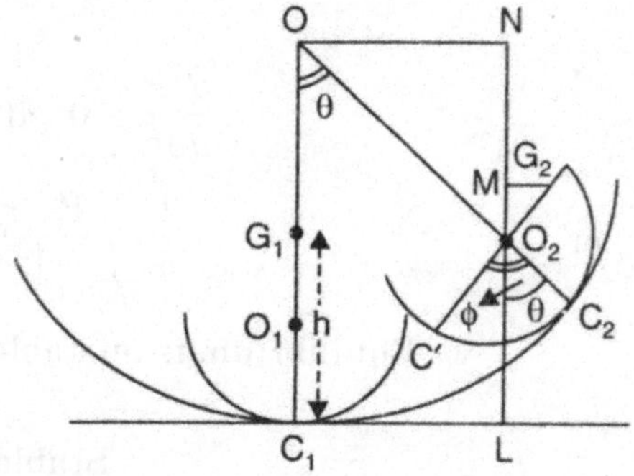

Fig. 5

$$\angle C'O_2C_2 = \phi$$

$$\therefore \quad \angle C'O_2L = \phi - \theta$$

Since the body rolls only

$$\therefore \quad \text{Arc } \widehat{C_1C_2} = \text{Arc } \widehat{C'C_2}$$

$$R\theta = r\phi$$

$$\phi = \frac{R\theta}{r} \qquad \text{...(1)}$$

If z denote the depth of G_2 below the centre O

$$z = MN = O_2N - O_2M = OO_2 \cos\theta - O_2G_2 \cos(\phi - \theta)$$

$$= (R - r)\cos\theta - (C'G_2 - C'O_2)\cos(\phi - \theta)$$

$$= (R - r)\cos\theta - (h - r)\cos(\phi - \theta) \qquad \because C_1G_1 = C'G_2 = h$$

Putting ϕ from (1), we get

$$z = (R - r)\cos\theta - (h - r)\cos\left(\frac{R - r}{r}\theta\right) \qquad \text{...(2)}$$

For the discussion of condition of equilibrium

We have to put $\frac{dz}{d\theta} = 0$ $\frac{d^2z}{d\theta^2} > 0$ for unstable equilibrium

$\frac{d^2z}{d\theta^2} < 0$ for stable equilibrium.

(Conditions are reverse in case of depth below the fixed plane)

Differentiating (2) w.r.t. θ

$$\frac{dz}{d\theta} = -(R - r)\sin\theta + (h - r)\left(\frac{R - r}{r}\right)\sin\left(\frac{R - r}{r}.\theta\right) = 0$$

which is satisfied by $\theta = 0$ (By Inspection)

$$\frac{d^2z}{d\theta^2} = -(R - r)\cos\theta + (h - r)\left(\frac{R - r}{r}\right)^2\cos\left(\frac{R - r}{r}.\theta\right) \qquad \text{...(3)}$$

Putting $\theta = 0$

$$\frac{d^2z}{d\theta^2} = -(R - r) + (h - r)\left(\frac{R - r}{r}\right)^2$$

$$= \left(\frac{R-r}{r}\right)^2 \left[h - r - \frac{r^2}{R-r}\right] = \left(\frac{R-r}{r}\right)^2 \left[h - \frac{rR}{R-r}\right] \quad ...(4)$$

$$\frac{d^2z}{d\theta^2} > 0 \quad \text{if } h > \frac{rR}{R-r}$$

or $$\frac{1}{h} < \frac{R-r}{rR} \quad \text{or} \quad \frac{1}{h} < \frac{1}{r} - \frac{1}{R}.$$

$\therefore$ Equilibrium is unstable if $\frac{1}{h} < \frac{1}{r} - \frac{1}{R}$.

Stable if $\frac{1}{h} > \frac{1}{r} - \frac{1}{R}$.

If $$\frac{d^2z}{d\theta^2} = 0 \quad i.e., \quad h = \frac{rR}{R-r}$$

We find, $$\frac{d^3z}{d\theta^3} = (R-r)\sin\theta - (h-r)\left(\frac{R-r}{r}\right)^3 \sin\left(\frac{R-r}{r}\theta\right)$$

$$\frac{d^4z}{d\theta^4} = (R-r)\cos\theta - (h-r)\left(\frac{R-r}{r}\right)^4 \cos\left(\frac{R-r}{r}\theta\right)$$

at $\theta = 0$ $\frac{d^3z}{d\theta^3} = 0$ and

$$\frac{d^4z}{d\theta^4} = R - r - \left(\frac{rR}{R-r} - r\right)\left(\frac{R-r}{r}\right)^4$$

$$= (R-r) - \left[\frac{rR - rR + r^2}{R-r}\right]\left(\frac{R-r}{r}\right)^4 = (R-r) - \frac{(R-r)^3}{r^2}$$

$$= (R-r)\left[1 - \left(\frac{R-r}{r}\right)^2\right] = (R-r)\left[1 - \left(\frac{R}{r} - 1\right)^2\right]$$

$$= (R-r)\left[1 - \left(1 - \frac{R}{r}\right)^2\right]$$

$$1 - \left(\frac{R}{r} - 1\right)^2 > 0 \quad \text{if} \quad 1 > \left(\frac{R}{r} - 1\right)^2$$

i.e., $R < 2r$

Equilibrium is unstable if $R < 2r$, $h = \frac{rR}{R-r}$

Stable if $R > 2r$

Thus we conclude that equilibrium is

Stable	**if**	$\frac{1}{h} > \frac{1}{r} - \frac{1}{R}$
Unstable	**if**	$\frac{1}{h} < \frac{1}{r} - \frac{1}{R}$

If $\quad \frac{1}{h} = \frac{1}{r} - \frac{1}{R}$

Stable if **R > 2r**

Unstable if **R< 2r** .

Theorem. A body rests in equilibrium upon another body, which is fixed and the portion of two bodies in contact have radii of curvature ρ_1 and ρ_2 respectively. The C.G. of the first body is at height h above the point of contact and the common normal makes an angle α with the vertical. The equilibrium is stable or unstable if

$$h < \text{or} > \frac{\rho_1 \rho_2}{\rho_1 + \rho_2} \cos \alpha$$

Proof. Let O and O_1 be the centres of curvature of the lower and upper bodies in the position of rest. C_1 be the point of contact. OO_1 is common normal. OP is vertical line through O . $\angle O_1OP = \alpha$. Let G_1 be the C.G. of upper body. C_1G_1 is vertical $\angle G_1C_1O_1 = \alpha$. When the upper body rolls on the lower the new point of contact is C_2 and C_1 takes new position as C′. C_1G_1 takes new position as $C'G_2$ such that $C_1G_1 = C'G_2 = h$. OO_2 is new common normal.

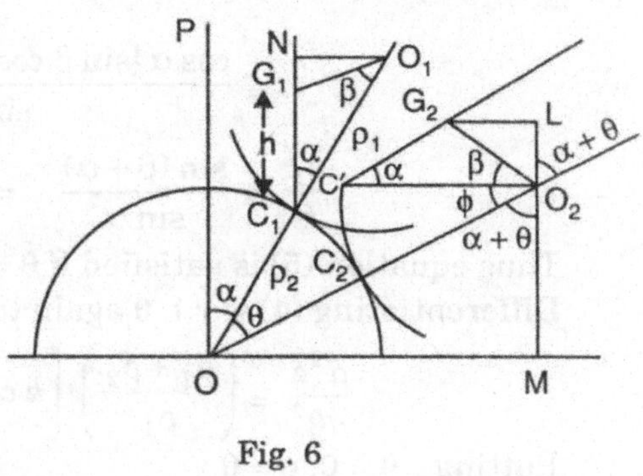

Fig. 6

$$\therefore \quad \angle G_1C_1O_1 = \angle G_2C'O_2 = \alpha$$

$$O_1C_1 = O_2C_2 = \rho_1, \; OC_1 = OC_2 = \rho_2$$

Let $\angle G_1O_1C_1 = \angle G_2O_2C' = \beta$

$$\text{Arc } \widehat{C_1C_2} = \text{Arc } \widehat{C'C_2} \Rightarrow \rho_2\theta = \rho_1\phi \qquad ...(1)$$

Now, If the height of G_2 above O is z

then $z = LM = LO_2 + O_2M$

$$= O_2G_2 \cos\{\pi - (\alpha + \theta + \beta + \phi)\} + OO_2 \cos(\alpha + \theta)$$

$$= -k \cos(\alpha + \theta + \beta + \phi) + (\rho_1 + \rho_2)\cos(\alpha + \theta) \qquad ...(2)$$

where $O_2G_2 = k$ (say)

Differentiating (1), we have $\frac{d\phi}{d\theta} = \frac{\rho_2}{\rho_1}$ ...(3)

To discuss the equilibrium, the necessary condition is $\frac{dz}{d\theta} = 0$

$$\therefore \quad \frac{dz}{d\theta} = k \sin(\alpha + \beta + \theta + \phi)\left(1 + \frac{d\phi}{d\theta}\right) - (\rho_1 + \rho_2)\sin(\alpha + \theta)$$

$$= k \sin(\alpha + \beta + \theta + \phi)\left(1 + \frac{\rho_2}{\rho_1}\right) - (\rho_1 + \rho_2)\sin(\alpha + \theta) \qquad \text{Using (3)}$$

$$= \left(\frac{\rho_1 + \rho_2}{\rho_1}\right)\{k \sin(\alpha + \beta + \theta + \phi) - \rho_1 \sin(\alpha + \theta)\} = 0 \qquad ...(4)$$

$$\Rightarrow \qquad k \sin(\alpha+\beta+\theta+\phi) = \rho_1 \sin(\alpha+\theta)$$

i.e., $$\frac{k}{\rho_1} = \frac{\sin(\alpha+\theta)}{\sin(\alpha+\beta+\theta+\phi)} \qquad \ldots(5)$$

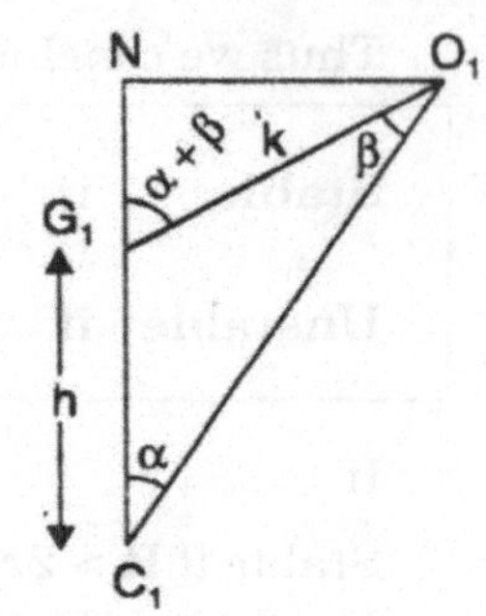

Fig. 7

Equation (4) is satisfied for $\theta = 0, \phi = 0$

As $$h = C_1G_1 = C_1N - NG_1$$
$$= \rho_1 \cos\alpha - k\cos(\alpha+\beta)$$

$$\frac{h}{k} = \frac{\rho_1}{k}\cos\alpha - \cos(\alpha+\beta) \qquad \ldots(6)$$

Also from sine rule, in $\Delta O_1C_1G_1$

$$\frac{h}{\sin\beta} = \frac{k}{\sin\alpha} \quad \Rightarrow \quad \frac{h}{k} = \frac{\sin\beta}{\sin\alpha}.$$

Putting in (6), we have

$$\frac{\sin\beta}{\sin\alpha} = \frac{\rho_1}{k}\cos\alpha - \cos(\alpha+\beta)$$

$$\therefore \quad \frac{\rho_1}{k}\cos\alpha = \frac{\sin\beta}{\sin\alpha} + \cos(\alpha+\beta) = \frac{\sin\beta + \sin\alpha\,[\cos\alpha\cos\beta - \sin\alpha\sin\beta]}{\sin\alpha}$$

$$= \frac{\sin\beta\,(1-\sin^2\alpha) + \sin\alpha\cos\alpha\cos\beta}{\sin\alpha}$$

$$= \frac{\cos\alpha\,[\sin\beta\cos\alpha + \cos\beta\sin\alpha]}{\sin\alpha}$$

$$\therefore \quad \frac{\rho_1}{k} = \frac{\sin(\beta+\alpha)}{\sin\alpha} \quad \Rightarrow \quad \frac{k}{\rho_1} = \frac{\sin\alpha}{\sin(\alpha+\beta)}. \qquad \ldots(7)$$

Thus equation (5) is satisfied if $\theta = 0, \phi = 0$

Differentiating (4) w.r.t. θ again to discuss stability.

$$\frac{d^2z}{d\theta^2} = \left(\frac{\rho_1+\rho_2}{\rho_1}\right)\left[k\cos(\alpha+\beta+\theta+\phi)\left(1+\frac{d\phi}{d\theta}\right) - \rho_1\cos(\alpha+\theta)\right]$$

Putting $\theta = 0, \phi = 0$

$$= \left(\frac{\rho_1+\rho_2}{\rho_1}\right)\left[k\cos(\alpha+\beta)\left(1+\frac{\rho_2}{\rho_1}\right) - \rho_1\cos\alpha\right]$$

$$= \left(\frac{\rho_1+\rho_2}{\rho_1^2}\right)[k(\rho_1+\rho_2)\cos(\alpha+\beta) - \rho_1^2\cos\alpha]$$

Putting $k\cos(\alpha+\beta)$ from equation (6)

$$\left.\frac{d^2z}{d\theta^2}\right]_{\substack{\theta=0\\ \phi=0}} = \frac{(\rho_1+\rho_2)}{\rho_1^2}\left[(\rho_1+\rho_2)(\rho_1\cos\alpha - h) - \rho_1^2\cos\alpha\right]$$

$$= \left(\frac{\rho_1+\rho_2}{\rho_1^2}\right)\left[\rho_1^2\cos\alpha + \rho_1\rho_2\cos\alpha - h(\rho_1+\rho_2) - \rho_1^2\cos\alpha\right]$$

$$= \left(\frac{\rho_1+\rho_2}{\rho_1^2}\right)[\rho_1\rho_2\cos\alpha - h(\rho_1+\rho_2)]$$

$$= \left(\frac{\rho_1 + \rho_2}{\rho_1}\right)^2 \left[\frac{\rho_1 \rho_2}{\rho_1 + \rho_2} \cos \alpha - h\right]$$

For stable equilibrium

$$\frac{d^2 z}{d\theta^2} > 0 \quad \Rightarrow \quad h < \frac{\rho_1 \rho_2}{\rho_1 + \rho_2} \cos \alpha$$

For unstable equilibrium

$$\frac{d^2 z}{d\theta^2} < 0 \quad \Rightarrow \quad h > \frac{\rho_1 \rho_2}{\rho_1 + \rho_2} \cos \alpha.$$

Hence Proved.

Particular case

If $\alpha = 0$, this result reduces to

$$h < \frac{\rho_1 \rho_2}{\rho_1 + \rho_2} \text{ for stable equilibrium}$$

i.e.,

$$\frac{1}{h} > \frac{1}{\rho_1} + \frac{1}{\rho_2} \text{ for stable equilibrium}$$

$$< \frac{1}{\rho_1} + \frac{1}{\rho_2} \text{ for unstable equilibrium}$$

This is the same result as of article 2.

Corrolary. A lamina in the form of a cycloid whose generating circle is of radius a, rest on the top of another cycloid whose generating circle is of radius b, their vertices being in contact and their axes vertical. If h be the height of the C.G. of upper cycloid above its vertex, show that the equilibrium is stable only if

$$h < \frac{4ab}{a+b} \text{ and unstable if } h \geq \frac{4ab}{a+b} \qquad (K.U.\ 2005)$$

Taking, $\rho_1 = a$ and $\rho_2 = b$ and $\alpha = 0$.

Example 1. *A cubical box of edge a is placed on the top of a fixed sphere. The centre of the face of the cube is in the contact with the highest point of the sphere. Find the least radius of the sphere for stable equilibrium.*

Sol. The height of C.G. above the point of contact $h = \frac{a}{2}$ radius of flat surface of upper body $r = \infty$. Let radius of lower sphere be R.

For stable equilibrium

$$\frac{1}{h} > \frac{1}{r} + \frac{1}{R}$$

$$\frac{2}{a} > 0 + \frac{1}{R} \quad \Rightarrow \quad R > \frac{a}{2}$$

$\therefore$ Least value of $R = \frac{a}{2}$

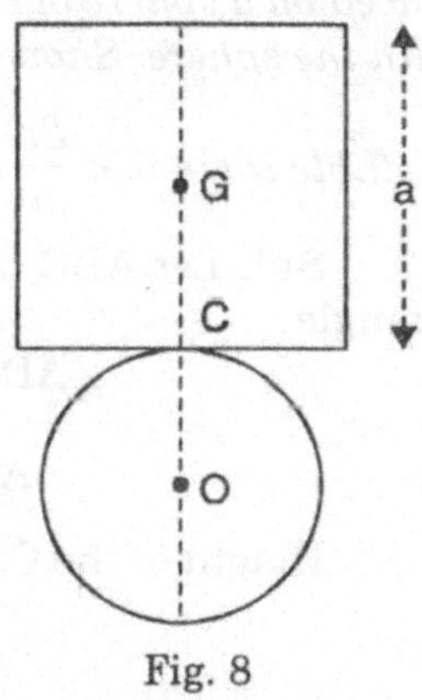

Fig. 8

Example 2. *A hemisphere rests in equilibrium on a sphere of equal radius. Show that the equilibrium is stable or unstable according to the flat or curved surface of hemisphere is in contact.*

Sol. Case (*i*). When flat surface rests on the sphere. C.G. of hemisphere divides the central radius in the ratio of 3 : 5.

Let radius = a

$$h = \text{CG} = \frac{3a}{8}$$

$$r = \infty$$

$$\text{R} = a$$

$$\frac{1}{h} = \frac{8}{3a},$$

$$\frac{1}{r} + \frac{1}{\text{R}} = 0 + \frac{1}{a}$$

$$\therefore \quad \frac{1}{h} > \frac{1}{r} + \frac{1}{\text{R}}$$

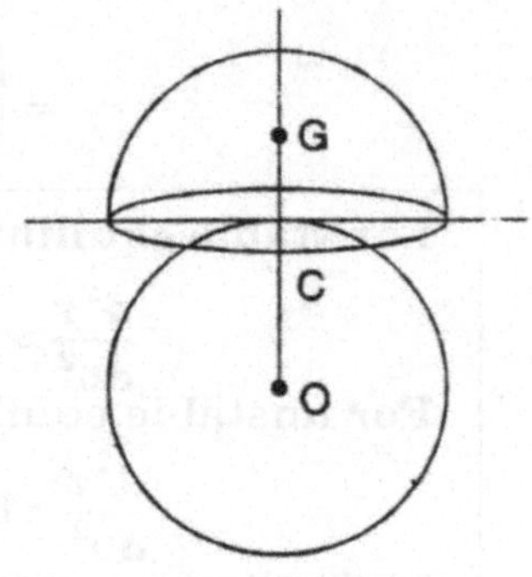

Fig. 9

$\Rightarrow$ Equilibrium is stable.

Case (*ii*). When curved surface lie on the sphere.

$$\text{OC} = a = \text{R} \qquad \text{(Radius of lower body)}$$

$$\text{CG} = h = \frac{5a}{8}$$

(height of C.G. above the point of contact)

$$\text{O}_1\text{C} = r = a \qquad \text{(Radius of upper body)}$$

$$\frac{1}{r} + \frac{1}{\text{R}} = \frac{1}{a} + \frac{1}{a} = \frac{2}{a} = \frac{10}{5a}$$

$$\frac{1}{h} = \frac{8}{5a}, \quad \frac{1}{h} < \frac{1}{r} + \frac{1}{\text{R}}$$

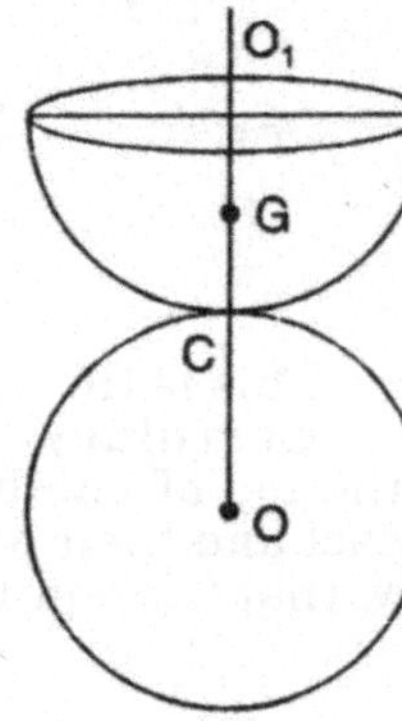

Fig. 10

$\therefore$ Equilibrium is unstable.

Example 3. *A Lamina in the form of an isosceles triangle whose vertical angle α is placed on a sphere of radius r so that its plane is vertical and one of the equal sides is in contact with the sphere. Show that if the triangle be slightly displaced in its own plane the equilibrium is stable if* $\sin \alpha < \frac{3r}{a}$ *where a is one of the equal sides.*

Sol. Let ABC be the triangle with AB = AC = a, $\angle$BAC = α. Let G be the C.G. of the triangle

$$\angle \text{ADB} = \angle \text{ADC} = \pi/2$$

$$\therefore \quad \text{AD} = a \cos \frac{\pi}{2}$$

Height of the C.G. above the point of contact is GH = h

$$h = \text{AG} \sin \frac{\alpha}{2} = \frac{2}{3} \text{AD} \sin \frac{\alpha}{2}$$

$$= \frac{2}{3} a \cos \frac{\alpha}{2} \sin \frac{\alpha}{2}$$

$$= \frac{1}{3} a \sin \alpha$$

R = radius of lower body = r

r = radius of upper body = ∞

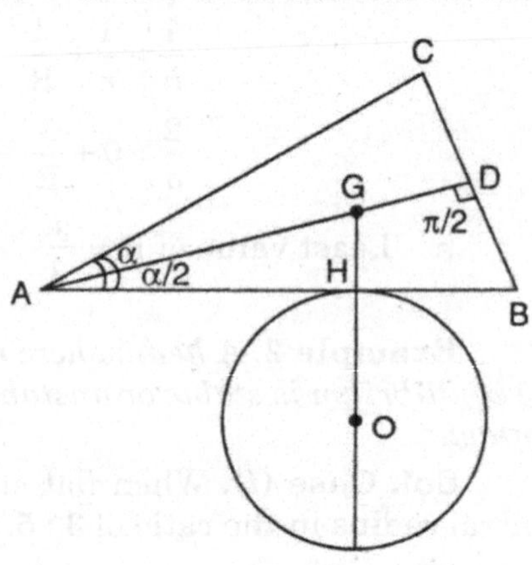

Fig. 11

For stable equilibrium

$$\frac{1}{h} > \frac{1}{r} + \frac{1}{R} \Rightarrow \frac{3}{a \sin \alpha} > \frac{1}{r} \quad \text{or} \quad \sin \alpha < \frac{3r}{a}.$$

Example 4. *A heavy hemispherical shell of radius r has a particle attached to a point on its rim and rests on a rough sphere of radius R at the highest point. If their curved surfaces are in contact then equilibrium is stable if $\frac{R}{r} > \sqrt{5} - 1$.*

Sol. Let the weight hemispherical shell be W acting at C.G. G′.

$\therefore$ $O'G' = \frac{r}{2}$ where O′ is the centre of the shell. Let ω be the weight of the particle attached at the rim. Let G be the C.G. of the hemisphere after attaching the weight at A.

The resultant of these two parallel forces lies on AG′ at G.

Let C the point of contact of hemisphere and sphere. Let ∠O′AG′ = θ

$$\tan \theta = \frac{O'G'}{O'A} = \frac{\frac{r}{2}}{r} = \frac{1}{2}, \ \sin \theta = \frac{1}{\sqrt{5}}$$

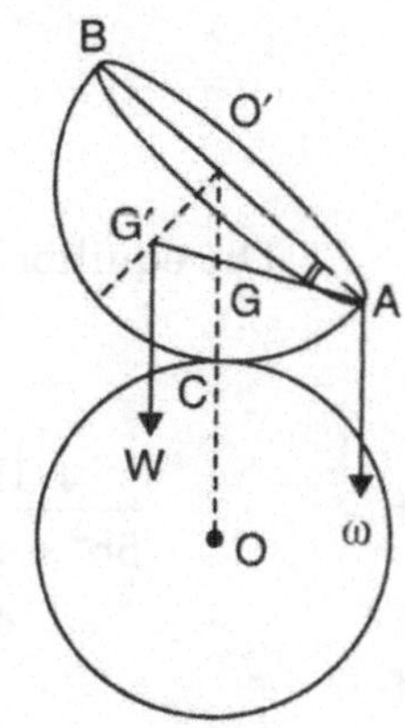

Fig. 12

Let height of G above C $= h$

$$h = O'C - O'G = r - r \sin \theta = r - r \cdot \frac{1}{\sqrt{5}}$$

The equilibrium is stable

if $$\frac{1}{h} > \frac{1}{r} + \frac{1}{R}$$

if $$h < \frac{rR}{r+R}$$

or $$r - \frac{r}{\sqrt{5}} < \frac{rR}{r+R}$$

or if $$\frac{r}{\sqrt{5}} > r - \frac{rR}{r+R}$$

or if $$\frac{1}{\sqrt{5}} > 1 - \frac{R}{r+R}$$

or if $$\frac{r+R}{\sqrt{5}} > r + R - R$$

or if $$\frac{r+R}{r} > \sqrt{5} \Rightarrow \frac{R}{r} > \sqrt{5} - 1$$

$\because$ The result is independent of the weight attached at the rim.

Therefore, the equilibrium is stable for any weight attached.

Example 5. *A body consists of a cone and a hemisphere on the same base, rests on a rough horizontal table with curved surface of hemisphere in contact with the table. Show that greatest height of the cone so that the equilibrium may be stable is $\sqrt{3}$ times the radius of hemisphere.*

Sol. Let H and r be the height and radius of base of the cone respectively. If C.G. of hemisphere is at G_1 and C.G. of cone is at G_2 then

$$OG_1 = \frac{3r}{8}, OG_2 = \frac{H}{4}.$$

C.G. of the combined body is given by

$$\bar{x} = \frac{w_1 \bar{x}_1 + w_2 \bar{x}_2}{w_1 + w_2}$$

Height of C.G. from the fixed plane h

$$h = \frac{\left(\frac{2}{3}\pi r^3\right).\frac{5r}{8} + \left(\frac{1}{3}\pi r^2 H\right)\left(r + \frac{H}{4}\right)}{\frac{2}{3}\pi r^3 + \frac{1}{3}\pi r^2 H}$$

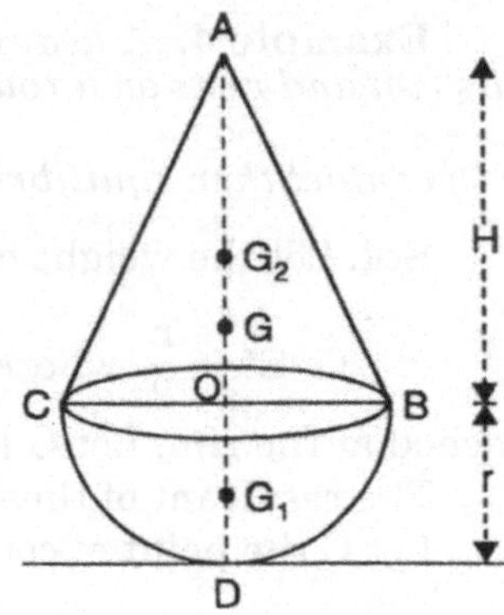

Fig. 13

$$= \frac{\frac{5r^2}{4} + Hr + \frac{H^2}{4}}{2r + H} = \frac{5r^2 + 4rH + H^2}{4(H + 2r)}$$

The equilibrium is stable if

$$\frac{1}{h} > \frac{1}{r} + \frac{1}{R} \qquad R = \infty, \qquad r = r$$

if $$\frac{4(H + 2r)}{5r^2 + 4rH + H^2} > \frac{1}{r}$$

if $$4Hr + 8r^2 > 5r^2 + 4rH + H^2$$

if $$H^2 < 3r^2$$

$$H < r\sqrt{3}$$

Hence the greatest height of the cone consistent with the stable equilibrium is $\sqrt{3}$ times the radius of the hemisphere.

Example 6. *A Lamina in the form of a cycloid whose generating circle is of radius a rests on the top of another cycloid whose generating circle is of radius b. Their vertices are in contact and axes vertical. If h is the height of upper cycloid above the vertex the equilibrium is stable or unstable if h < or >* $\frac{4ab}{a+b}$. (K.U. 2005)

Sol. The intrinsic equation of the cycloid

$$s = 4a \sin \psi$$

radius of curvature $\rho = \frac{ds}{d\psi} = 4a \cos \psi$

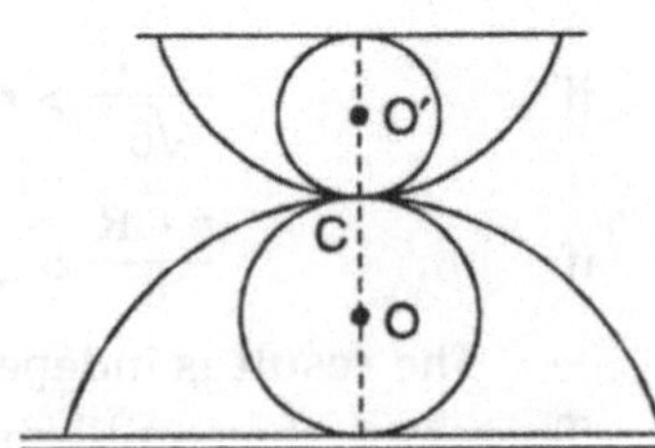

Fig. 14

At the vertex $\psi = 0$ and

so $\rho = 4a$

In the present case $\rho_1 = 4a$, $\rho_2 = 4b$, $h = h$.

The equilibrium is stable

if $$\frac{1}{h} > \frac{1}{\cos\alpha}\left(\frac{1}{\rho_1} + \frac{1}{\rho_2}\right) \qquad \alpha = 0 \text{ at the highest point.}$$

if $$\frac{1}{h} > \frac{1}{4a} + \frac{1}{4b} = \frac{a+b}{4ab}$$

if $$h < \frac{4ab}{a+b}$$

and unstable if $h > \dfrac{4ab}{a+b}$.

Example 7. *A solid hemisphere rests on a plane inclined to the horizon at an angle $\sin^{-1}\left(\frac{3}{8}\right)$. The plane is rough enough to prevent sliding. Find the position of equilibrium and show that it is stable.*

Sol. Let O and G be the centre and C.G. of the hemisphere of radius r

$$OC = OD = r$$

$$OG = \frac{3r}{8}$$

Let α be the inclination of the plane with horizontal

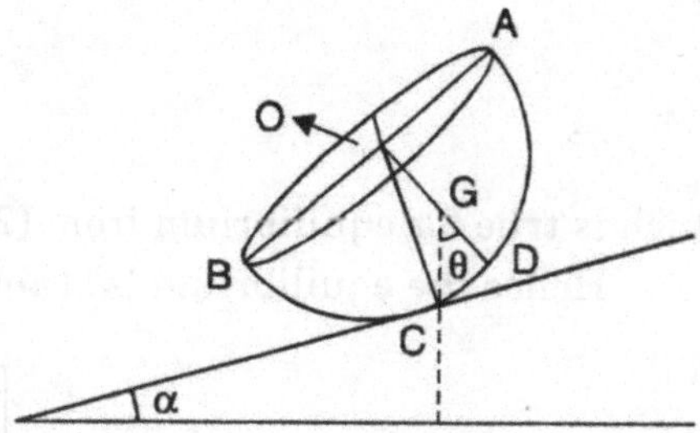

Fig. 15

$$\therefore \quad \sin\alpha = \frac{3}{8}$$

$$\angle OCG = \alpha, \quad \text{Let } \angle CGD = \theta$$

From ΔOGC

$$\frac{OG}{\sin\alpha} = \frac{OG}{\sin(\pi-\theta)} \Rightarrow \frac{3r}{8\sin\alpha} = \frac{r}{\sin\theta}$$

$$\therefore \quad \sin\theta = \frac{8}{3}\sin\alpha \qquad ...(1)$$

$\because$ $\sin\theta < 1 \Rightarrow$ **Equilibrium is possible if**

$$\frac{8}{3}\sin\alpha < 1$$

or $$\sin\alpha < \frac{3}{8} \quad \text{or} \quad \alpha < \sin^{-1}\left(\frac{3}{8}\right) \qquad ...(2)$$

For stability of equilibrium we find height of C.G. above the point of contact.

Let $CG = h$

From ΔOGC $$\frac{OG}{\sin\alpha} = \frac{CG}{\sin(\theta-\alpha)}$$

$$\frac{3r}{8\sin\alpha} = \frac{h}{\sin(\theta-\alpha)} \Rightarrow h = \frac{3r\sin(\theta-\alpha)}{8\sin\alpha} \qquad ...(3)$$

For stability, $$\frac{1}{h} > \left(\frac{1}{\rho_1} + \frac{1}{\rho_2}\right)\frac{1}{\cos\alpha} \qquad \rho_2 = \infty$$

$$\Rightarrow \quad \frac{1}{h} > \frac{1}{r\cos\alpha} \qquad \rho_1 = r$$

$$\Rightarrow \qquad h < r \cos \alpha$$

Putting the value of h from (3), we have

$$\Rightarrow \qquad \frac{3r \sin(\theta - \alpha)}{8 \sin \alpha} < r \cos \alpha$$

$$\Rightarrow \qquad 3 \sin(\theta - \alpha) < 8 \sin \alpha \cos \alpha$$

$$\Rightarrow \qquad 3 [\sin \theta \cos \alpha - \cos \theta \sin \alpha] < 8 \sin \alpha \cos \alpha$$

$$\Rightarrow \qquad 3\left[\frac{8}{3} \sin \alpha \cos \alpha - \sqrt{1 - \frac{64}{9}\sin^2 \alpha} \sin \alpha\right] < 8 \sin \alpha \cos \alpha \qquad \text{Using (1)}$$

$$\Rightarrow \qquad -3 . \sqrt{9 - 64 \sin^2 \alpha} . \sin \alpha < 0 \qquad \Rightarrow \sin^2 \alpha (9 - 64 \sin^2 \alpha) > 0$$

$$\Rightarrow \qquad 64 \sin^2 \alpha < 9 \qquad \Rightarrow \sin \alpha < \frac{3}{8}$$

$$\Rightarrow \qquad \alpha < \sin^{-1}\left(\frac{3}{8}\right)$$

which is true for equilibrium from (2).

Hence the equilibrium is stable.

EXERCISE 1

1. A uniform beam of thickness $2b$ rests symmetrically on a perfect rough horizontal cylinder of radius a. Show that equilibrium is stable or unstable according as b is less than or greater than a.
2. A heavy uniform cube of side $\frac{\pi r}{2}$ balances on the highest point of sphere of radius r. If there is no sliding show that cube can rock through a right angle without falling.
3. Two equal particles are connected by a light inextensible string which is slung over the top of a smooth vertical circle. If the length of the string is less than half the circumference of the circle then show that the equilibrium is unstable.
4. An elliptic cylinder rests on another equal elliptic cylinder such that their lengths are horizontal and minor axis of cross sections in the same vertical plane. Prove that for stability, $e > \frac{1}{\sqrt{2}}$, where e is eccentricity of cross-section.
5. A solid sphere of radius r rests inside a fixed rough hemisphere of radius $2r$. Show that the equilibrium is stable for any weight attached at the highest point of the sphere.
6. A sphere of weight W and radius a lies within a fixed sphere of radius b. A particle of weight w is fixed to the upper end of the vertical diameter. Prove that equilibrium is stable if $\frac{W}{w} \geq \frac{b - 2a}{a}$.
7. A solid ellipsoid whose axes are $2a$, $2b$ and $2c$ with the axis vertical rests on a rough horizontal plane. The C.G. is on the vertical axis at a distance h from the bottom . Show that equilibrium is stable if $h < \frac{a^2}{c}$ and $\frac{b^2}{c}$ both.

8. A solid body consists of a hemisphere of radius a and a right circular cone of same material on its base. The convex side of the hemisphere rests on the highest point of the sphere of radius b. Show that greatest height of the cone for the equilibrium to be stable is $\frac{a}{a+b}[\sqrt{\{(3b+a)(b-a)-2a\}}]$.

SOME MORE ILLUSTRATIVE EXAMPLES

Example 1. *A uniform beam of length 2a rests on two smooth inclined planes of inclinations α and β with the horizontal respectively. Show that inclination θ of the beam to horizontal in the equilibrium position is given by* $\tan\theta = \frac{1}{2}(\cot\alpha \sim \cot\beta)$. *Also show that equilibrium is unstable.* (K.U. 2005)

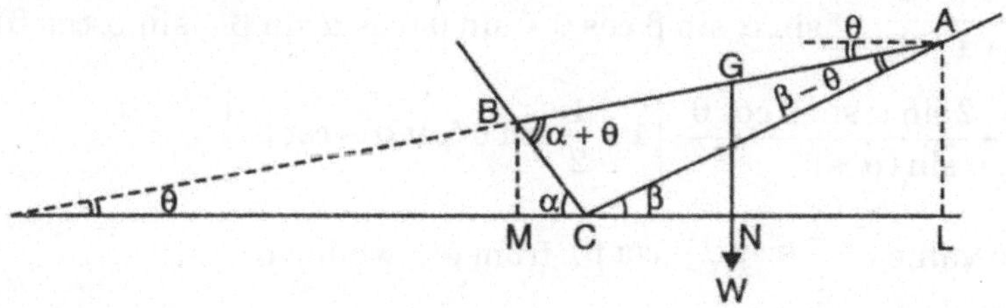

Fig. 16

Sol. Let the rod AB with C.G. at G rests on the planes CB and CA

Let $\angle BCM = \alpha$, $\angle ACL = \beta$

Let the rod AB makes an angle θ with the horizon, where AL and BM are perpendiculars from A and B on the horizontal plane.

The height of C.G. of rod is at height z = GN

$$z = \frac{1}{2}(AL + BM) = \frac{1}{2}(AC\sin\beta + BC\sin\alpha) \quad \text{...(1)}$$

In ΔABC

$$\frac{BC}{\sin(\beta-\theta)} = \frac{AC}{\sin(\alpha+\theta)} = \frac{AB}{\sin\{\pi-(\alpha+\beta)\}}$$

or

$$\frac{BC}{\sin(\beta-\theta)} = \frac{AC}{\sin(\alpha+\theta)} = \frac{2a}{\sin(\alpha+\beta)} \quad \text{...(2)}$$

Putting the values of BC and AC from (2) in (1)

$$z = \frac{1}{2}\left[\frac{2a\sin(\alpha+\theta)}{\sin(\alpha+\beta)}\sin\beta + \frac{2a\sin(\beta-\theta)}{\sin(\alpha+\beta)}\sin\alpha\right]$$

$$= \frac{a}{\sin(\alpha+\beta)}[\sin(\alpha+\theta)\sin\beta + \sin(\beta-\theta)\sin\alpha]$$

For equilibrium

$$\frac{dz}{d\theta} = \frac{a}{\sin(\alpha+\beta)}[\cos(\alpha+\theta)\sin\beta - \cos(\beta-\theta)\sin\alpha] = 0 \quad \text{...(3)}$$

or $(\cos\alpha\cos\theta - \sin\alpha\sin\theta)\sin\beta - (\cos\beta\cos\theta + \sin\beta\sin\theta)\sin\alpha = 0$

or $2\sin\theta\sin\alpha\sin\beta = \cos\theta(\cos\alpha\sin\beta - \sin\alpha\cos\beta)$

or $$\tan\theta = \frac{1}{2}[\cot\alpha - \cot\beta] \qquad ...(4)$$

This gives the position of equilibrium.

For stability or unstability we find $\dfrac{d^2z}{d\theta^2}$

$$\frac{d^2z}{d\theta^2} = \frac{a}{\sin(\alpha+\beta)}[-\sin(\alpha+\theta)\sin\beta - \sin(\beta-\theta)\sin\alpha]$$

$$= \frac{-a}{\sin(\alpha+\beta)}[(\sin\alpha\cos\theta + \cos\alpha\sin\theta)\sin\beta + (\sin\beta\cos\theta - \cos\beta\sin\theta)\sin\alpha]$$

$$= \frac{-a}{\sin(\alpha+\beta)}[2\sin\alpha\sin\beta\cos\theta + \sin\theta(\cos\alpha\sin\beta - \sin\alpha\cos\beta)]$$

$$= \frac{-a\,.\,2\sin\alpha\sin\beta\cos\theta}{\sin(\alpha+\beta)}\left[1+\frac{1}{2}\tan\theta(\cot\alpha-\cot\beta)\right]$$

Putting the value of $\frac{1}{2}(\cot\alpha - \cot\beta)$ from (4), we have

$$\frac{d^2z}{d\theta^2} = \frac{-2a\sin\alpha\sin\beta\cos\theta}{\sin(\alpha+\beta)}[1+\tan^2\theta]$$

$\because$ α, β, θ are acute angles and $(\alpha+\beta) < 180°$

$\therefore$ $$\frac{d^2z}{d\theta^2} < 0$$

Therefore, the equilibrium is unstable.

Example 2. *A heavy uniform bar AB of weight W can move freely in a vertical plane about a hinge at A. A string is attached to the end B and after passing over a smooth pulley at C vertically above A carried a weight P at the other end. Show that the position of equilibrium in which AB is inclined to the vertical is unstable.*

Sol. Let the length of string be l.

Let $BC = r$, $AC = b$, $AB = 2a$ then $PC = l - r$.

Let G be the C.G. of rod. The height of G above is $AD = a\cos\theta$ and the height of P above $A = b - (l - r)$.

Let z be the height of C.G. of P and W above A

$$z = \frac{W\,.\,a\cos\theta + P\{b-(l-r)\}}{W+P} \quad \left(\bar{z} = \frac{W_1\bar{z}_1 + W_2\bar{z}_2}{W_1+W_2}\right) \qquad ...(1)$$

Differentiating w.r.t. θ

$$\frac{dz}{d\theta} = \frac{1}{W+P}\left[-Wa\sin\theta + P\,.\frac{dr}{d\theta}\right] \qquad ...(2)$$

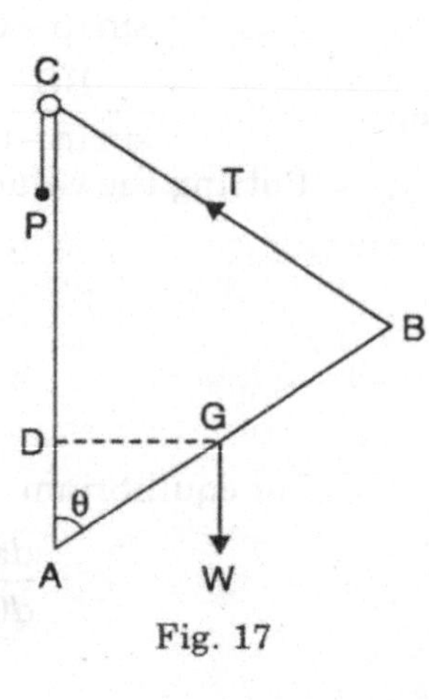

Fig. 17

From Δ ABC using cosine formula

$$\cos\theta = \frac{AB^2 + AC^2 - BC^2}{2\,AB\,.\,AC} = \frac{4a^2 + b^2 - r^2}{2\,.\,2a\,.\,b}$$

$$\Rightarrow \quad r^2 = 4a^2 + b^2 - 4ab\cos\theta$$

$$\Rightarrow \quad 2r\frac{dr}{d\theta} = 4ab\sin\theta$$

$$\Rightarrow \quad \frac{dr}{d\theta} = \frac{2ab}{r}\sin\theta \qquad ...(3)$$

Putting value of $\frac{dr}{d\theta}$ from (3) in (2)

$$\frac{dz}{d\theta} = \frac{1}{W+P}\left[-W a\sin\theta + \frac{P\,.\,2ab\sin\theta}{r}\right] = 0 \qquad ...(4)$$

$$\Rightarrow \quad \sin\theta = 0 \quad \text{or} \quad r = \frac{2Pb}{W} \qquad ...(5)$$

But since rod is inclined to the vertical therefore, $\sin\theta \neq 0$

∴ Equilibrium is possible if $r = \frac{2bP}{W}$

Further differentiating (4)

$$\frac{d^2z}{d\theta^2} = \frac{a}{W+P}\left[-W\cos\theta + \frac{2Pb}{r}\cos\theta - \frac{2Pb}{r^2}\sin\theta\frac{dr}{d\theta}\right]$$

$$= \frac{a}{W+P}\left[-W\cos\theta + \frac{rW\cos\theta}{r} - \frac{2Pb}{r^2}\,.\sin\theta\,.\frac{2ab\sin\theta}{r}\right]$$

$$= \frac{-4a^2b^2P\sin^2\theta}{r^3} < 0$$

∴ z is maximum.

Hence equilibrium is unstable.

Example 3. *A square lamina rests with its plane perpendicular to a smooth wall. One corner of the square is attached to a point in the wall by the light string of length equal to the side of the square. Show that the equilibrium is stable.* (K.U. 2004)

Sol. Let ABCD be the square of side $2a$

$$BE = AB = 2a$$

Where BE is the string attached at B and E.

Let $\angle BAE = \theta = \angle BEA$ (∵ AB = BE)

Let z be the depth of C.G. below E

$$z = EL + MG \qquad ...(1)$$

From ΔBEL, $EL = 2a\cos\theta$

From ΔBMG,

$$MG = BG\sin(\theta + 45°) \qquad ...(2)$$

$BG^2 + AG^2 = AB^2$ From right angle triangle ABG

$$2BG^2 = 4a^2$$

$$BG = a\sqrt{2}$$

Putting in (2), we have

$$MG = a\sqrt{2}\,\sin(\theta + 45°)$$

$$= a\sqrt{2}\,[\sin\theta\cos 45° + \cos\theta\sin 45°]$$

$$= a(\sin\theta + \cos\theta)$$

$$\therefore \quad z = 2a\cos\theta + a(\sin\theta + \cos\theta)$$

$$= 3a\cos\theta + a\sin\theta.$$

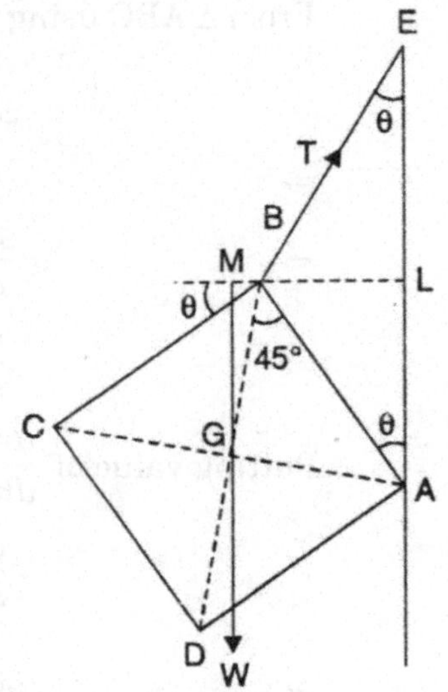

Fig. 18

For equilibrium $\dfrac{dz}{d\theta} = -3a\sin\theta + a\cos\theta = 0$

$\Rightarrow \quad \tan\theta = \dfrac{1}{3} \quad$ or $\quad \theta = \tan^{-1}\dfrac{1}{3}$.

For stability $\dfrac{d^2z}{d\theta^2} = -3a\cos\theta - a\sin\theta$

$$= -3a\cdot\frac{3}{\sqrt{10}} - a\cdot\frac{1}{\sqrt{10}} = -a\sqrt{10} < 0$$

$\because$ Depth of C.G. is maximum from a fixed point.

$\therefore$ The equilibrium is stable.

Example 4. *A smooth ellipse is fixed with its axis vertical and in it is placed a beam with its ends resting on the arc of the ellipse. If length of the beam be not less than the latus rectum of the ellipse, show that when it is in stable equilibrium, it will pass through the forces.*

Sol. Referred to S as pole the equation of ellipse is

$$\frac{l}{r} = 1 + e\cos\theta$$

Let ZZ′ be the directrix.

By the property of conic

$$AS = eAL,\ BS = eBM,$$

where AB is the rod resting in the ellipse S is focus

Height of C.G. above the directrix is

$$GN = \frac{1}{2}(AL + BM)$$

$$= \frac{1}{2}\frac{(AS + BS)}{e}$$

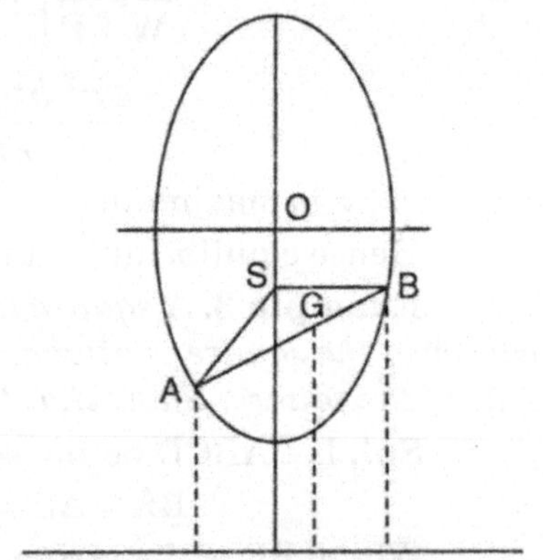

Fig. 19

For stability GN should be minimum for which AS + BS is minimum

or A, B, S is in the same line $\Rightarrow$ rod AB passes through S

In such case the length of rod = AB = AS + BS = $\dfrac{l}{1 + e\cos\theta} + \dfrac{l}{1 + e\cos(\pi - \theta)}$

$$= \frac{2l}{1 - e^2\cos^2\theta}$$

For minimum length, $\theta = \pi/2$

Length of rod $= 2l =$ Latus rectum.

Example 5. *A heavy body the section of which is a cycloid, rests on a rough horizontal plane and has its C.G. at the centre of curvature of the curve at the point of contact. Show that the equilibrium is unstable.* (M.D.U. 2005)

Sol. Equation of cycloid is $s = 4a \sin \psi$

$$\rho = \frac{ds}{d\psi} = 4a \cos \psi$$

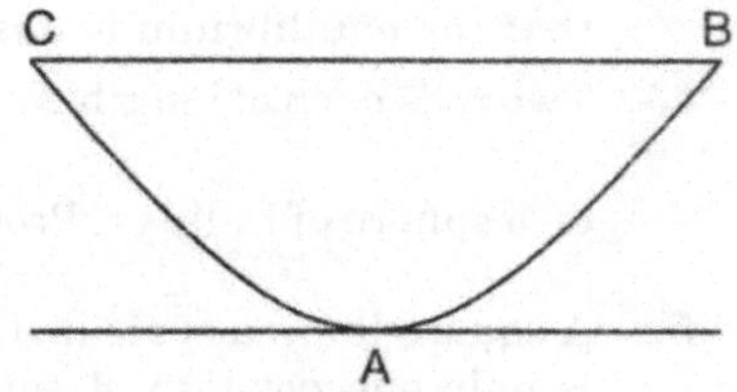

Fig. 20

At the point of contact $\psi = 0 \quad \therefore \quad \rho = 4a$

The equilibrium is stable or unstable

$$h < \text{or} > \frac{\rho_1 \rho_2 \cos \alpha}{\rho_1 + \rho_2}$$

In this case $\alpha = 0$

$\frac{1}{h} >$ or $< \frac{1}{\rho_1} + \frac{1}{\rho_2}$ lower surface is plane $\therefore \rho_1 = \infty$

$$\frac{1}{4a} > \text{or} < \frac{1}{\infty} + \frac{1}{4a} \Rightarrow \frac{1}{4a} > \text{or} < \frac{1}{4a}$$

The expression is an equality, therefore it does not give any result. In such case the equilibrium is stable or unstable if $\frac{d}{ds}\left(\frac{1}{\rho_1}\right) + \frac{d}{ds}\left(\frac{1}{\rho_2}\right) >$ or < 0

$$= \frac{d}{ds}(0) + \frac{d}{ds}\left(\frac{1}{4a}\right) = 0 + 0 \qquad \rho_1 = \infty,\ \rho_2 = 4a$$

$\therefore$ This test also fails.

Then the equilibrium is stable or unstable if

$$\frac{d^2}{ds^2}\left(\frac{1}{\rho_1}\right) + \frac{d^2}{ds^2}\left(\frac{1}{\rho_2}\right) + \frac{(\rho_1 + \rho_2)(\rho_2 + 2\rho_1)}{\rho_1^2 \rho_2^3} \quad \text{is} < \text{or} > 0$$

$$= 0 + 0 + \frac{\left(1 + \frac{\rho_2}{\rho_1}\right) + \left(2 + \frac{\rho_2}{\rho_1}\right)}{\rho_2^3} = \frac{2}{\rho_2^3} > 0 \quad \rho_1 \to \infty$$

$\therefore$ Equilibrium is unstable.

EXERCISE 2

1. A uniform beam of length $2a$ rests against a smooth vertical wall and upon a peg at a distance b from the wall. Show that in the position of equilibrium the inclination of the beam to the vertical is $\sin^{-1}\left(\frac{b}{a}\right)^{\frac{1}{3}}$ and show that it is unstable. *(K.U. 2007)*

2. One end of uniform rod AB of length $2a$ and weight W is hinged at A, while B is tied to a light string which passed over a smooth pulley at distance d vertically above A and carries a weight

$\frac{W}{4}$. If $2a < d < 4a$, show that the system is in stable equilibrium when AB is vertical. There is also a position of equilibrium in which rod is inclined to the vertical.

3. A uniform rod AB of weight W rests with one end A on a smooth horizontal plane AC and the other end against a smooth vertical wall BC. The end A is connected by a string which passes over a smooth pulley at C and carries a weight W′. If A, B, C are in same vertical plane, show that the equilibrium is unstable.
4. Two rods each of length $2a$ have their ends united at an angle α and are placed in a vertical plane on a sphere of radius r. Prove that equilibrium is stable or unstable according as $\sin\alpha >$ or $< \frac{2r}{a}$.
5. A square lamina rests in a vertical plane on two smooth pegs at the same level. Show that there is only one position of equilibrium unless the distance between the pegs is greater than one fourth of the diagonal of the square. If this condition is satisfied, there may be three positions of equilibrium out of which only one is stable.
6. A uniform isosceles triangular lamina ABC rests in equilibrium with its equal sides AB and AC in contact with two smooth pegs in the same horizontal line at a distance C apart. If length of perpendicular AD from A on BC is h, show that there are three positions of equilibrium, out of which one is stable when AD is vertical and other two are unstable, if $h < 3\,C\,\text{cosec}\,A$; whilest if $h \geq 3C\,\text{cosec}\,A$, there is only one position of equilibrium which is unstable.
7. A uniform rod of length $2a$ rests partly within and partly outside a fixed hemispherical bowl of radius r. The rim of the bowl is horizontal. If in the position of equilibrium the rod is inclined at an angle θ with the horizon show that $2r\cos 2\theta = a\cos\theta$. Show also that the equilibrium is stable.

SECTION III

8

Radial and Transverse Velocities

1. Introduction

Consider a particle moving along a plane curve. Let P and Q be the positions of the particle at time t and $t + \delta t$. Consider O as origin and OX as initial line.

Let OP = r and OQ = $r + \delta r$, $\angle XOP = \theta$ and $\angle XOQ = \theta + \delta\theta$.

The resolved part of velocity along the radius vector OP in the sense of r increasing is called the **radial velocity.**

The resolved part of velocity at P at right angle to OP in the sense of increasing θ is called the **transverse velocity.**

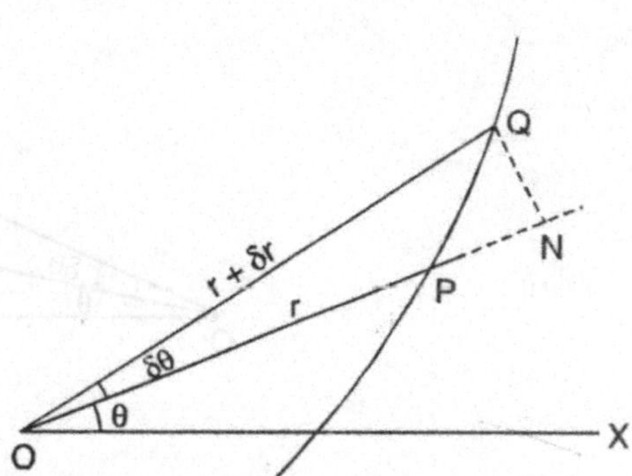

Fig. 1

Draw QN ⊥ OP

∴ **Radial velocity at P**

$$= \lim_{\delta t \to 0} \frac{\text{displacement along OP in time } \delta t}{\delta t}$$

$$= \lim_{\delta t \to 0} \frac{PN}{\delta t} = \lim_{\delta t \to 0} \frac{ON - OP}{\delta t} = \lim_{\delta t \to 0} \frac{(r + \delta r)\cos\delta\theta - r}{\delta t}$$

$$= \lim_{\delta t \to 0} \frac{(r + \delta r)\left\{1 - \frac{\delta\theta^2}{2!} + \frac{\delta\theta^4}{4!} - \ldots\ldots\right\} - r}{\delta t}$$

$$= \lim_{\delta t \to 0} \frac{(r + \delta r).1 - r}{\delta t} \qquad \text{Neglecting higher powers of } \delta\theta$$

$$= \lim_{\delta t \to 0} \frac{\delta r}{\delta t} = \frac{dr}{dt} = \dot{r} \text{ in the direction of OP.}$$

Transverse velocity at P

$$= \lim_{\delta t \to 0} \frac{\text{displacement perpendicular to OP}}{\delta t}$$

$$= \lim_{\delta t \to 0} \frac{QN}{\delta t} = \lim_{\delta t \to 0} \frac{(r + \delta r)\sin\delta\theta}{\delta t} = \lim_{\delta t \to 0} \frac{(r + \delta r)\left\{\delta\theta - \frac{\delta\theta^3}{3!} + \ldots\ldots\right\}}{\delta t}$$

$$= \lim_{\delta t \to 0} r\frac{\delta\theta}{\delta t} = r\frac{d\theta}{dt} = r\dot{\theta} \text{ in the sense of } \theta \text{ increasing.}$$

Thus **radial velocity = $\dot{r}$**

Transverse velocity = $r\dot{\theta}$

2. Radial and Transverse Velocities and Accelerations

Let u and v be the radial and transverse velocities at P(r, θ) and $u + \delta u$, $v + \delta v$ be the radial and transverse velocities at Q $(r + \delta r, \theta + \delta\theta)$.

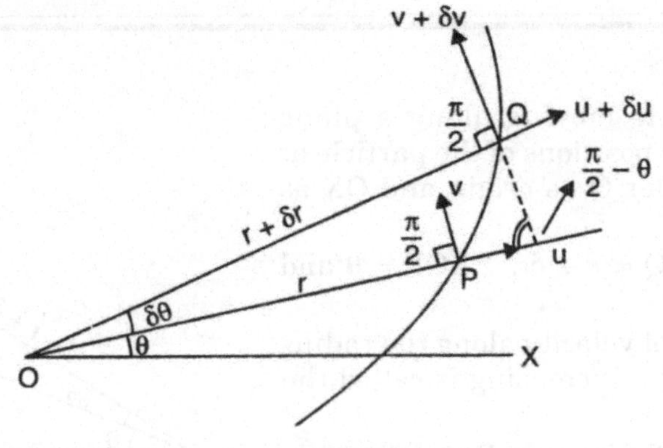

Fig. 2

Radial acceleration at P

$$= \lim_{\delta t \to 0} \frac{\text{change of velocity along OP}}{\delta t}$$

$$= \lim_{\delta t \to 0} \frac{(u + \delta u)\cos\delta\theta - (v + \delta v)\cos\left(\frac{\pi}{2} - \delta\theta\right) - u}{\delta t}$$

$$= \lim_{\delta t \to 0} \frac{(u + \delta u)\cos\delta\theta - (v + \delta v)\sin\delta\theta - u}{\delta t}$$

$$= \lim_{\delta t \to 0} \frac{(u + \delta u) - (v + \delta v)\,\delta\theta - u}{\delta t}$$

neglecting square and higher powers of $\delta\theta$

$$= \lim_{\delta t \to 0} \frac{\delta u - v\,\delta\theta}{\delta t} = \frac{du}{dt} - v\frac{d\theta}{dt} \qquad u = \frac{dr}{dt},\ v = r\frac{d\theta}{dt}$$

$$= \frac{d^2r}{dt^2} - r\left(\frac{d\theta}{dt}\right)^2$$

Radial acceleration = $\ddot{r} - r\dot{\theta}^2$ in the sense of r increasing.

Transverse acceleration at P

$$= \lim_{\delta t \to 0} \frac{\text{change in velocity} \perp \text{to OP}}{\delta t}$$

$$= \lim_{\delta t \to 0} \frac{(u+\delta u)\sin\delta\theta + (v+\delta v)\sin\left(\frac{\pi}{2}-\delta\theta\right) - v}{\delta t}$$

$$= \lim_{\delta t \to 0} \frac{(u+\delta u)\sin\delta\theta + (v+\delta v)\cos\delta\theta - v}{\delta t}$$

Expanding $\sin\delta\theta$ and $\cos\delta\theta$ and neglecting higher powers of $\delta\theta$, we have

$$= \lim_{\delta t \to 0} \frac{(u+\delta u)\,\delta\theta + (v+\delta v).1 - v}{\delta t} = u\frac{d\theta}{dt} + \frac{dv}{dt}$$

$$= \frac{dr}{d\theta}.\frac{d\theta}{dt} + \frac{d}{dt}\left(r\frac{d\theta}{dt}\right) = \frac{dr}{dt}.\frac{d\theta}{dt} + \frac{dr}{dt}.\frac{d\theta}{dt} + r\frac{d^2\theta}{dt^2}$$

$$= 2\frac{dr}{dt}.\frac{d\theta}{dt} + r\frac{d^2\theta}{dt^2} = \frac{1}{r}\frac{d}{dt}\left(r^2\frac{d\theta}{dt}\right)$$

Transverse acceleration = $r\ddot{\theta} + 2\dot{r}\dot{\theta}$ or $\frac{1}{r}\frac{d}{dt}(r^2\dot{\theta})$.

3. Expressions for radial and transverse velocities and accelerations (vector method)

Let $\vec{OP} = \vec{r} = |\vec{r}|\hat{r}$

Let $\angle POX = \theta$.

If $\hat{i}$ and $\hat{j}$ are unit vector along OX and OY then

$$\hat{r} = \cos\theta\,\hat{i} + \sin\theta\,\hat{j} \quad ...(1)$$

and $$\hat{n} = \cos\left(\frac{\pi}{2}+\theta\right)\hat{i} + \sin\left(\frac{\pi}{2}+\theta\right)\hat{j}$$

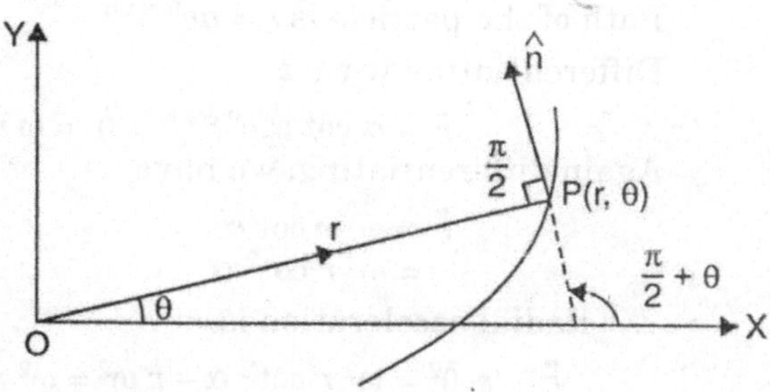

Fig. 3

where $\hat{n}$ is the unit vector perpendicular to $\vec{OP}$

i.e., $$\hat{n} = -\sin\theta\,\hat{i} + \cos\theta\,\hat{j} \quad ...(2)$$

Now, $$\frac{d}{dt}\hat{r} = -\sin\theta\frac{d\theta}{dt}\hat{i} + \cos\theta\frac{d\theta}{dt}\hat{j} = (-\sin\theta\,\hat{i} + \cos\theta\,\hat{j})\dot{\theta} = \hat{n}\,\dot{\theta} \quad ...(3)$$

and $$\frac{d}{dt}\hat{n} = -\cos\theta\,\hat{i}\frac{d\theta}{dt} - \sin\theta\,\hat{j}\frac{d\theta}{dt} = -(\cos\theta\,\hat{i} + \sin\theta\,\hat{j})\dot{\theta} = -\hat{r}\,\dot{\theta} \quad ...(4)$$

$\because$ $\vec{OP} = \vec{r} = r\hat{r}$ where $\vec{r}$ is position vector of P.

If velocity of P is $\vec{V}$

then $$\vec{V} = \frac{d\vec{r}}{dt} = \frac{d}{dt}(r\hat{r}) = \frac{dr}{dt}\hat{r} + r\frac{d}{dt}\hat{r}$$

$$\vec{V} = \dot{r}\hat{r} + r\dot{\theta}\hat{n} \quad ...(5) \quad \text{using (3)}$$

Thus component of linear velocity V along $\hat{r}$ *is* $\dot{r}$ which is **radial velocity** and $r\dot{\theta}$ is in the direction of $\hat{n}$ which is **transverse velocity.**

Differentiating (5) w.r.t. t, we get the acceleration

$$\vec{a} = \frac{d}{dt}(\dot{r}\,\hat{r}) + \frac{d}{dt}(r\,\dot{\theta}\,\hat{n})$$

$$= \ddot{r}\hat{r} + \dot{r}\frac{d}{dt}\hat{r} + \frac{dr}{dt}\dot{\theta}\,\hat{n} + r\,\ddot{\theta}\,\hat{n} + r\,\dot{\theta}\frac{d}{dt}\hat{n}$$

$$= \ddot{r}\hat{r} + \dot{r}\,\hat{n}\,\dot{\theta} + \dot{r}\,\dot{\theta}\,\hat{n} + r\,\ddot{\theta}\,\hat{n} + r\,\dot{\theta}(-\dot{\theta}\,\hat{r}) \qquad \text{using (3) and (4)}$$

$$\vec{a} = (\ddot{r} - r\dot{\theta}^2)\hat{r} + (2\dot{r}\,\dot{\theta} + r\,\ddot{\theta})\hat{n} \qquad \text{...(6)}$$

Hence **radial acceleration is $\ddot{r} - r\,\dot{\theta}^2$** and **transverse acceleration is $2\dot{r}\dot{\theta} + r\ddot{\theta}$**.

Example 1. *If the particle moves along the equiangular spiral $r = a\,e^{\theta\cot\alpha}$ with constant angular velocity, show that the resultant acceleration of the particle makes an angle 2α with the radius vector and is of magnitude $\frac{V^2}{r}$, where V is the speed of the particle.*

Sol. The angular velocity $\frac{d\theta}{dt} = \dot{\theta} = \omega$ (constant)

Path of the particle is $r = ae^{\theta\cot\alpha}$. ...(1)

Differentiating w.r.t. t

$$\therefore \quad \dot{r} = a\cot\alpha\, e^{\theta\cot\alpha}\,.\,\dot{\theta} = \omega\, r\cot\alpha \qquad \text{...(2)}$$

Again differentiating, we have

$$\ddot{r} = \dot{r}\,\omega\cot\alpha \qquad \text{using (2)}$$

$$= \omega^2 r\cot^2\alpha \qquad \text{...(3)}$$

$\therefore$ Radial acceleration is

$$\ddot{r} - r\,\dot{\theta}^2 = \omega^2 r\cot^2\alpha - r\,\omega^2 = \omega^2 r(\cot^2\alpha - 1) \qquad \text{...(4)}$$

Transverse acceleration is

$$= \frac{1}{r}\frac{d}{dt}(r^2\,\dot{\theta}) = \frac{1}{r}\frac{d}{dt}(\omega r^2) = \frac{1}{r}\,2r\,\omega\,\dot{r}$$

$$= 2\,\omega\,(\omega\, r\cot\alpha) \qquad \text{using (2)}$$

$$= 2\,\omega^2 r\cot\alpha \qquad \text{...(5)}$$

If β be the angle which the resultant acceleration makes with the radius vector then

$$\tan\beta = \frac{2\omega^2 r\cot\alpha}{\omega^2 r(\cot^2\alpha - 1)}$$

$$= \frac{2\cos\alpha}{\sin\alpha}\cdot\frac{\sin^2\alpha}{\cos^2\alpha - \sin^2\alpha} = \frac{\sin 2\alpha}{\cos 2\alpha} = \tan 2\alpha$$

$$\therefore \quad \beta = 2\alpha.$$

Also, $\quad V^2 = \dot{r}^2 + (r\,\dot{\theta})^2 = \omega^2 r^2\cot^2\alpha + \omega^2 r^2 = \omega^2 r^2\operatorname{cosec}^2\alpha$

$$V = \omega\, r\operatorname{cosec}\alpha \qquad \text{...(6)}$$

Resultant acceleration

$$= \sqrt{(\text{radial acceleration})^2 + (\text{transverse acceleration})^2}$$

$$= \sqrt{\omega^4 r^2(\cot^2\alpha - 1)^2 + 4\,\omega^4 r^2\cot^2\alpha}$$

$$= \omega^2 r\sqrt{(\cot^2\alpha - 1)^2 + 4\cot^2\alpha} = \omega^2 r(\cot^2\alpha + 1) = \omega^2 r \operatorname{cosec}^2\alpha$$

$$= \frac{(\omega r \operatorname{cosec}\alpha)^2}{r} = \frac{V^2}{r}. \qquad \text{Hence Proved.}$$

Example 2. *The velocities of a particle along and perpendicular to the radius vector are λr and $\mu\theta$. Find the path of the particle and show that the radial and transverse acceleration are $\lambda^2 r - \mu^2\theta^2/r$ and $\mu\theta(\lambda + \mu/r)$.*

Sol. Given $\dot{r} = \lambda r$ ...(1) $\quad r\dot{\theta} = \mu\theta$ ...(2)

Differentiating (1) w.r.t. t

$$\ddot{r} = \lambda\dot{r} = \lambda^2 r \qquad \text{...(3)}$$

Dividing (1) by (2), we have

$$\frac{dr}{r d\theta} = \frac{\lambda r}{\mu\theta} \quad \text{or} \quad \frac{d\theta}{\theta} = \frac{\mu}{\lambda}\frac{dr}{r^2}$$

Integrating, we get

$$\log\theta = -\frac{\mu}{\lambda}\frac{1}{r} + \log c$$

or $$\log\frac{\theta}{c} = -\frac{\mu}{\lambda r} \quad \text{or} \quad \theta = ce^{-\mu/\lambda r}$$

which is the required path of the particle.

Also, Radial acceleration

$$= \ddot{r} - r\dot{\theta}^2 = \lambda^2 r - r.\frac{\mu^2\theta^2}{r^2} = \lambda^2 r - \frac{\mu^2\theta^2}{r}$$

Transverse acceleration

$$= \frac{1}{r}\frac{d}{dt}(r^2\dot{\theta}) = \frac{1}{r}\frac{d}{dt}\left(r^2\frac{\mu\theta}{r}\right) = \frac{1}{r}\frac{d}{dt}(\mu r\theta)$$

$$= \frac{\mu}{r}[r\dot{\theta} + \dot{r}\theta] = \frac{\mu}{r}[\mu\theta + \lambda r\theta]$$

$$= \mu\theta\left[\frac{\mu}{r} + \lambda\right]. \qquad \text{Hence proved.}$$

Example 3. *A particle describes an equiangular spiral $r = ae^{\theta}$ in such a manner that its acceleration has no radial component. Prove that its angular velocity is constant and the magnitude of the velocity and acceleration is each proportional to r.* (KU 2004)

Sol. Path of the particle is $r = ae^{\theta}$ ...(1)

$$\therefore \quad \dot{r} = ae^{\theta}\dot{\theta} = r\dot{\theta}$$

$$\ddot{r} = r\ddot{\theta} + \dot{r}\dot{\theta} = r\ddot{\theta} + r\dot{\theta}^2$$

or $$\ddot{r} - r\dot{\theta}^2 = r\ddot{\theta}$$

$\because$ Radial acceleration is zero $\quad \therefore \quad \ddot{r} - r\dot{\theta}^2 = 0$

$\Rightarrow \quad r\ddot{\theta} = 0, \quad r \neq 0$

$\Rightarrow \quad \ddot{\theta} = 0 \quad \Rightarrow \quad \dot{\theta} = \text{constant} = \omega \text{ (say)}$ ...(2)

Resultant velocity $= \sqrt{\dot{r}^2 + (r\dot{\theta})^2} = \sqrt{(r\dot{\theta})^2 + (r\dot{\theta})^2}$

$= \sqrt{2r^2\omega^2}$ using (1) and (2)

$= \sqrt{2}\, r\omega$

$\therefore$ Velocity $\propto r$

Resultant acceleration

$= \sqrt{(\text{radial acceleration})^2 + (\text{transverse acceleration})^2}$

$= \sqrt{0 + (2\dot{r}\dot{\theta} + r\ddot{\theta})^2} = 2\, r\omega\omega + 0 \qquad \because \ddot{\theta} = 0$

$= 2\,\omega^2 r$

$\therefore$ Acceleration $\propto r$. Hence proved.

Example 4. *A straight smooth tube revolves with angular velocity ω in a horizontal plane about one extremity which is fixed. If at zero time a particle inside it be at a distance a from the fixed end and moving with constant velocity V along the tube, show that at time t its distance is*

$$a \cosh \omega t + \frac{V}{\omega} \sinh \omega t.$$

Sol. Let OX be the initial position of the tube. O is fixed end of the tube and at time t particle is at P, such that OP $= r$ and $\angle$POX $= \theta$.

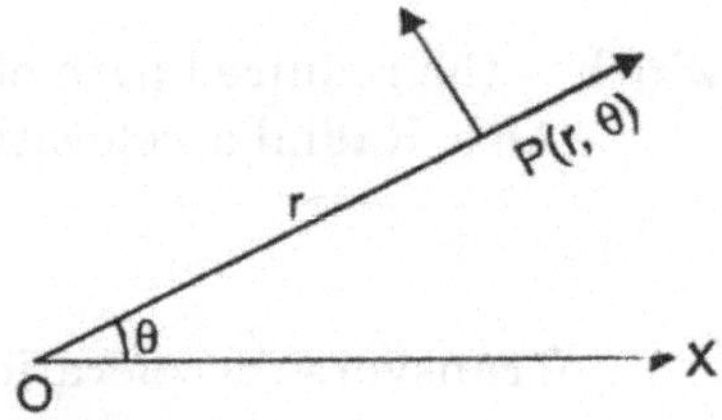

Fig. 4

Since the particle is moving with constant velocity V along the radius vector, therefore radial acceleration is zero.

$$\therefore \qquad \ddot{r} - r\dot{\theta}^2 = 0$$

or
$$\frac{d^2r}{dt^2} - \omega^2 r = 0 \qquad ...(1)$$

Its solution is $r = C_1 \cosh \omega t + C_2 \sinh \omega t$ ...(2)

At $t = 0, r = a$

$\therefore$ $a = C_1$

Differentiating $\dfrac{dr}{dt} = \dot{r} = C_1\, \omega \sinh \omega t + C_2\, \omega \cosh \omega t$

at $t = 0$, $\dot{r} = V$

$$\therefore \qquad V = 0 + C_2\,\omega \Rightarrow C_2 = \frac{V}{\omega}$$

Putting the values of C_1 and C_2 in (2)

$$r = a \cosh \omega t + \frac{V}{\omega} \sinh \omega t$$

which is the required distance at time t from O.

Example 5. *A particle moves along a circle of radius 'a' with a uniform speed V. Show that the radial and transverse accelerations are* $-\dfrac{V^2}{a}\cos\theta$ *and* $-\dfrac{V^2}{a}\sin\theta$, *if a diameter is taken as initial line and one end of this diameter as pole.*

Sol. Let the diameter OA be the initial line with O as pole. Let C be the centre. The polar equation of the circle is

$$r = 2a \cos \theta \qquad ...(1)$$

Let P be the position of the particle at any time and V be the velocity of P along the tangent at P.

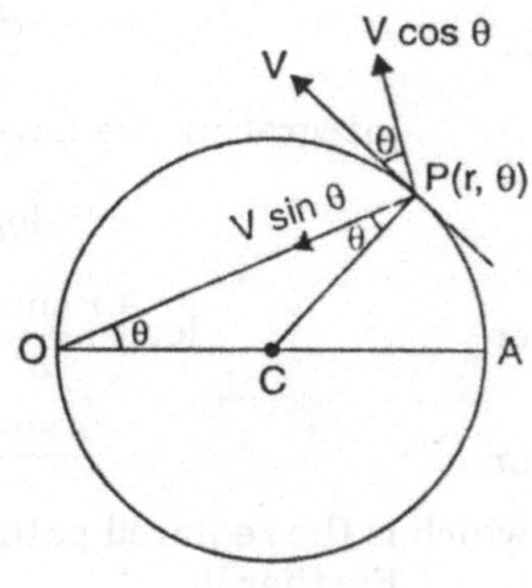

Fig. 5

∴ Resolved parts of V along OP and perpendicular to OP are – V sin θ and V cos θ

$$\therefore \qquad \dot{r} = -V \sin \theta \qquad ...(2)$$

$$r\dot{\theta} = V \cos \theta \qquad ...(3)$$

Differentiating (2), we get

$$\ddot{r} = -V \cos \theta \, \dot{\theta}$$

$$= -V \cos \theta . \frac{V \cos \theta}{r} \qquad \text{(using 3)}$$

$$\ddot{r} = -\frac{V^2 \cos^2 \theta}{r}$$

∴ Radial acceleration at P is

$$\ddot{r} - r\dot{\theta}^2 = -\frac{V^2 \cos^2 \theta}{r} - r\left(\frac{V \cos \theta}{r}\right)^2 = -\frac{2V^2 \cos^2 \theta}{2a \cos \theta} \qquad \text{From (1)}$$

$$= -\frac{V^2 \cos \theta}{a}$$

Transverse acceleration at P is

$$\frac{1}{r}\frac{d}{dt}(r^2 \dot{\theta}) = \frac{1}{r}\frac{d}{dt}\left(r^2 \frac{V \cos \theta}{r}\right) = \frac{V}{r}\frac{d}{dt}(r \cos \theta) = \frac{V}{r}[\dot{r} \cos \theta - r \sin \theta \, \dot{\theta}]$$

$$\text{From (2) and (3)} \qquad = \frac{V}{2a \cos \theta}[-V \sin \theta \cos \theta - V \sin \theta \cos \theta]$$

$$= -\frac{V^2 \sin \theta}{a}$$

Example 6. *A boat which is rowed with constant velocity u starts from a point on the bank of a river which flows with a constant velocity v and it points always towards a fixed point B on the other bank opposite to A. Find the equation of the path of the boat. If V = u, show that the path is a parabola.*

Sol. Consider B as pole and BX as initial line. Let P be the position of the boat at time *t*.

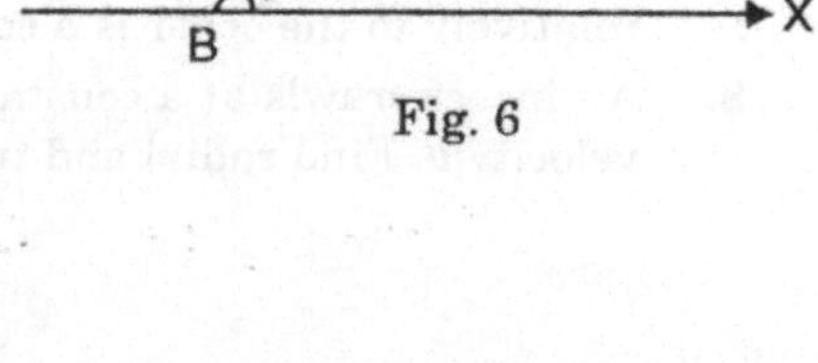

Fig. 6

P has two velocities *v* along the current and *u* towards B.

∴ Radial velocity of boat at P

$$\frac{dr}{dt} = \dot{r} = v \cos \theta - u \qquad ...(1)$$

and transverse velocity

$$r\frac{d\theta}{dt} = r\dot{\theta} = -v \sin \theta \qquad ...(2)$$

in the direction of increasing r and θ

Dividing (1) by (2)

$$\frac{dr}{r\,d\theta} = \frac{v\cos\theta - u}{-v\sin\theta}$$

or
$$\frac{dr}{r} = \frac{u}{v}\ \text{cosec}\,\theta - \cot\theta$$

Integrating, we have

$$\log r = \frac{u}{v}\ \log\tan\theta/2 - \log\sin\theta + \log c$$

or
$$\log\left(\frac{r\sin\theta}{c}\right) = \frac{u}{v}\ \log\tan\theta/2$$

or
$$\frac{r\sin\theta}{c} = (\tan\theta/2)^{u/v}$$

which is the required path.

Further if $v = u$

$$\frac{r\sin\theta}{c} = \tan\theta/2$$

or
$$\frac{c}{r} = \frac{\sin\theta}{\tan\theta/2} = \frac{2\sin\theta/2\cos\theta/2}{\sin\theta/2} \times \cos\theta/2 = 2\cos^2\theta/2$$

or
$$\frac{c}{r} = 1 + \cos\theta$$

which is the polar equation of parabola.

EXERCISE 1

1. A particle describes an equiangular spiral $r = ae^{m\theta}$ with constant angular velocity. Find its velocity and acceleration.
2. A point P describes an equiangular spiral $r = a\,e^{\theta\cot\alpha}$ with constant angular velocity about the pole O. Find its acceleration and show that its direction makes the same angle with the tangent at P as the radius vector OP makes with the tangent.
3. A particle moves along a circle $r = 2a\cos\theta$ in such a manner that radial component of acceleration is always zero. Show that transverse acceleration is proportional to fifth power of cosec θ.
4. If angular velocity of a point moving in a plane curve be constant about a fixed origin, show that the transverse acceleration varies as the radial velocity.
5. A ring which can slide on a thin long smooth rod rests at a distance d from one end O. The rod is then set revolving uniformly about O in a horizontal plane. Show that ring describes the curve $r = d\cosh\theta$.
6. The velocities of a particle (radial and transverse) are λr^2 and $\mu\theta^2$, where λ and μ are constants. Find the equation of the path of the particle. Also find radial and transverse accelerations.
7. A vessel steams at a constant speed v along a straight line, whilest another streaming at a constant speed V, keeps the first always exactly a beam. Show that the path of either vessel relatively to the other is a conic section of ecentricity v/V.
8. An insect crawls at a constant rate u along the spoke of a cart wheel of radius a, moving with velocity v. Find radial and transverse accelerations. *(M.D.U. 2005)*

9. A small bead slides with a constant speed v on a smooth wire in the shape of a Cardiod $r = a\,(1 + \cos\,\theta)$. Show that the angular velocity is $\dfrac{v}{2a}$ sec $\theta/2$ and radial acceleration is constant.

10. A point moves along a parabola $\dfrac{2a}{r} = 1 + \cos\theta$ such that transverse component is constant. Show that acceleration of the point is constant in magnitude.

11. A point starts from origin in the direction of initial line with velocity f/ω and moves with constant angular velocity ω about origin with constant radial acceleration $-f$. Show that rate of growth of the radial velocity is never positive but tends to the limit zero, and equation of the path is $\omega^2 r = f\,(1 - e^{-\theta})$.

12. The acceleration of a point moving in a plane curve is resolved into two components one parallel to the initial line and other along the radius vector. Prove that these components are $-\dfrac{1}{r\sin\theta}\dfrac{d}{dt}(r^2\,\dot{\theta})$ and $\dfrac{\cot\theta}{r}\dfrac{d}{dt}(r^2\,\dot{\theta}) + \ddot{r} - r\dot{\theta}^2$.

13. A particle moves in a circular paths of radius a so that its angular velocity ω about a fixed point in the circumference is constant. Show that resultant acceleration of the particle at every point is $4a\omega^2$. *(K.U. 2007)*

14. A small ring is at rest on a smooth straight horizontal rod of length a at a distance b from one end O of the rod. The rod is then suddenly set rotating in a horizontal plane about O with constant angular velocity ω. Prove that the ring will leave the rod with velocity $\omega\sqrt{2a^2 - b^2}$ after a time $\dfrac{1}{\omega}\cosh^{-1} a/b$.

15. A point moves on a parabola $2a = r\,(1 + \cos\theta)$ in such a manner that the component of velocity at right angles to the radius vector from the focus is constant. Show that the acceleration of the point is constant in magnitude.

16. If the path of a particle is $r = a\tan\theta$ and acceleration is directed towards the origin, show that the acceleration in terms of r is $\dfrac{k^2}{r^2}\left[3 + \dfrac{2a^2}{r^2}\right]$, where $h = r^2\dfrac{d\theta}{dt}$.

17. If the angular velocity ω of a particle about origin is constant and the rate of change of acceleration is directed wholly along the radius vector, prove that $d^2r/dt^2 = \omega^2 r/3$.

18. A particle is moving in a plane under the attraction of magnitude $2/r^3$ per unit mass towards a fixed point in the plane. Initially $r = 2$ and radial and transverse velocities are $\sqrt{3}/2$ and 1 respectively. Show that $\ddot{r} = \dfrac{2}{r^3}$.

ANSWERS

1. $\sqrt{m^2 + 1}\,\omega r,\ \omega^2 r(m^2 + 1)$

6. $2\lambda^2 r^3 - \dfrac{\mu^2\theta^4}{r},\ \dfrac{2\mu^2\theta^3}{r} + \lambda\mu r\theta^2$

8. $-ut\ v^2/a^2$, $2uv/a$

4. Tangential acceleration

Suppose a particle is moving along a plane curve. Let A be the fix point on the curve. Let the particle be at P and Q at time t and $t + \delta t$. Let $\widehat{AP} = s$, $\widehat{PQ} = \delta s$. Let velocity of the particle at P and Q be v and $v + \delta v$ along the tangents at these points. Let the tangents PT and QS make angle ψ and $\psi + \delta\psi$ with OX.

Tangential acceleration at P

$$= \lim_{\delta t \to 0} \frac{\text{change of velocity along the tangent at P in time } \delta t}{\delta t}$$

$$= \lim_{\delta t \to 0} \frac{(v + \delta v)\cos \delta\psi - v}{\delta t}$$

$$= \lim_{\delta t \to 0} \frac{(v + \delta v)\left(1 - \frac{\delta\psi^2}{2!} + \ldots\ldots\right) - v}{\delta t}$$

$$= \lim_{\delta t \to 0} \frac{\delta v}{\delta t} = \frac{dv}{dt}$$

neglecting terms of 2nd and higher orders.

$$= \frac{d}{dt}\left(\frac{ds}{dt}\right) = \frac{d^2 s}{dt^2}$$

or $$\frac{dv}{dt} = \frac{dv}{ds}\frac{ds}{dt} = v\frac{dv}{ds}$$

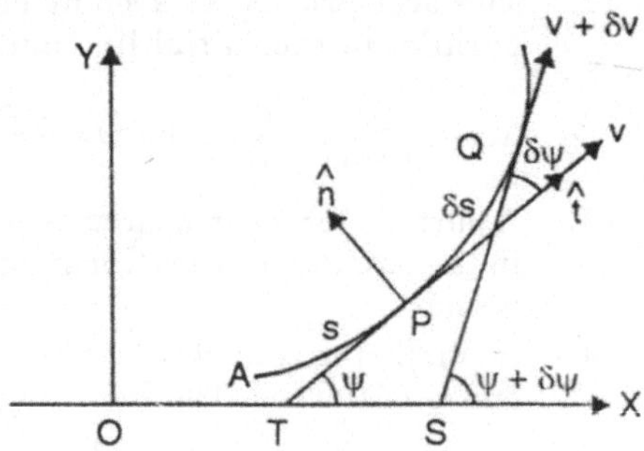

Fig. 7

Thus tangential acceleration is $\frac{dv}{dt}$, $v\frac{dv}{ds}$ or $\frac{d^2s}{dt^2}$

Normal acceleration at P

$$= \lim_{\delta t \to 0} \frac{(v + \delta v)\sin \delta\psi - v\cos 90}{\delta t} = \lim_{\delta t \to 0} \frac{(v + \delta v)\left(\delta\psi - \frac{\delta\psi^3}{3!} + \ldots\ldots\right)}{\delta t}$$

$$= v\frac{d\psi}{dt} \quad \text{neglecting higher order terms.}$$

Normal acceleration $= v\frac{d\psi}{ds}\frac{ds}{dt} = \frac{v^2}{\rho}$

where $\rho = \frac{ds}{d\psi}$ is the radius of curvature.

$\therefore$ Resultant Acceleration

$$= \sqrt{(\text{tangential acceleration})^2 + (\text{normal acceleration})^2}$$

Resultant acceleration $= \sqrt{\left(\frac{dv}{dt}\right)^2 + \left(\frac{v^2}{\rho}\right)^2}$

Direction of the resultant acceleration with the tangent is given by **$\tan\alpha = \frac{v^2/\rho}{dv/dt}$**.

5. Tangential and Normal Acceleration (Vector Method) (M.D.U. 2005)

Refer Fig. 7.

Let the particle be at P at time t and at Q at time $t + \delta t$. Let $\hat{t}$ and $\hat{n}$ be the unit vector along the tangent and inward normal.

$$\hat{t} = \cos\psi\,\hat{i} + \sin\psi\,\hat{j}$$

$$\hat{n} = \cos\left(\frac{\pi}{2}+\psi\right)\hat{i} + \sin\left(\frac{\pi}{2}+\psi\right)\hat{j}$$

$$\therefore \quad = -\sin\psi\,\hat{i} + \cos\psi\,\hat{j}$$

$$\frac{d\hat{t}}{dt} = -\sin\psi\,\dot{\psi}\hat{i} + \cos\psi\,\dot{\psi}\hat{j}$$

$$= (-\sin\psi\,\hat{i} + \cos\psi\,\hat{j})\dot{\psi}$$

$$\frac{\mathbf{d\hat{t}}}{\mathbf{dt}} = \mathbf{\hat{n}}\,\dot{\psi}$$

$$\frac{d}{dt}\hat{n} = -\cos\psi\,\dot{\psi}\,\hat{i} - \sin\psi\,\dot{\psi}\,\hat{j}$$

$$= -(\cos\psi\,\hat{i} + \sin\psi\,\hat{j})\,\dot{\psi}$$

$$\frac{\mathbf{d\hat{n}}}{\mathbf{dt}} = -\mathbf{\hat{t}}\,\dot{\psi}$$

Velocity of P is given as $\vec{V} = v\hat{t}$ where $v = \dfrac{ds}{dt}$.

Acceleration vector $\vec{a}$ at P

$$\vec{a} = \frac{d}{dt}(v\hat{t}) = \frac{dv}{dt}\hat{t} + v\frac{d}{dt}\hat{t} = \frac{dv}{dt}\hat{t} + v\,\dot{\psi}\,\hat{n} \qquad \because \frac{d}{dt}\hat{t} = \dot{\psi}\,\hat{n}$$

$$= \frac{dv}{dt}\hat{t} + v\frac{d\psi}{dt}\hat{n} = \frac{dv}{dt}\hat{t} + v\frac{d\psi}{ds}\frac{ds}{dt}\hat{n}$$

$$\vec{a} = \left(\frac{dv}{dt}\right)\hat{t} + \left(\frac{v^2}{\rho}\right)\hat{n} = \dot{v}\hat{t} + kv^2\hat{n} \text{ where } k = \frac{1}{\rho} \text{ (curvature)} \qquad \ldots(1)$$

Thus **tangential acceleration** = $\mathbf{\dfrac{dv}{dt}}$

and **normal acceleration** = $\mathbf{\dfrac{v^2}{\rho}}$

where ρ is the radius of curvature.

Example 1. *A particle moves along a plane curve such that its tangential and normal acceleration are equal and the angular velocity is constant. Find the path of the particle.*

Sol. Tangential acceleration = Normal acceleration.

$$v\frac{dV}{ds} = \frac{v^2}{\rho} \qquad \ldots(1)$$

$$\frac{dv}{v} = \frac{ds}{\rho} \qquad \left(\rho = \frac{ds}{d\psi}\right)$$

$$\therefore \quad \frac{dv}{v} = d\psi$$

Integrating, $\log v = \psi + \log c$

or $\log \frac{v}{c} = \psi$ or $v = ce^{\psi}$

or $\frac{ds}{dt} = ce^{\psi}$ or $\frac{ds}{d\psi}\frac{d\psi}{dt} = ce^{\psi}$ $\frac{d\psi}{dt} = \omega$ (constant) given.

or $ds = \frac{c}{\omega} e^{\psi} d\psi$

Integrating again, we get $s = \frac{c}{\omega} e^{\psi} + d.$ ...(2)

Example 2. *A particle moves along the curve $x = 4t$, $y = 6t - t^2$. Find tangential and normal acceleration at $t = 3$.*

Sol. $x = 4t \quad , \quad y = 6t - t^2$ ...(1)

$$\dot{x} = 4 \quad , \quad \dot{y} = 6 - 2t$$

$$\ddot{x} = 0 \quad , \quad \ddot{y} = -2$$

Radius of curvature

$$\rho = \frac{(\dot{x}^2 + \dot{y}^2)^{3/2}}{\dot{x}\ddot{y} - \dot{y}\ddot{x}}$$

$$V = \sqrt{\dot{x}^2 + \dot{y}^2}$$

$$\therefore \quad \rho = \frac{V^3}{4 \times (-2) - 0} = \frac{V^3}{-8}$$

$$V = \sqrt{16 + 4(3-t)^2} = 2\sqrt{t^2 - 6t + 13}$$

$$\text{Tangential acceleration} = \frac{dV}{dt} = 2 \times \frac{1}{2} \frac{2t - 6}{\sqrt{t^2 - 6t + 13}}$$

At $t = 3$ Tangential acceleration = 0

$t = 3$ $V = 2\sqrt{9 - 18 + 13} = 4.$

$$\text{Normal acceleration} = \frac{V^2}{\rho} = \frac{V^2 \times 8}{-V^3} = \frac{-8}{V} = -2 = -2 \text{ unit/sec}^2.$$

Example 3. *A particle describes a cycloid with uniform speed. Prove that the normal acceleration at any point varies inversely as the square root of the distance from the base of the cycloid.*

Sol. The intrinsic equation of the cycloid is $s = 4a \sin \psi$

$$v = \frac{ds}{dt} = \text{constant} \qquad \text{(given)}$$

$$\therefore \quad \frac{dv}{dt} = 0 \Rightarrow \text{Tangential acceleration} = 0$$

$\therefore$ Normal acceleration $= \dfrac{v^2}{\rho}$

where $\rho = \dfrac{ds}{d\psi} = 4a \cos\psi$

$$\therefore \text{ Normal acceleration} = \frac{v^2}{4a\cos\psi} = \frac{v^2}{4a\sqrt{1-\sin^2\psi}}$$

$$= \frac{v^2}{\sqrt{16a^2 - (4a\sin\psi)^2}} = \frac{v^2}{\sqrt{16a^2 - s^2}} \quad \text{...(2)}$$

We know that if distance of any point on the cycloid from the tangent at the vertex be y, then the distances of the point measured from the vertex along the arc is given by $s^2 = 8\,ay$.

$$\therefore \text{ Normal acceleration} = \frac{v^2}{\sqrt{16a^2 - 8ay}} = \frac{v^2}{\sqrt{8a\,(2a-y)}}.$$

$$\therefore \text{ Normal acceleration} \propto \frac{1}{\sqrt{2a-y}}$$

where $2a - y$ is the distance of the point from the base.

Example 4. *A point moves in a plane curve so that its tangential acceleration is constant and the magnitude of the tangential velocity and normal acceleration are in a constant ratio. Find the intrinsic equation of the curve.*

Sol. Tangential acceleration

$$v\frac{dv}{ds} = k \quad \text{...(1)}$$

$$\frac{v}{\left(\dfrac{v^2}{\rho}\right)} = \lambda \quad \text{...(2)}$$

From (1) $\quad v\,dv = k\,ds$

Integrating, $\quad \dfrac{v^2}{2} = ks + c \quad$ or $\quad v = \sqrt{2ks + 2c}$ ...(3)

From (2), we have $\quad v\lambda = \rho$

or $\quad \dfrac{ds}{d\psi} = \lambda\sqrt{2ks + 2c}$ $\quad$ using (3)

$$\therefore \quad \frac{ds}{\sqrt{2ks+2c}} = \lambda\, d\psi$$

Integrating, we get

$$\frac{\sqrt{2ks+2c}}{k} = \lambda\psi + \alpha$$

or $\quad \sqrt{2ks+2c} = k\lambda\psi + k\alpha \quad$ or $\quad 2ks + 2c = k^2(\lambda\psi + \alpha)^2$

or $\quad ks + c = \dfrac{k^2}{2}(\lambda^2\psi^2 + 2\lambda\psi\alpha + \alpha^2)$

or $$s = \frac{k\lambda^2}{2}\psi^2 + k\lambda\alpha\psi + \frac{k\alpha^2}{2} - \frac{c}{k}$$

$$s = A\psi^2 + B\psi + D$$

where $A = \frac{k\lambda^2}{2}$, $B = k\lambda\alpha$, $D = \frac{k\alpha^2}{2} - \frac{C}{k}$.

Example 5. *A particle projected with a velocity u is acted on by a force which produces a constant acceleration f in the plane of the motion inclined at a constant angle α with the direction of motion. Find the intrinsic equation of the curve described, and the particle will be moving in the opposite direction to that of projection at time* $\frac{u}{f \cos \alpha}(e^{\pi \cot \alpha} - 1)$.

Sol. Since acceleration is inclined at an angle α with the direction of motion

$$\therefore \quad v\frac{dv}{ds} = f\cos\alpha \quad \text{...(1)}$$

$$\frac{v^2}{\rho} = f\sin\alpha \quad \text{...(2)}$$

Integrating (1), we get

$$\frac{v^2}{\rho} = fs\cos\alpha + C$$

Initially at $s = 0, v = u$

$$\therefore \quad \frac{u^2}{2} = C.$$

Substituting the value of C, we have

$$v^2 = 2fs\cos\alpha + u^2 \quad \text{...(3)}$$

Putting the value of v^2 from (2) into (3)

$$\rho f \sin\alpha = 2fs\cos\alpha + u^2$$

or $$\frac{ds}{d\psi} f\sin\alpha = 2fs\cos\alpha + u^2$$

or $$\frac{f\sin\alpha\, ds}{u^2 + 2fs\cos\alpha} = d\psi$$

Integrating both sides

$$\frac{\sin\alpha}{2\cos\alpha}\int\frac{2f\cos\alpha}{u^2 + 2fs\cos\alpha}ds = \int d\psi + C'$$

$$\frac{1}{2}\tan\alpha\log(u^2 + 2fs\cos\alpha) = \psi + C' \quad \text{...(4)}$$

Let at $s = 0, \psi = 0$

$$\frac{1}{2}\tan\alpha\log u^2 = C'$$

Putting C′ in (4), we have

$$\frac{1}{2}\tan\alpha\log(u^2 + 2fs\cos\alpha) = \psi + \frac{1}{2}\tan\alpha\log u^2$$

or $$\log(u^2 + 2fs\cos\alpha) - \log u^2 = 2\psi\cot\alpha$$

or $$\log\frac{(u^2 + 2fs\cos\alpha)}{u^2} = 2\psi\cot\alpha$$

or $$\frac{u^2 + 2fs\cos\alpha}{u^2} = e^{2\psi\cot\alpha} \quad ...(5)$$

or $$\frac{2fs\cos\alpha}{u^2} = (e^{2\psi\cot\alpha} - 1)$$

or $$s = \frac{u^2}{2f\cos\alpha}(e^{2\psi\cot\alpha} - 1)$$

which is the required equation of the path.

Further $$\frac{v^2}{\rho} = f\cos\alpha$$

Integrating $$v = ft\cos\alpha + c''$$

At $$s = 0\ v = u \Rightarrow c'' = u$$

Hence $$v = u + ft\cos\alpha \quad ...(6)$$

From (3) $$v = \sqrt{u^2 + 2fs\cos\alpha}$$

But $$u^2 + 2fs\cos\alpha = u^2 e^{2\psi\cot\alpha} \quad \text{from (5)}$$

Hence $$v = \sqrt{u^2 e^{2\psi\cot\alpha}} \quad \text{or} \quad v = ue^{\psi\cot\alpha} \quad ...(7)$$

Comparing (6) and (7)

$$u + ft\cot\alpha = ue^{\psi\cot\alpha} \quad \text{or} \quad t = \frac{u(e^{\psi\cot\alpha} - 1)}{f\cos\alpha}$$

When the particle is moving in a direction opposite to the direction of projection we have $\psi = \pi$.

$$\therefore \quad t = \frac{u(e^{\psi\cot\alpha} - 1)}{f\cos\alpha}.$$ Hence proved.

EXERCISE 2

1. A particle describes a circle of radius r with a uniform speed V. Show that its acceleration at any point of the path is $\frac{V^2}{r}$ and is always directed towards the centre of the circle.

2. Prove that acceleration of a particle moving in a curve with uniform speed is $\rho\dot{\psi}^2$.

3. If tangential and normal acceleration of a moving point are equal then prove that the velocity varies as e^{ψ}. *(K.U. 2005)*

4. A particle describes a plane curve with a constant speed and its acceleration is constant in magnitude. Prove that its path is a circle.

5. A particle describes a curve for which s and ψ vanish simultaneously with uniform speed V. If acceleration at any point be $v^2c/(s^2 + c^2)$ find the equation of the curve.

6. A particle moves in a catenary $s = c \tan \psi$. The direction of its acceleration is equally inclined with the tangent and normal to the path. The speed at the vertex ($\psi = 0$) is u. Show that the velocity and acceleration at any point ψ are ue^{ψ} and $\frac{\sqrt{2}}{c} u^2 e^{2\psi} \cos^2 \psi$ respectively.

7. If speed of a particle moving along the circle of radius a at certain time is u and tangential acceleration is λ times the normal acceleration, then prove that the particle will return to the same point after a time $(a/\lambda u)(1 - e^{-2\pi\lambda})$.

8. A curve is described by a particle having constant acceleration in a direction inclined at a constant angle α to the tangent. Show that the curve is an equiangular spiral.

9. A particle is moving in a parabola $p^2 = ar$ with uniform angular velocity about the focus. Prove that its normal acceleration at any point is proportional to the radius of curvature of the path at that point.

10. A particle describes a plane curve such that tangential and normal accelerations are each constant during the motion. Prove that the angle ψ through which the direction of motion turns in time t is given by $\psi = A \log (1 + Bt)$.

11. One point describes the diameter AB of the circle with constant velocity and another the semi-circumference AB from rest with constant tangential acceleration. They start together from A and arrive together at B. Show that velocities at B are $\pi : 1$.

12. A particle moves in a plane such that its tangential and normal accelerations are equal and its velocity varies as $e^{(\tan^{-1} s/c)}$, s being the length of the arc of the curve measured from the fixed point on the curve. Show that the path of the curve is $s = c \tan \psi$.

9

Simple Harmonic Motion

Definition. If the motion of a particle is such that the acceleration is always directed towards a fixed point and is proportional to the distance of the particle from the fixed point then the motion is called simple harmonic motion.

1. Expression for Acceleration, Velocity and Displacement of a Particle Executing S.H.M.

(K.U. 2004 ; M.D.U. 2005)

Let O be the fixed point. A particle starts from rest at A (OA = a) and moves towards O. Let at time t the particle be at P such that OP = x. Since the particle executes S.H.M. its acceleration is μx.

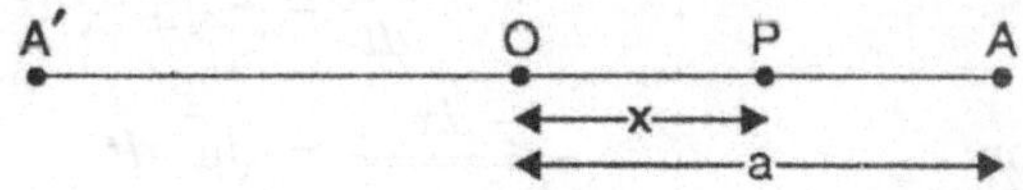

Fig. 1

$\therefore$ Equation of motion is

$$v\frac{dv}{dx} = \frac{d^2x}{dt^2} = -\mu x \quad \text{...(1)}$$

Negative sign on the R.H.S. is taken because $\frac{d^2x}{dt^2}$ is the acceleration in the direction of x increasing while μx is the acceleration towards O

$$vdv = -\mu x\, dx$$

Integrating both sides, we have

$$\frac{v^2}{2} = \frac{-\mu x^2}{2} + C \quad \text{...(2)}$$

$\because$ Particles starts from A with velocity zero

$\therefore$ At $x = a \quad v = 0$

$$\therefore \quad C = \frac{\mu a^2}{2}$$

Putting the value of C in (2), we have

$$\boxed{\mathbf{v^2 = \mu (a^2 - x^2)}} \quad \text{...(3)}$$

or

$$v = \pm \sqrt{\mu}\ \sqrt{a^2 - x^2}\,.$$

Case. (*i*) When time is measured from the fixed point. Particle moves away from O and t increases implies x increases.

$\therefore \quad \frac{dx}{dt}$ is (+) ve.

$$\frac{dx}{dt} = \sqrt{\mu}\ \sqrt{a^2 - x^2}$$

or $$\frac{dx}{\sqrt{a^2 - x^2}} = \sqrt{\mu}\ dt$$

Integrating both sides

$$\sin^{-1}\frac{x}{a} = \sqrt{\mu}\, t + C_1$$

At $t = 0\ x = 0 \quad \Rightarrow \quad C_1 = 0$

$$\therefore \quad \boxed{\mathbf{x = a \sin \sqrt{\mu}\ t}}$$

Case. (*ii*) When time is measured from A. At $x = a$, $v = 0$, $t = 0$. **Particle moves towards O and** x **decreases with the increase of** t **hence** $\frac{dx}{dt}$ **is negative.**

$$\frac{dx}{dt} = -\sqrt{\mu}\ \sqrt{a^2 - x^2}$$

or $$\frac{-dx}{\sqrt{a^2 - x^2}} = \sqrt{\mu}\ dt$$

Integrating, we get

$$\cos^{-1}\frac{x}{a} = \sqrt{\mu}\, t + C_2$$

At $x = a, \quad t = 0 \quad \Rightarrow \quad C_2 = 0$

We have $$\boxed{\mathbf{x = a \cos \sqrt{\mu}\ t}}$$

Maximum velocity of the particle is at $x = 0$

$$V_{max} = \sqrt{\mu}\ a$$

Minimum velocity is at $x = a$ *i.e.*, $V_{min} = 0$.

Maximum displacement a is called the amplitude of S.H.M.

Maximum acceleration μa is at the extreme positions where particle is at rest.

2. S.H.M. is a Periodic Motion

A function $f(x)$ is periodic if $f(x) = f(t + x)$, t is called the period.

We know that displacement x in S.H.M. is given by

$$x = a \sin \sqrt{\mu}\, t = a \sin (\sqrt{\mu}\, t + 2\pi)$$

$$= a \sin \sqrt{\mu}\left(t + \frac{2\pi}{\sqrt{\mu}}\right)$$

$$\therefore \quad \boxed{\mathbf{T} = \frac{2\pi}{\sqrt{\mu}}}$$ is the time period of S.H.M.

$\therefore$ The particle occupies the same position and moves in the same direction after $\frac{2\pi}{\sqrt{\mu}}$ seconds.

Frequency of S.H.M. is the number of complete oscillations in a second

$$\therefore \quad n = \frac{1}{T} \quad \text{or} \quad \boxed{n = \frac{\sqrt{\mu}}{2\pi}}$$

Discussion of the motion. The particle starts from A with velocity zero and moves towards O with acceleration μa. Its velocity increases and attains maximum velocity $a\sqrt{\mu}$ at O and crosses O and moves along OA′. Since acceleration is always towards O therefore the velocity decreases and vanishes at A′ such that OA = OA′. It then retraces its path and passes through O and again it is at rest at A. Thus the particle oscillates between A and A′.

Note. When we have to find time the particle takes from O to some position P or *vice versa* we use $x = a \sin \sqrt{\mu}\, t$. when we have to find time from some position P to extreme position A or from A to P, we use $x = a \cos \sqrt{\mu}\, t$. x is always measured from O.

Example 1. *A particle moves in a straight line with S.H.M. of period 2 secs. If it starts from rest at a distance of 13 cm. from the centre of the path, show that the greatest velocity and the velocity acquired by it, when it has described 8 cm are 13 π cm / sec and 12 π cm / sec respectively.*

Sol. Time period $\frac{2\pi}{\sqrt{\mu}} = 2 \Rightarrow \mu = \pi^2$

$\because$ Particles starts from rest at a distance 13 cm. from centre hence amplitude $a = 13$.

Maximum velocity = $a\sqrt{\mu}$ = 13 π cm/sec.

$\because$ Particle travelled 8 cm. from extreme position its displacement from centre is $x = 13 - 8 = 5$ cm.

$$\therefore \quad V = \sqrt{\mu}\sqrt{a^2 - x^2} = \pi\sqrt{169 - 24} = 12\,\pi \text{ cm/sec.}$$

Example 2. *A mass of one gramme vibrates through a millimetre on each side of the middle point of its path 256 times per second.*

Assuming its motion to be S.M.M. ; show that maximum force upon the particle is $\frac{1}{10}(512\,\pi)^2$ *dynes.*

Sol. Particle oscillates 1 mm on both sides of the centre hence its amplitude

$$a = 1 \text{ mm.} = 0.1 \text{ cm.}$$

Frequency $n = 256 = \frac{\sqrt{\mu}}{2\pi}$

$\therefore \quad \mu = (512\,\pi)^2$

Maximum acceleration on the particle = $a\mu = 0.1(512\,\pi)^2$

$\therefore$ Max. force = (mass)(max. accel.)

$$= 1 \times 0.1(512\,\pi)^2 = \frac{1}{10}(512\,\pi)^2.$$

Example 3. *At the end of three successive seconds, the distances of a particles moving with S.H.M. from mean position are* x_1, x_2 *and* x_3. *Show that the time period of S.H.M is* $\frac{2\pi}{\theta}$ *sec where* $\cos\theta = \frac{x_1 + x_3}{2x_2}$.

Sol. Let x_1, x_2, x_3 be the displacements from the mean position at t, $t + 1$, $t + 2$ sec. respectively.

Hence,
$$x_1 = a \sin \sqrt{\mu}\, t \qquad ...(1)$$
$$x_2 = a \sin \sqrt{\mu}\,(t + 1) \qquad ...(2)$$
$$x_3 = a \sin \sqrt{\mu}\,(t + 2) \qquad ...(3)$$

$$\frac{x_1 + x_3}{2x_2} = \frac{a \sin \sqrt{\mu}\, t + a \sin \sqrt{\mu}\,(t+2)}{2a \sin \sqrt{\mu}\,(t+1)}$$

$$= \frac{a\left[2 \sin \frac{\sqrt{\mu}\, t + \sqrt{\mu}\,(t+2)}{2} . \cos \frac{\sqrt{\mu}\, t - \sqrt{\mu}\,(t+2)}{2}\right]}{2a \sin \sqrt{\mu}\,(t+1)}$$

$$\frac{x_1 + x_3}{2x_2} = \cos \sqrt{\mu}\,. \qquad \text{But } \cos\theta = \frac{x_1 + x_3}{2x_2} \quad \text{(given)}$$

$$\therefore \qquad \cos\theta = \cos\sqrt{\mu} \quad \Rightarrow \quad \sqrt{\mu} = \theta$$

$$\therefore \quad \text{Time period} \qquad T = \frac{2\pi}{\sqrt{\mu}} = \frac{2\pi}{\theta} \text{ sec.}$$

Example 4. *A particle rests in equilibrium under the attraction of two centres of force which attract directly as the distance, their intensity being* μ *and* μ'. *The particle is displaced slightly towards one of them. Show that time of a small oscillation is* $\frac{2\pi}{\sqrt{\mu + \mu'}}$.

Sol. Let A and A′ be the two centres of forces and the particle of mass m is at the point B, where AB = x and A′ B = y.

∵ Particle is in equilibrium at B, the two forces at B balance each other

i.e., $$m\mu x = m\mu' y \qquad ...(1)$$

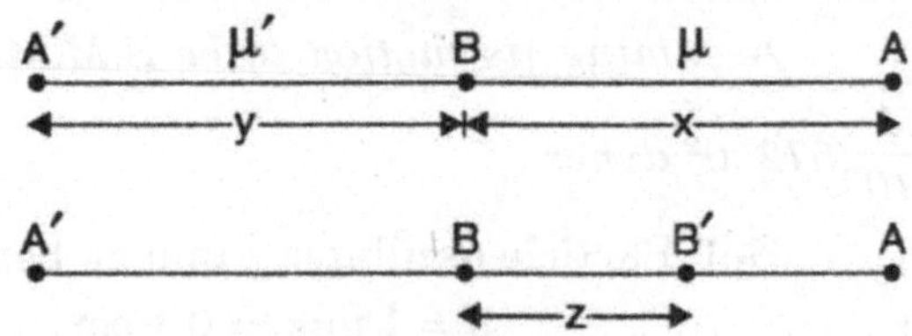

Fig. 2

Let the particle be slightly displaced to B′ towards A such that BB′ = z. Taking the direction towards A as positive, forces at B′ are $m\mu'(y + z)$ towards A′ and $m\mu(x - z)$ towards A.

∴ Resultant force towards A′ is
$$= m\mu'(y + z) - m\mu\,(x - z)$$
$$= m(\mu + \mu')\, z + m\mu'\, y - m\mu\, x = m(\mu + \mu')\, z \qquad \text{From (1)}$$

$\therefore$ Equation of motion is

$$m\frac{d^2z}{dt^2} = -m(\mu+\mu')\,z \quad \text{or} \quad \frac{d^2z}{dt^2} = -(\mu+\mu')\,z$$

Comparing with the equation of motion of S.H.M., we get

$$T = \frac{2\pi}{\sqrt{\mu+\mu'}}.$$

Example 5. *If in a simple harmonic motion u, v, w be the velocities at distances a, b, c from the fixed point on the straight line which is not the centre of the force, show that the period T is given by the equation*

$$\frac{4\pi^2}{T^2}(a-b)(b-c)(c-a) = \begin{vmatrix} u^2 & v^2 & w^2 \\ a & b & c \\ 1 & 1 & 1 \end{vmatrix}.$$

Sol. Let O be the centre of the force and A be the fixed point such that AB = a, AC = b, AD = c and let OA = s and amplitude be A.

$\therefore$ OB = $s + a$, OC = $s + b$ and OD = $s + c$

Velocities at B, C and D are u, v, w respectively.

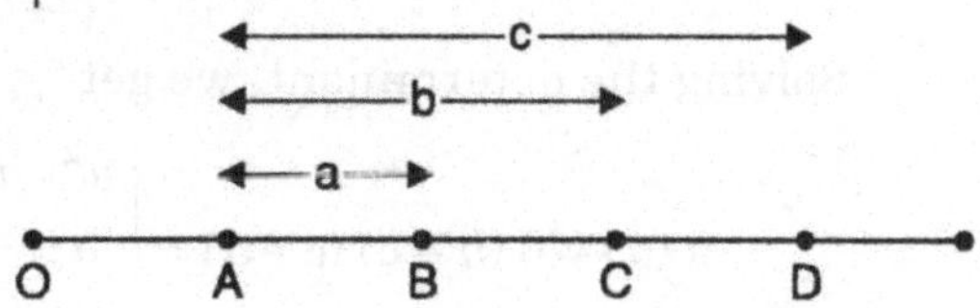

Fig. 3

$$\therefore \quad u^2 = \mu\,[A^2 - (s+a)^2] \quad \text{...(1)}$$
$$v^2 = \mu\,[A^2 - (s+b)^2] \quad \text{...(2)}$$
$$w^2 = \mu\,[A^2 - (s+c)^2] \quad \text{...(3)}$$

or

$$\frac{u^2}{\mu} = (A^2 - s^2) - 2as - a^2, \quad \frac{v^2}{\mu} = (A^2 - s^2) - 2bs - b^2$$

$$\frac{w^2}{\mu} = (A^2 - s^2) - 2cs - c^2$$

or

$$\left(\frac{u^2}{\mu} + a^2\right) + 2as + s^2 - A^2 = 0 \quad \text{...(4)}$$

$$\left(\frac{v^2}{\mu} + b^2\right) + 2bs + s^2 - A^2 = 0 \quad \text{...(5)}$$

$$\left(\frac{w^2}{\mu} + b^2\right) + 2cs + s^2 - A^2 = 0 \quad \text{...(6)}$$

From (4), (5) and (6) eliminating s and $s^2 - A^2$ using determinants, we get

$$\begin{vmatrix} \frac{u^2}{\mu} + a^2 & a & 1 \\ \frac{v^2}{\mu} + b^2 & b & 1 \\ \frac{w^2}{\mu} + c^2 & c & 1 \end{vmatrix} = 0$$

or $$\begin{vmatrix} \frac{u^2}{\mu} & a & 1 \\ \frac{v^2}{\mu} & b & 1 \\ \frac{w^2}{\mu} & c & 1 \end{vmatrix} + \begin{vmatrix} a^2 & a & 1 \\ b^2 & b & 1 \\ c^2 & c & 1 \end{vmatrix} = 0 \quad \text{Property of determinant}$$

$$\begin{vmatrix} a^2 & a & 1 \\ b^2 & b & 1 \\ c^2 & c & 1 \end{vmatrix} = -\frac{1}{\mu}\begin{vmatrix} u^2 & a & 1 \\ v^2 & b & 1 \\ w^2 & c & 1 \end{vmatrix}$$

$$-\mu\begin{vmatrix} 1 & a & a^2 \\ 1 & b & b^2 \\ 1 & c & c^2 \end{vmatrix} = -\begin{vmatrix} u^2 & v^2 & w^2 \\ a & b & c \\ 1 & 1 & 1 \end{vmatrix}$$

Solving the determinant, we get

$$\mu\,(a-b)\,(b-c)\,(c-a) = \begin{vmatrix} u^2 & v^2 & w^2 \\ a & b & c \\ 1 & 1 & 1 \end{vmatrix} \qquad \text{...(7)}$$

But $$T = \frac{2\pi}{\sqrt{\mu}} \Rightarrow \mu = \frac{4\pi^2}{T^2}$$

Putting μ in (7), we get

$$\frac{4\pi^2}{T^2}(a-b)(b-c)(c-a) = \begin{vmatrix} u^2 & v^2 & w^2 \\ a & b & c \\ 1 & 1 & 1 \end{vmatrix}.$$

EXERCISE 1

1. Show that the particle executing S.H.M. requires 1/6th of its period to move from the position of maximum displacement to one in which the displacement is half the amplitude.
2. At what distance from the centre will the velocity be half of the maximum ?
3. The maximum velocity of a body moving with S.H.M. is 2 cm/sec and its period is $\frac{1}{5}$ sec. Find the amplitude.
4. A particle moving with S.H.M. of period 12 sec. travels 10 cm. from the position of rest in 2 sec. Find velocity at the end of 2 sec.
5. At the end of three successive seconds, the distances of moving point with S.H.M. from its mean position measured in the same direction are 1, 5, 5. Show that the period of complete oscillation is $\frac{2\pi}{\theta}$ sec., where $\cos\theta = 3/5$. *(K.U. 2005)*
6. A particle moving with S.H.M. travels a distance a in first second from rest and distance b in the next second. Prove that amplitude of the motion is $\frac{2a^2}{3a-b}$.

7. A particle is performing S.H.M. of period T about a centre O and it passes through the position P(OP = b) with velocity v in the direction OP. Prove that the time that elapses before it returns to P is $\frac{T}{\pi}\tan^{-1}\left(\frac{vT}{2\pi b}\right)$. *(K.U. 2007)*

8. A horizontal shelf is moved up and down with S.H.M. of period $\frac{1}{2}$ sec. What is the amplitude admissible in order that a weight placed on the shelf may be Jerked off ?

9. Show that in a simple harmonic motion of amplitude a and period T, the velocity v at a distance x from the centre is given by the relation $v^2T^2 = 4\pi^2(a^2 - x^2)$.

10. A particle executes S.H.M. such that in two of its positions the velocities are u and v and accelerations are α and β. Show that distance between the positions is $\frac{v^2 - u^2}{\alpha + \beta}$ and amplitude is

$$\frac{[(v^2 - u^2)(\alpha^2 v^2 - \beta^2 u^2)]^{\frac{1}{2}}}{\beta^2 - \alpha^2}.$$

11. A body moving in a straight line OAB with S.H.M. has zero velocity at A and B. Whose distances from O are a and b respectively. If v is the velocity half way between A and B then time period of one oscillation is $\frac{\pi(b-a)}{v}$.

12. If the displacement of a moving point at any time t is given by $x = a\cos nt + b\sin nt$. Show its motion is simple harmonic with amplitude $\sqrt{a^2 + b^2}$ and time period $\frac{2\pi}{n}$. Also find maximum velocity and acceleration.

13. A point moves in a straight line with S.H.M., the centre of motion being O and extreme position A. A point is taken on OA such that $2\,OQ^2 = OA^2$. Show that the time taken from A to Q is same as that from Q to O.

14. The motion of a particle is S.H.M. of period T and amplitude a. Find velocity in terms of a, T and x and prove that

$$\int_0^T V^2\,dt = \frac{2\pi^2 a^2}{T}.$$

15. A particle is attached to one end of an extensible string and the other end moves in a vertical line with S.H.M. of amplitude a taking n revolutions per second. Show that the string will not remain tight during the motion unless $n^2 < g/(4\pi^2 a)$.

16. The speed v of a point moving in a line is given by $v^2 = a + 2bx - cx^2$, where x is the distance from a fixed point on its path. Show that the motion is simple harmonic if a, b, c are constant and $c > 0$. Find its period.

ANSWERS

2. $\frac{a\sqrt{3}}{2}$ 3. $\frac{1}{5\pi}$ 4. $\frac{5\pi}{\sqrt{3}}$ cm/sec.

8. $\frac{g}{16\pi^2}$ 16. $\frac{2\pi}{\sqrt{c}}$.

10

Elastic Strings

When some tensile force is applied on an elastic string it produces some extension in the string. It is found experimentally that within elastic limits, the extension is proportional to the tensile force. This law is known as Hooke's Law. If l is the natural length and ε is the extension produced by a tensile force T, then $T = \frac{\lambda\varepsilon}{l}$ where λ is a constant called modulus of the string. The value of λ depends on the material and geometry of the string. If we have a steel wire of uniform cross-sectional area A and natural length l and a tensile force T produces an extension ε then $\frac{T}{A} = \frac{E\varepsilon}{l}$ where E is called Young's modulus. T/A is called stress and ε/l is called strain in the wire.

1. Work done by an Elastic String *(K.U. 2007)*

Let a tensile force T be applied to a string of natural length l, modulus λ produces an extension ε, $T = \lambda\, \varepsilon/l$. If a further small extension $\delta\varepsilon$ is given to the string work done by the force is $\delta w = \frac{\lambda\varepsilon}{l}\delta\varepsilon$.

Total work done in giving extension from ε_1 to ε_2 is $W = \int_{\varepsilon_1}^{\varepsilon_2} \frac{\lambda\varepsilon}{l}\, d\varepsilon$

$$= \frac{\lambda}{l}\left(\frac{\varepsilon_2^{\,2} - \varepsilon_1^{\,2}}{2}\right) = \lambda\,\frac{(\varepsilon_2 + \varepsilon_1)}{2l}\,(\varepsilon_2 - \varepsilon_1)$$

$$= \left(\frac{\lambda\varepsilon_2}{l} + \frac{\lambda\varepsilon_1}{l}\right)\frac{(\varepsilon_2 - \varepsilon_1)}{2} = \left(\frac{T_1 + T_2}{2}\right)(\varepsilon_2 - \varepsilon_1)$$

$$= \text{(Mean tension) (Net extension)}$$

2. Horizontal Elastic String

An elastic string is fixed to a point on a smooth horizontal table and the other end A is attached to a particle of mass m. The end A is stretched to a point B and then released. Discuss the motion. *(K.U. 2005)*

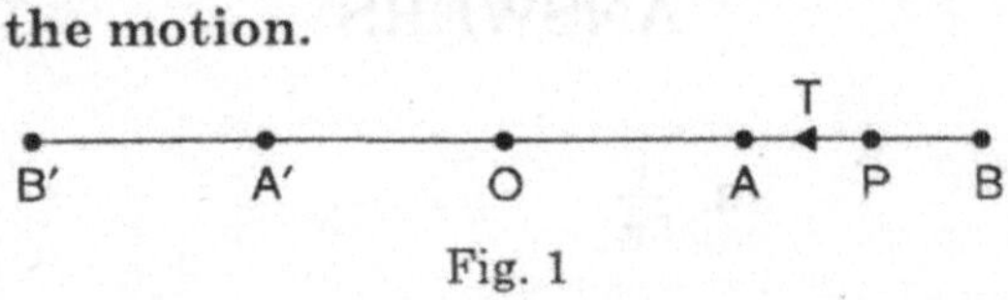

Fig. 1

Let mass m be attached to A.

Let OA = l be the natural length of the string. Let the string be stretched to B and then released. Let it be at P at time t such that AP = x.

$$T = \frac{\lambda x}{l}, \quad AB = a$$

∴ Equation of motion of the particle is

$$m\frac{d^2x}{dt^2} = -\frac{\lambda x}{l} \qquad ...(1)$$

or
$$\frac{d^2x}{dt^2} = -\left(\frac{\lambda}{ml}\right)x$$

or
$$\frac{d^2x}{dt^2} = -\mu x, \text{ where } \mu = \frac{\lambda}{ml} \qquad ...(2)$$

Thus when the string is in the stretched position its motion is simple harmonic, about A. Also equation of motion (2) can be written as

$$v\frac{dv}{dx} = -\mu x.$$

Integrating, we get

$$\frac{v^2}{2} = -\frac{\mu x^2}{2} + c$$

When particle is at B, velocity, is zero

∴ $$x = a,\ v = 0 \quad \Rightarrow \quad c = \frac{\mu a^2}{2}$$

Substituting the value of c

$$v^2 = \mu(a^2 - x^2) \qquad ...(3)$$

At $\quad x = 0 \quad v = a\sqrt{\mu}$

Thus when particle reaches A, string is in the natural length, there is no tension in the string. Therefore, by Newton's first law there is no force to disturb the motion and particle moves left to O with uniform velocity till it reaches A′ where OA′ = l. When particle moves beyond A′, the string is again stretched and tension comes into action similar to as it was on the right hand side of A and comes to instantaneous rest at B′ and retraces the former path.

Time period of one oscillation

= 2 [Time from B to B′]

= 2 [Time from B to A + Time from A to A′ + Time from A′ to B′]

$$= 2\left[\frac{1}{4}\frac{2\pi}{\sqrt{\mu}} + \frac{2l}{v} + \frac{1}{4}\frac{2\pi}{\sqrt{\mu}}\right] = 2\left[\frac{\pi}{\sqrt{\mu}} + \frac{2l}{a\sqrt{\mu}}\right] = 2\sqrt{\frac{ml}{\lambda}}\left(\pi + \frac{2l}{a}\right)$$

∴ $$T = 2\sqrt{\frac{ml}{\lambda}}\left(\pi + \frac{2l}{a}\right) \qquad ...(4)$$

3. Vertical Elastic String

A particle of mass m hangs in equilibrium from a fixed point O by an elastic string of natural length l, modulus λ. The string is extended by an amount e. If the string is further extended by an amount $a < e$ and then let go ; discuss the motion.

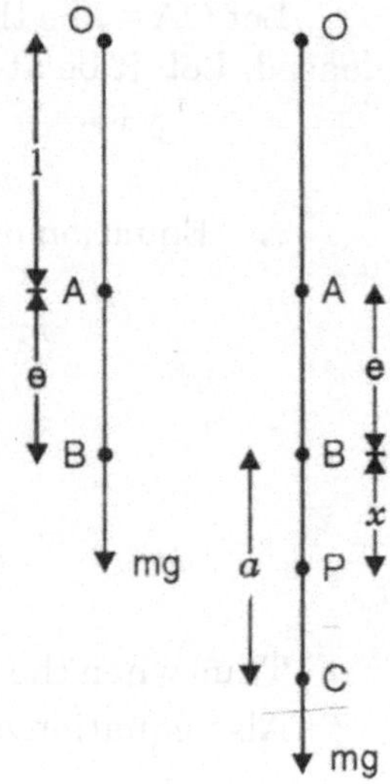

Fig. 2

Let the string of length l be fixed at O and mass m is suspended from A freely. String is extended upto B due to weight mg. Let AB = e

$$\therefore \qquad T_0 = mg = \frac{\lambda e}{l} \text{ (Hooke's Law)}$$

or
$$\lambda = \frac{mgl}{e} \qquad \text{...(1)}$$

Let the mass be pulled upto C such that BC = a and let go. Let at any instant the particle be at P such that BP = x and Tension in the string due to extension be T

The equation of motion is

$$m\frac{d^2x}{dt^2} = mg - T$$

$$= mg - \lambda\frac{AP}{l}$$

$$= mg - \lambda\frac{(e+x)}{l}$$

Putting the value of λ

$$m\frac{d^2x}{dt^2} = mg - \frac{mgl}{e}\frac{(e+x)}{l}$$

or
$$m\frac{d^2x}{dt^2} = -\frac{mg\,x}{e}$$

$$\therefore \qquad \frac{d^2x}{dt^2} = -\left(\frac{g}{e}\right)x = -\mu x \qquad \text{where } \mu = g/e$$

Thus the motion is simple harmonic with centre of oscillation is at B, the position of equilibrium. Since BC = a and $a < e$ mass does not go above A.

$\therefore$ Period of oscillation is

$$T = \frac{2\pi}{\sqrt{g/e}} \quad \text{or} \quad T = 2\pi\sqrt{\frac{e}{g}} \qquad \because \lambda = \frac{mg\,l}{e}$$

$$T = 2\pi\sqrt{\frac{ml}{\lambda}}.$$

Case. If string is extended to c such that BC = a and $a > e$. In this case the amplitude of S.H.M. is $a > e$. The particle rises above A. String becomes slack and the mass moves freely under gravity under retardation g till it comes to instanteneous rest. Then particle falls under gravity and returns to A with same velocity and then due to tension in the string it moves again with S.H.M.

Example 1. *A light elastic string of modulus λ is stretched to double of its natural length and is tied to two fixed points distance 2l apart. A particle of mass m, tied to its middle point is displaced in the line of the string through a distance equal to half its distance from the fixed point and released. Find the time period of complete oscillation and maximum velocity acquired.*

Sol. Let string of natural length l be stretched to length $2\,l$ and tied between A and B.

Fig. 3

Let O be the middle point such that OA = OB = l. Particle of mass m is displaced upto C, OC $= \dfrac{l}{2}$.

Let P be the position of the mass at same instant such that OP = x.

Then tension in the string AP = T_1

$$T_1 = \lambda \frac{\left[(l+x) - \dfrac{l}{2}\right]}{\dfrac{l}{2}}$$

T_2 = Tension in the string PB

$$T_2 = \frac{\lambda}{\dfrac{l}{2}} \left[(l-x) - \frac{l}{2}\right]$$

$\therefore$ The equation of motion is

$$m\frac{d^2x}{dt^2} = T_2 - T_1 = \frac{2\lambda}{l}\left[l - x - \frac{l}{2} - l - x + \frac{l}{2}\right]$$

$$\therefore \quad m\frac{d^2x}{dt^2} = -\frac{4\lambda x}{l}$$

or
$$\frac{d^2x}{dt^2} = -\left(\frac{4\lambda}{ml}\right)x = -\mu x \text{ where } \mu = \frac{4\lambda}{ml}$$

$\therefore$ Time period is

$$T = \frac{2\pi}{\sqrt{\dfrac{4\lambda}{ml}}} = \pi\sqrt{\frac{ml}{\lambda}}$$

$\because$ Amplitude of S.H.M. $= \dfrac{l}{2}$

$\therefore$ Maximum velocity $= \dfrac{l}{2}\sqrt{\dfrac{4\lambda}{ml}} = \sqrt{\dfrac{\lambda\, l}{m}}$.

Example 2. *An elastic string of natural length 2l can support a certain weight when it is stretched till its length is 3l. One end of the string is now attached to a fixed point on a smooth horizontal table and same weight is attached to the other end. Prove that if the weight be pulled to any distance and let go, the string will become slack after a time* $\dfrac{\pi}{2}\sqrt{l/g}$.

Sol. When mass m hangs vertically extension is $3l - 2l = l$

$$\therefore \quad T = mg = \frac{\lambda l}{2l} \Rightarrow \lambda = 2mg \qquad ...(1)$$

Let the string be attached to O and mass m is attached to A and pulled upto B and released. Let P be the position at time t such that AP = x.

$\therefore$ Equation of motion is

$$m\frac{d^2x}{dt^2} = -\frac{\lambda x}{2l}.$$

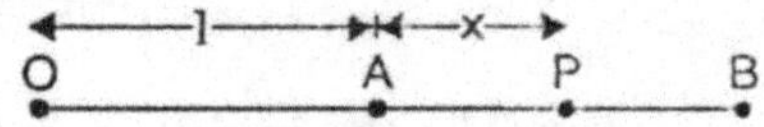

Fig. 4

Putting the value of $\lambda = 2\ mg$, we get

$$\frac{d^2x}{dt^2} = -\left(\frac{g}{l}\right)x = -\mu x, \quad \text{where } \mu = g/l.$$

$\therefore$ Motion is simple harmonic about the centre at A. The string becomes slack when particle crosses A.

$\because$ Required time $T = \frac{1}{4}\frac{2\pi}{\sqrt{\mu}} = \frac{\pi}{2}\sqrt{\frac{l}{g}}$.

Example 3. *A particle of mass 4 kg suspended from an elastic string of natural length 3 m extends it 2 m. One end of the string is fixed and a mass of 2 kg is attached to the other. The mass is held so that the string is just unstretched and is then let go. Find the period, amplitude and maximum velocity of the motion.*

Sol. Length of string OA = 3 m.

When 4 kg mass is suspended string is extended by 2m.

By Hook's Law :

$$4\,g = \lambda\,.\,\frac{2}{3} \quad \Rightarrow \quad \lambda = 6\,g$$

When 2 kg mass is suspended, extension AB is given by Hooke's Law

$$2\,g = \lambda\,.\,\frac{AB}{3} = 6\,g\,.\,\frac{AB}{3}$$

or $\qquad AB = 1$ m.

Let P be the position of particle when it is released from B, BP = x

Equation of motion is

$$2\frac{d^2x}{dt^2} = 2g - T,$$

where T is the tension in the string.

or $$2\frac{d^2x}{dt^2} = 2g - T. = 2g - \lambda : \frac{AP}{3}$$

$$= 2g - 6g\,.\,\frac{(1+x)}{3} = -\,2gx.$$

or $$\frac{d^2x}{dt^2} = -gx = -\mu x, \quad \text{where } \mu = g$$

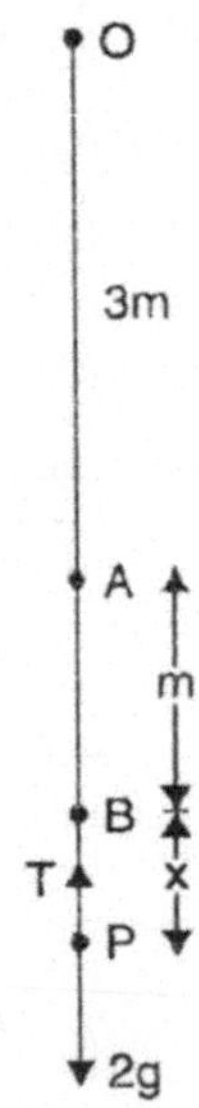

Fig. 5

$\therefore$ Time period of S.H.M. $= \frac{2\pi}{\sqrt{\mu}} = \frac{2\pi}{\sqrt{g}}$

$\because$ Particle is initially at rest at A

$\therefore$ Amplitude AC = 1 m.

Maximum velocity $= a\sqrt{\mu} = 1\sqrt{g} = \sqrt{g}$ m/sec.

Example 4. *A mass m hangs from a fixed point by a light elastic string and is given a small vertical displacement. Show that the motion is simple harmonic. If l is the length of string when the system is in equilibrium and makes n oscillations per second, show that the natural length of the string is* $l - g/(4\pi^2 n^2)$. (KU 2004)

Sol. Let O be the fixed point and mass m be attached at A such that OA = a (natural length)

Let the string be stretched to B and OB = l

$\therefore$ By Hooke's Law

$$mg = \frac{\lambda(l-a)}{a} \quad \ldots(1)$$

Let the particle be pulled to C and then released. Let P be some position of the particle at time T such that BP = x.

Tension T in the string in this position is

$$T = \lambda \frac{(l+x-a)}{a}$$

The equation of motion is

$$m\frac{d^2x}{dt^2} = mg - T$$

or
$$m\frac{d^2x}{dt^2} = mg - \lambda\frac{(l+x-a)}{a}$$

$$= mg - \frac{\lambda(l-a)}{a} - \frac{\lambda x}{a}.$$

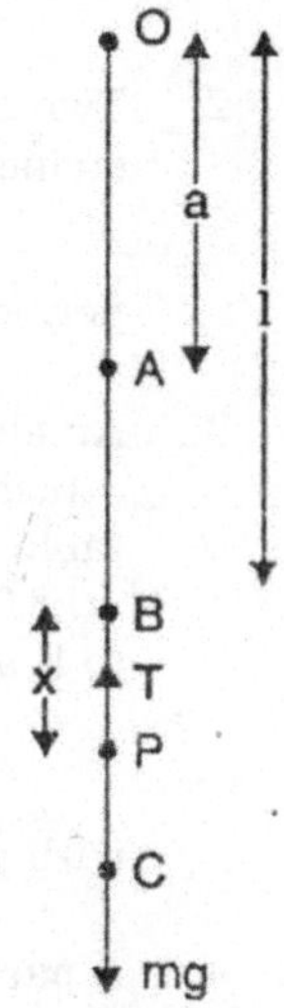

Fig. 6

From (1)

$$m\frac{d^2x}{dt^2} = -\frac{\lambda x}{a}$$

or
$$\frac{d^2x}{dt^2} = \frac{-\lambda x}{ma} = -\mu x, \quad \text{where } \mu = \frac{\lambda}{ma} \quad \ldots(2)$$

Frequency
$$n = \frac{1}{T} = \frac{\sqrt{\mu}}{2\pi} = \frac{1}{2\pi}\sqrt{\frac{\lambda}{ma}}$$

or
$$4\pi^2 n^2 = \frac{\lambda}{ma}$$

Putting the value of λ from (1), we get

$$4\pi^2 n^2 = \frac{a\,mg}{l-a} \times \frac{1}{ma} = \frac{g}{l-a}$$

or
$$l - a = \frac{g}{4\pi^2 n^2} \qquad \text{or} \quad a = l - \frac{g}{4\pi^2 n^2}.$$

EXERCISE 1

1. One end of an elastic string, modulus λ natural length l is fixed to a point on a smooth horizontal table and the other end is tied to a particle of mass m on the table. The particle is pulled to a distance such that the string becomes equal to twice its natural length and then released. Show that the time of one complete oscillation is

$$2(\pi + 2)\sqrt{\frac{ml}{\lambda}}.$$

2. Two elastic strings are fastened to a particle of mass m and their other ends are fixed so that strings are taught. The modulus of each is λ, the tension T, and lengths a and b. Show that the period of oscillation is $2\pi\sqrt{\frac{mab}{(T+\lambda)(a+b)}}$.

3. An elastic string of natural length $(a + b)$ where $a > b$ and modulus λ has a particle of mass m attached to it at a distance a from one end A, which is fixed to a point on a smooth horizontal table. The other end is fixed to point B and the string is just unstretched. If the particle be pulled to B and then released show that it will oscillate to and fro through a distance $\frac{b(\sqrt{a}+\sqrt{b})}{\sqrt{a}}$ with periodic time $\pi(\sqrt{a} + \sqrt{b})\sqrt{\frac{m}{\lambda}}$.

4. A particle is performing S.H.M. in the line joining two points A and B on a smooth plane and is connected with these points by elastic strings of length a and a' and moduli λ and λ' respectively. Show that the time period of one oscillation is $2\pi\,[m/\{\lambda/a + \lambda'/a'\}]^{1/2}$.

5. A light elastic string of length 4 m has one end fixed at A and the other end is attached a mass, which in equilibrium extends the string by 1 m. If the stone be dropped from A find the maximum extension produced.

6. A weight is attached to the lower end of a light elastic string whose upper end is fixed and released. If it oscilates through a distance $\frac{1}{2}$ m then find the period of oscillation.

7. If t_1 and t_2 be periods corresponding to two different attached weights and c_1, c_2 be the statical extensions due to these weights prove that $g = \frac{4\pi^2(c_1 - c_2)}{t_1^2 - t_2^2}$.

8. A light elastic string of natural length l has one end fixed at O and the other end is attached to a particle, the weight of which in equilibrium extends it to a length l_1. Show that if particle be released from O it will come to instantaneous rest at a depth $\sqrt{l_1^2 - l^2}$ below the equilibrium position.

ANSWERS

5. 8m

6. $\pi/\sqrt{g}$.

11

Motion in a Resisting Medium

In the discussion of motion of the particle in the previous chapters, we have not taken the account of the resistance of the medium in which the particle travels. Though there is no definite rule for the resistance of the medium yet it has been observed that resistance is some function of velocity of the particle and the density of the medium. The resistance force always acts in the direction opposite of the motion and is non conservative. Hence the principle of conservation of energy is not applicable. It has been observed that the projectiles moving in air with velocity less than 250 m/sec, the resistance varies as the square of the velocity and varies as cube of the velocity when velocity is between 250 m/sec and 400 m/sec. approx.

Terminal velocity. In the case of the particle falling under gravity in a resisting medium the velocity will never exceed some definite quantity. For suppose the law of resistance is proportional to v^n. The downward acceleration is then $g - k\,v^n$. When the acceleration vanishes $v = \left(\frac{g}{k}\right)^{1/n}$. This is the maximum velocity which the body can attain. It is called the **limiting** or **terminal velocity.**

Thus **the velocity at which the downward acceleration is zero, is called the terminal velocity.**

1. Motion of a Falling Body Under Gravity in Medium Whose Resistance is Proportional to the Square of Velocity (KU 2004 ; M.D.U. 2005)

Let *a* particle falls vertically from rest under gravity from some point A. Let P be its position when it has described a distance x in time t. The forces acting on the particle at P are

(*i*) The weight mg of the particle acting vertically downward.

(*ii*) The force of resistance $mk\,v^2$ acting vertically upward.

Hence the equation of motion is

$$mv\frac{dv}{dx} = mg - mk\,v^2$$

or
$$v\frac{dv}{dx} = g - kv^2 \qquad \text{...(1)}$$

Let V be the terminal velocity, then from (1)

$$g - kV^2 = 0 \quad \Rightarrow \quad V^2 = g/k$$

∴ Equation of motion can be written as

$$v\frac{dv}{dx} = g\left(1 - \frac{k}{g}v^2\right)$$

or
$$v\frac{dv}{dx} = g\left(1 - \frac{v^2}{V^2}\right) \qquad \text{...(2)}$$

or
$$\frac{-2v\,dv}{V^2 - v^2} = \frac{-2g}{V^2}\,dx$$

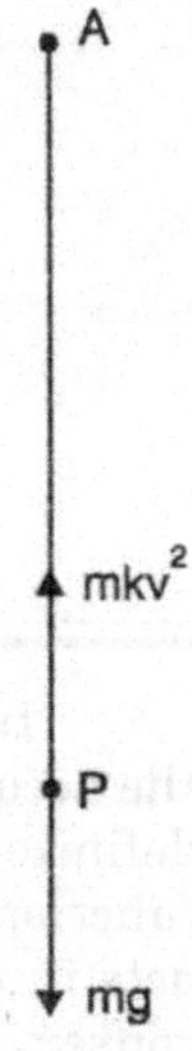

Fig. 1

Integrating both sides, we get

$$\log (V^2 - v^2) = \frac{-2gx}{V^2} + C$$

Applying the initial condition $x = 0, v = 0$

We get $C = \log V^2$

Hence
$$\log (V^2 - v^2) = \frac{-2gx}{V^2} + \log V^2$$

or
$$\log\left(\frac{V^2 - v^2}{V^2}\right) = \frac{-2gx}{V^2}$$

or
$$1 - \frac{v^2}{V^2} = e^{\frac{2gx}{V^2}}$$

or
$$\boxed{v^2 = V^2\left(1 - e^{\frac{-2gx}{V^2}}\right)} \qquad \text{...(3)}$$

This gives a relation between v and x. Further equation (2) can be written as

$$\frac{dv}{dt} = g\left(\frac{V^2 - v^2}{V^2}\right) \qquad \text{or} \qquad \frac{dv}{V^2 - v^2} = \frac{g}{V^2}\,dt$$

Integrating, we get

$$\frac{1}{2V}\log\frac{V+v}{V-v} = \frac{g}{V^2}t + C'$$

At $t = 0 \quad v = 0 \quad \Rightarrow \quad C' = 0$

Hence
$$\log\frac{V+v}{V-v} = \frac{2g}{V}t$$

or
$$\frac{V+v}{V-v} = e^{\frac{2g}{V}t} = \frac{e^{\frac{g}{V}t}}{e^{\frac{-g}{V}t}}$$

Applying componendo and dividendo

$$\frac{v}{V} = \frac{e^{\frac{g}{V}t} - e^{\frac{-g}{V}t}}{e^{\frac{g}{V}t} + e^{\frac{-g}{V}t}}$$

or
$$\frac{v}{V} = \tanh\left(\frac{g}{V}t\right)$$

or
$$\boxed{v = V\tanh\left(\frac{g}{V}t\right)} \qquad \text{...(4)}$$

Equation (4) gives the velocity at time t. Further

$$\frac{dx}{dt} = V \tanh\left(\frac{g}{V}t\right)$$

Integrating again, we get

$$x = \frac{V^2}{g} \log \cosh\left(\frac{g}{V}t\right) + C''$$

At $t = 0, x = 0$ gives $c'' = 0$

$$\therefore \quad x = \frac{V^2}{g} \log \cosh\left(\frac{g}{V}t\right) \quad ...(5)$$

Form equation (3) it is clear that when $x \to \infty$, $v \to V$. Hence the particle would not actually acquire the terminal velocity until it had fallen an infinite distance.

2. Motion of a Particle Projected Vertically Upward in Medium Whose Resistance Varies as Square of the Velocity

Let the particle be projected from A with initial velocity u, vertically upward under gravity. Let P be its position after time t such that AP = x.

The forces acting on the particle at P are :

(*i*) The weight mg acting vertically downwards.

(*ii*) The force of resistance $mk\,v^2$ acting vertically downwards.

The equation of motion is

$$mv\frac{dv}{dx} = -(mg + mk\,v^2)$$

or

$$v\frac{dv}{dx} = -g\left(1 + \frac{k}{g}v^2\right) \quad ...(1)$$

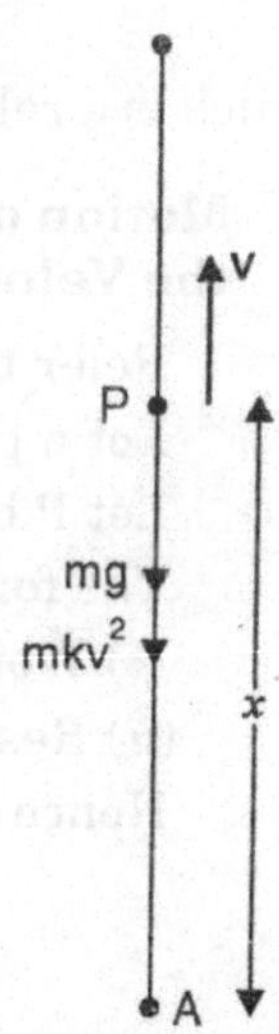

Fig. 2

Let V be the terminal velocity *i.e.*, the velocity when the downward acceleration is zero, then from equation (1) of article 1

$$g - kV^2 = 0$$

or

$$V^2 = g/k \quad ...(2)$$

Substituting (2) in (1)

$$v\frac{dv}{dx} = -g\left(1 + \frac{v^2}{V^2}\right) \quad ...(3)$$

or

$$\frac{v\,dv}{v^2 + V^2} = -\frac{g}{V^2}dx$$

or

$$\frac{2v\,dv}{v^2 + V^2} = -\frac{2g}{V^2}dx$$

Integrating both sides, we get

$$\log(v^2 + V^2) = -\frac{2gx}{V^2} + C$$

At $x = 0$, $v = u \Rightarrow C = \log(u^2 + V^2)$

$$\therefore \quad \log (v^2 + V^2) = -\frac{2gx}{V^2} + \log (u^2 + V^2)$$

or

$$\log\left(\frac{u^2 + V^2}{v^2 + V^2}\right) = \frac{2gx}{V^2} \qquad \text{...(4)}$$

This is a relation between x and v.

Further equation (3) can be written as

$$\frac{dv}{dt} = -g\left(1 + \frac{v^2}{V^2}\right)$$

$$\frac{dv}{v^2 + V^2} = -\frac{g}{V^2}\,dt$$

On integration, we get

$$\frac{1}{V}\tan^{-1}\frac{v}{V} = \frac{-g}{V^2}t + C'$$

At $t = 0, v = u,$ we get $C' = \frac{1}{V}\tan^{-1}\frac{u}{V}$

$$\therefore \quad \frac{1}{V}\tan^{-1}\frac{v}{V} = \frac{-g}{V^2}t + \frac{1}{V}\tan^{-1}\frac{u}{V}$$

or

$$t = \frac{V}{g}\left[\tan^{-1}\frac{u}{V} - \tan^{-1}\frac{v}{V}\right]$$

which is a relation between v and t.

3. Motion of a Body Falling Under Gravity in a Medium Whose Resistance Varies as the Velocity

Refer to Fig. 1.

Let a particle of mass m falls from A.

Let P be the position when it has described a distance x in time t.

The forces acting on the particle at P are

(*i*) Weight mg of the particle acting vertically downwards.

(*ii*) Resistance mkv acting vertically upwards.

Hence the equation of motion is

$$mv\frac{dv}{dx} = mg - mkv$$

or

$$v\frac{dv}{dx} = g - kv \qquad \text{...(1)}$$

If V be the terminal velocity then

$$g - kV = 0 \quad \Rightarrow \quad V = \frac{g}{k}$$

Therefore (1) can be written as

$$v\frac{dv}{dx} = g\left(1 - \frac{k}{g}v\right) = g\left(1 - \frac{v}{V}\right)$$

$$v\frac{dv}{dx} = g\left(\frac{V - v}{V}\right) \qquad \text{...(2)}$$

or
$$\frac{vdv}{V - v} = \frac{g}{V}dx$$

or
$$\frac{g}{V}dx = \left(-1 + \frac{V}{V - v}\right)dv$$

Integrating

$$\frac{gx}{V} = -v - V\log(V - v) + C$$

At $x = 0, \quad v = 0 \quad \Rightarrow \quad C = V\log V$

$$\therefore \quad \frac{gx}{V} = -v - V\log(V - v) + V\log V$$

or
$$\frac{gx}{V} = -v + V\log\frac{V}{V - v}$$

or
$$x = \frac{-vV}{g} + V^2\log\frac{V}{V - v} \qquad \text{...(3)}$$

Equation (3) is relation between v and x.

If $v \to V$ then $x \to \infty$. Therefore terminal velocity is never attained.

Further equation (2) can be written as

$$\frac{dv}{dt} = \frac{g(V - v)}{V}$$

or
$$dt = \frac{V}{g}\left(\frac{dv}{V - v}\right)$$

Integrating,

$$t = \frac{-V}{g}\log(V - v) + C'.$$

At $t = 0, \quad v = 0 \quad \Rightarrow \quad C' = \frac{V}{g}\log V$

Hence
$$t = \frac{V}{g}[\log V - \log(V - v)] = \frac{V}{g}\log\frac{V}{V - v}$$

or
$$\frac{gt}{V} = \log\frac{V}{V - v}$$

or
$$\frac{V - v}{V} = e^{-(g/V)t}$$

$$1 - \frac{v}{V} = e^{-(g/V)t}$$

or
$$v = V[1 - e^{-(g/V)t}] \qquad \text{...(4)}$$

Equation (4) is a relation between t and v.

As $t \to \infty \quad v \to V$ (Terminal velocity)

4. Motion of a Body Moving Upward in a Medium Whose Resistance Varies as the Velocity : *(K.U. 2007)*

Refer to Fig. 2.

Let the particle be projected from A with initial velocity u, vertically upward under gravity. Let P be its position after time t such that AP = x.

The forces on the particle at P are :

(*i*) The weight mg acting vertically downwards.

(*ii*) The force of resistance mkv acting vertically downwards.

The equation of motion is

$$mv\frac{dv}{dx} = -(mg + mkv)$$

or

$$\frac{vdv}{dx} = -(g + kv) \quad ...(1)$$

Let V be the terminal velocity, *i.e.*, the velocity when the downward acceleration is zero

$$\therefore \quad g - k\mathrm{V} = 0 \quad \Rightarrow \quad \mathrm{V} = \frac{g}{k} \quad ...(2)$$

Substituting in (1)

$$\frac{vdv}{dx} = -g\left(1 + \frac{v}{\mathrm{V}}\right) \quad ...(3)$$

$$\frac{vdv}{\mathrm{V} + v} = \frac{-g}{\mathrm{V}}dx$$

or

$$\left(1 - \frac{\mathrm{V}}{\mathrm{V} + v}\right)dv = \frac{-g}{\mathrm{V}}dx$$

Integrating $v - \mathrm{V}\log(\mathrm{V} + v) = \dfrac{-gx}{\mathrm{V}} + \mathrm{C}$

At $x = 0, \quad v = u \quad \Rightarrow \quad \mathrm{C} = u - \mathrm{V}\log \mathrm{V}$

$$\therefore \quad v - \mathrm{V}\log(\mathrm{V} + v) = \frac{-gx}{\mathrm{V}} + u - \mathrm{V}\log \mathrm{V}$$

or

$$\frac{gx}{\mathrm{V}} = u - v + \mathrm{V}\log\left(\frac{\mathrm{V} + v}{\mathrm{V}}\right)$$

or

$$x = \frac{\mathrm{V}}{g}(u - v) + \frac{\mathrm{V}^2}{g}\log\left(\frac{\mathrm{V} + v}{\mathrm{V}}\right) \quad ...(4)$$

Equation (4) gives the relation between velocity and displacement.

Further equation (3) can be written as

$$\frac{dv}{dt} = -g\left(1 + \frac{v}{\mathrm{V}}\right) = -g\left(\frac{\mathrm{V} + v}{\mathrm{V}}\right)$$

or

$$dt = -\frac{\mathrm{V}}{g}\frac{dv}{(\mathrm{V} + v)}$$

Integrating

$$t = -\frac{\mathrm{V}}{g}\log(\mathrm{V} + v) + \mathrm{C}'$$

At $t = 0,\ v = u \quad \Rightarrow \quad C' = \frac{V}{g}\log(V + u)$

Hence $$t = \frac{V}{g}\log(V+u) - \frac{V}{g}\log(V+v)$$

$$t = \frac{V}{g}\log\left(\frac{V+u}{V+v}\right) \qquad ...(5)$$

(5) gives relation between v and t.

Further (5) can be written as

$$\frac{V+v}{V+u} = e^{-gt/V}$$

$$V + v = (V+u)\,e^{-gt/V}$$

or $$v = -V + (V+u)\,e^{-gt/V}$$

or $$\frac{dx}{dt} = -V + (V+u)\,e^{-gt/V}$$

Integrating $$x = -Vt - (V+u)\frac{V}{g}e^{-gt/V} + C''$$

At $t = 0,\ x = 0 \quad \Rightarrow \quad C'' = (V+u)\frac{V}{g}$

$$x = -Vt + (V+u)\left[1 - \frac{V}{g}e^{-gt/V}\right] \qquad ...(6)$$

Equation (6) gives a relation between x and t.

Example 1. *A particle of mass m is falling under gravity through a medium whose resistance is μv. If particle be released from rest, show that the distance fallen through in time t is*

$$\frac{gm^2}{\mu}\left[e^{\frac{-\mu t}{m}} - 1 + \frac{\mu t}{m}\right].$$ (K.U. 2005)

Sol. Let v the velocity of the particle at some instant. Forces on the particle are weight mg acting vertically downward and resistance force μv acting vertically up.

∴ The equation of motion is

$$m\frac{dv}{dt} = mg - \mu v$$

or $$\frac{dv}{dt} + \frac{\mu v}{m} = g \qquad ...(1)$$

This is a linear differential equation of Ist order

Its integrating factor $= e^{\int \frac{\mu}{m}dt} = e^{\frac{\mu t}{m}}$

Its solution is $$v \,.\, e^{\frac{\mu t}{m}} = \int g.e^{\frac{\mu t}{m}}\,dt + C = g\frac{m}{\mu}.e^{\frac{\mu t}{m}} + C$$

Applying initial condition $t = 0, v = 0,$ $\Rightarrow$ $C = \frac{-mg}{\mu}$. Substituting C

$$\therefore \quad v e^{\frac{\mu t}{m}} = g\frac{m}{\mu} e^{\frac{\mu t}{m}} - g\frac{m}{\mu}$$

or
$$v = \frac{mg}{\mu}\left(1 - e^{\frac{-\mu t}{m}}\right) \qquad ...(2)$$

or
$$\frac{dx}{dt} = \frac{mg}{\mu}\left[1 - e^{\frac{-\mu t}{m}}\right]$$

Integrating again, we get

$$x = \frac{mg}{\mu}\left[t + \frac{me^{\frac{-\mu t}{m}}}{\mu}\right] + C'$$

Initially at $t = 0, x = 0,$ $\Rightarrow$ $C' = -\frac{m^2 g}{\mu^2}$

Substituting the value of C′, we get

$$x = \frac{mg}{\mu}\left[t + \frac{m}{\mu}e^{\frac{-\mu t}{m}}\right] - \frac{m^2 g}{\mu^2}$$

or
$$x = \frac{m^2 g}{\mu^2}\left[\frac{\mu t}{m} + e^{\frac{-\mu t}{m}} - 1\right].$$ Hence proved.

Example 2. *A particle falls from rest under gravity through a distance x in a medium whose resistance varies as the square of the velocity. If v the velocity actually acquired by it and v_0 be the velocity it would have acquired, had there been no resisting medium, and V be the terminal velocity, show that*

$$\frac{v^2}{{v_0}^2} = 1 - \frac{1}{2}\cdot\frac{{v_0}^2}{V^2} + \frac{1}{2.3}\cdot\frac{{v_0}^4}{V^4} - \frac{1}{2.3.4}\frac{{v_0}^6}{V^6} + ...$$

Sol. Particle is falling under gravity and the resistance kv^2.

$\therefore$ Equation of motion is

$$mv\frac{dv}{dx} = mg - mk\,v^2$$

or
$$v\frac{dv}{dx} = g\left(1 - \frac{k}{g}v^2\right) \qquad ...(1)$$

If V is the terminal velocity, then $V^2 = \frac{g}{k}$

$$\therefore \quad v\frac{dv}{dx} = g\left(1 - \frac{v^2}{V^2}\right) \qquad ...(2)$$

or $$\frac{v\,dv}{V^2 - v^2} = \frac{g}{V^2}\,dx$$

or $$\frac{-2v\,dv}{V^2 - v^2} = \frac{-2g}{V^2}\,dx$$

Integrating both sides

$$\log (V^2 - v^2) = \frac{-2gx}{V^2} + C$$

At $x = 0,\ v = 0 \Rightarrow C = \log V^2$

Hence $$\log (V^2 - v^2) = \frac{-2gx}{V^2} + \log V^2$$

or $$\log \frac{(V^2 - v^2)}{V^2} = \frac{-2gx}{V^2} \quad \text{or} \quad 1 - \frac{v^2}{V^2} = e^{\frac{-2gx}{V^2}}$$

or $$v^2 = V^2 \left(1 - e^{\frac{-2gx}{V^2}}\right) \qquad ...(3)$$

If the particle falls under gravity when there is no resisting medium velocity acquired by falling a distance x from rest,

$$v_0^2 = 0 + 2gx$$

Putting the value of $2gx$ in equation (3), we get

$$v^2 = V^2 \left(1 - e^{\frac{-v_0^2}{V^2}}\right) = V^2\left[1 - \left\{1 - \frac{v_0^2}{V^2} + \frac{v_0^2}{2!\,V^4} - \frac{v_0^6}{3!\,V^6} + \ldots\ldots\right\}\right]$$

$$\frac{v^2}{v_0^2} = 1 - \frac{v_0^2}{2V^2} + \frac{v_0^4}{2.3.\,V^4} - \ldots\ldots$$ Hence Proved.

Example 3. *A heavy particle is projected upwards with velocity u in a medium whose resistance is $gu^{-2}\tan^2\alpha$ times the square of the velocity, α being a constant. Show that the particle will return to the point of projection with velocity $u \cos \alpha$, after a time*

$$ug^{-1} \cot \alpha \left[\alpha + \log \frac{\cos \alpha}{1 - \sin \alpha}\right].$$

Sol. Since the particle is projected upwards, therefore mg and resistance both act vertically downwards. Hence the equation of motion is

$$mv\frac{dv}{dx} = -mg - mg\,u^{-2}\tan^2\alpha\, v^2$$

or $$v\frac{dv}{dx} = \frac{-g}{u^2}\,[u^2 + v^2\tan^2\alpha] \qquad ...(1)$$

or $$\frac{v\,dv}{u^2 + v^2\tan^2\alpha} = \frac{-g}{u^2}\,dx$$

Multiplying both sides by $2\tan^2\alpha$ and integrating

$$\int \frac{2v\tan^2\alpha\,dv}{u^2 + v^2\tan^2\alpha} = -\int \frac{2g\tan^2\alpha}{u^2}\,dx + C$$

$$\log (u^2 + v^2 \tan^2 \alpha) = \frac{-2g \tan^2 \alpha}{u^2} x + C \qquad ...(2)$$

At $\quad x = 0,\ v = u \ \Rightarrow\ C = \log (u^2 \sec^2 \alpha)$

Putting in equation (2)

$$\log (u^2 + v^2 \tan^2 \alpha) = \frac{-2gx}{u^2} \tan^2 \alpha + \log u^2 \sec^2 \alpha$$

$$\log \left(\frac{u^2 + v^2 \tan^2 \alpha}{u^2 \sec^2 \alpha} \right) = \frac{-2gx}{u^2} \tan^2 \alpha$$

If h is the maximum height attained then at $x = h,\ v = 0$,
we get,

$$\log \left(\frac{u^2}{u^2 \sec^2 \alpha} \right) = \frac{-2gh}{u^2} \tan^2 \alpha$$

or
$$-\log \sec^2 \alpha = \frac{-2gh}{u^2} \tan^2 \alpha$$

$\therefore$
$$h = \frac{u^2}{2g} \cot^2 \alpha \log \sec^2 \alpha \qquad ...(3)$$

Equation (1) can be written as

$$\frac{dv}{dt} = \frac{-g}{u^2} (u^2 + v^2 \tan^2 \alpha)$$

or
$$dt = -\frac{u^2}{g} \left(\frac{dv}{u^2 + v^2 \tan^2 \alpha} \right).$$

Let t be the time to reach the highest point, then

$\therefore$
$$t = \int_u^0 \frac{-u^2}{g} \left(\frac{dv}{u^2 + v^2 \tan^2 \alpha} \right)$$

$$= \frac{-u^2}{g} \cdot \frac{1}{u \tan \alpha} \tan^{-1} \frac{v \tan \alpha}{u} \Bigg]_u^0 = \frac{u}{g \tan \alpha} [\alpha] \qquad ...(4)$$

From the highest point particle starts moving downwards under the forces weight mg acting vertically downwards and resistance upwards.

$\therefore$ Equation of motion is

$$mv \frac{dv}{dx} = mg - mg\, u^{-2} \tan^2 \alpha\, v^2$$

or
$$v \frac{dv}{dx} = \frac{g}{u^2} [u^2 - v^2 \tan^2 \alpha]$$

$$\frac{-2v\, dv}{u^2 - v^2 \tan^2 \alpha} = \frac{-2g\, dx}{u^2}$$

Integrating we get,

$$\frac{\log (u^2 - v^2 \tan^2 \alpha)}{\tan^2 \alpha} = \frac{-2gx}{u^2} + C' \qquad ...(5)$$

At $x = 0,\ v = 0 \Rightarrow C' = \cot^2 \alpha \log u^2$

$\therefore$ Substituting the value of C′ in (5), we get

$$\frac{2gx}{u^2} = \cot^2 \alpha \log \frac{u^2}{u^2 - v^2 \tan^2 \alpha} \quad \text{...(6)}$$

If v_1 is the velocity acquired when the particle returns to the point of projection, it can be obtained by putting $x = h$ in (6)

i.e.
$$\frac{2gh}{u^2} = \cot^2 \alpha \log \frac{u^2}{u^2 - v_1^2 \tan^2 \alpha}$$

Putting the value of $\frac{2gh}{u^2}$ from (3), we get

$$\log \sec^2 \alpha = \log \frac{u^2}{u^2 - v_1^2 \tan^2 \alpha}$$

or
$$\sec^2 \alpha = \frac{u^2}{u^2 - v_1^2 \tan^2 \alpha}$$

or
$$u^2 \sec^2 \alpha - v_1^2 \tan^2 \alpha \sec^2 \alpha = u^2$$

or
$$v_1^2 \tan^2 \alpha \sec^2 \alpha = u^2 (\sec^2 \alpha - 1)$$

or
$$v_1 = u \cos \alpha \quad \text{...(7)}$$

$\therefore$ Particle will return to point of projection with velocity $u \cos \alpha$. Let t_2 be the time to return from highest point.

The equation of motion can be written as

$$\frac{dv}{dt} = \frac{g}{u^2}(u^2 - v^2 \tan^2 \alpha)$$

$\therefore$ Time of return t' can be obtained by integrating

$$t' = \int dt = \int_0^{u \cos \alpha} \frac{u^2}{g} \frac{dv}{u^2 - v^2 \tan^2 \alpha}$$

$$= \frac{u^2}{g} \cdot \frac{1}{2u \tan \alpha} \log \left[\frac{u + v \tan \alpha}{u - v \tan \alpha}\right]_0^{u \cos \alpha}$$

$$= \frac{u}{2g \tan \alpha}\left[\log\left(\frac{1 + \sin \alpha}{1 - \sin \alpha}\right) - \log 1\right]$$

$$= \frac{u}{2g \tan \alpha}\left[\log \frac{(1 - \sin^2 \alpha)}{(1 - \sin \alpha)^2}\right]$$

$$= \frac{u}{2g \tan \alpha}\left[2 \log \frac{\cos \alpha}{1 - \sin \alpha}\right]$$

$$= \frac{u}{g \tan \alpha} \log \frac{\cos \alpha}{1 - \sin \alpha} \quad \text{...(7)}$$

$\therefore$ Total time from point of projection to highest point and back

$$t + t' = \frac{u\alpha}{g\tan\alpha} + \frac{u}{g\tan\alpha}\log\frac{\cos\alpha}{1-\sin\alpha}$$

$$= \frac{u}{g\tan\alpha}\left[\alpha + \log\frac{\cos\alpha}{1-\sin\alpha}\right].$$ Hence proved .

Example 4. *A particle is projected with velocity u along a smooth horizontal plane in a medium whose resistance per unit mass is μ times the cube of the velocity. Show that if v is the velocity at time t at distance s, then* $v = \frac{u}{1+\mu us}$ *and* $t = \frac{s}{u} + \frac{1}{2}\mu s^2$.

Sol. Since the particle is moving on a smooth horizontal plane, therefore only force acting on the particle is $m\mu v^3$.

The equation of motion of the particle is

$$mv\frac{dv}{ds} = -m\mu v^3$$

or $$\frac{dv}{ds} = -\mu v^2$$

or $$\frac{-dv}{v^2} = +\mu ds \quad ...(1)$$

Integrating

$$\frac{1}{v} = \mu s + C$$

At $s = 0, \quad v = u \quad \Rightarrow \quad C = \frac{1}{u}$

$\therefore$ $$\frac{1}{v} = \mu s + \frac{1}{u}$$

or $$v = \frac{u}{1+\mu us} \quad ...(2)$$

Further $$\frac{ds}{dt} = \frac{u}{1+\mu us}$$

or $$\left(\frac{1+\mu us}{u}\right)ds = dt$$

Integrating $$t = \frac{1}{u}\left[s + \mu u\frac{s^2}{2}\right] + C'$$

At $t = 0, \quad s = 0 \quad \Rightarrow \quad C' = 0$

$\therefore$ $$t = \frac{s}{u} + \frac{\mu s^2}{2}.$$ Hence Proved.

Example 5. *A body of mass m is projected vertically upwards in a medium for which the resistance is* mk^2v^2. *If initial velocity is* v_0; *show that* $v_1^2 = \frac{g v_0^2}{g + k^2 v_0^2}$, *where* v_1 *is the velocity with which the particle returns to the point of projection.* (K.U. 2007)

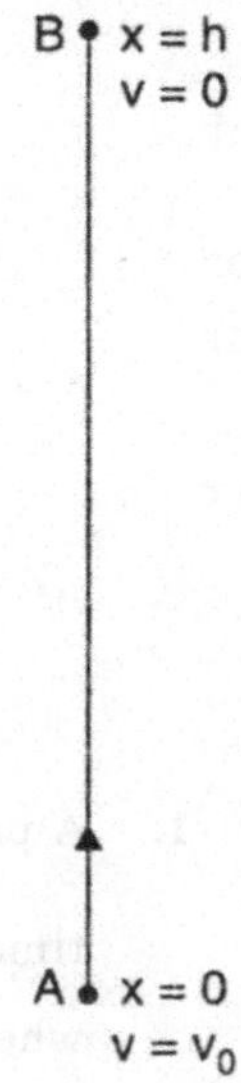

Fig. 3

Sol. When the particle is projected up with air resistance $mk^2 v^2$, the equation of motion is
$$mv \frac{dv}{dx} = -mg - mk^2v^2$$

or
$$v \frac{dv}{dx} = -g - k^2v^2$$

or
$$\frac{v\,dv}{g + k^2 v^2} = -dx$$

Integrating both sides, we get
$$\frac{1}{2k^2} \log (g + k^2v^2) = -x + C_1$$

At $x = 0$, $\quad v = v_0 \Rightarrow C_1 = \dfrac{1}{2k^2} \log (g + k^2{v_0}^2)$

$$\therefore \quad \frac{1}{2k^2} \log (g + k^2v^2) = -x + \frac{1}{2k^2} \log (g + k^2{v_0}^2)$$

At the highest point B, $x = h$, $v = 0$
$$\therefore \quad \frac{1}{2k^2} \log g = -h + \frac{1}{2k^2} \log (g + k^2{v_0}^2)$$

or
$$h = \frac{1}{2k^2} \log \left[\frac{g + k^2{v_0}^2}{g}\right] \qquad ...(1)$$

Now from the highest point B, particle starts downwards with velocity zero. Its equation of motion is refer to highest point as origin is
$$mv \frac{dv}{dx} = mg - mk^2v^2 \quad \text{or} \quad \frac{v\,dv}{g - k^2v^2} = dx$$

Integrating both sides, we get
$$-\frac{1}{2k^2} \log (g - k^2v^2) = x + C_2$$

At $x = 0$, $x = 0 \Rightarrow C_2 = -\dfrac{1}{2k^2} \log g$

Putting the value of C_2, we get
$$x = \frac{1}{2k^2} [\log g - \log (g - k^2v^2)] = \frac{1}{2k^2}\left[\log \frac{g}{g - k^2v^2}\right]$$

When particles returns to the point of projection, it has travelled a distance h from the highest point. Let the velocity acquired be v_1
$$\therefore \quad h = \frac{1}{2k^2}\left[\log \frac{g}{g - k^2{v_1}^2}\right] \qquad ...(2)$$

Comparing (1) and (2),
$$\frac{1}{2k^2} \log \left[\frac{g + k^2{v_0}^2}{g}\right] = \frac{1}{2k^2}\left[\log \frac{g}{g - k^2{v_1}^2}\right]$$

or $$\frac{g + k^2 v_0^{\,2}}{g} = \frac{g}{g - k^2 v_1^{\,2}}$$

or $$g^2 - k^2 g v_1^{\,2} + k^2 g v_0^{\,2} - k^4 v_0^{\,2} v_1^{\,2} = g^2$$

or $$(k^2 g + k^4 v_0^{\,2})\, v_1^{\,2} = k^2 g v_0^{\,2}$$

or $$v_1^{\,2} = \frac{g\, v_0^{\,2}}{g + k^2 v_0^{\,2}}$$

EXERCISE 1

1. A particle of mass m, is projected vertically under gravity, the resistance of the air being mk times the velocity. Show that the greatest height attained by the particle is $\frac{V^2}{g}\,[\lambda - \log(1 + \lambda)]$, where V is the terminal velocity and λV is the velocity of projection.
2. A particle is projected with velocity v along a smooth horizontal plane in a medium whose resistance per unit mass is μv^3. Show that the distance it has described in time t is

   $\frac{1}{\mu v}\left[\sqrt{1 + 2\mu v^2 t} - 1\right]$ and its velocity then is $\frac{v}{\sqrt{1 + 2\mu v^2\, t}}$.
3. A heavy particle is projected vertically upwards with a velocity u in a medium, the resistance of which varies as the cube of the velocity. Find the height to which the particle will ascend.
4. A particle is projected with velocity u along a smooth horizontal plane in a medium whose resistance per unit mass is k(velocity) show that the velocity after time t and distance s in that time are given by

   $$v = ue^{-kt} \quad \text{and} \quad s = \frac{u}{k}(1 - e^{-kt}).$$
5. A particle moves in a straight line in a medium whose resistance per unit mass is kv^2. If its initial velocity be u, show that space described in time t is $\frac{1}{k}\log(1 + kut)$.
6. A particle falls from rest in a medium in which the resistance is kv^2 per unit mass. Show that the distance fallen in time t is $\frac{1}{k}\left[\log\cosh\left(\frac{t}{\sqrt{gk}}\right)\right]$.
7. If the resistance vary as the fourth power of the velocity, the energy of mass m at a depth x below the heighest point when moving in a vertical line under gravity will be E tan (mgx/E), when rising and E tanh (mgx/E) when falling, where E is the terminal energy.
8. A particle of unit mass is projected vertically upwards with velocity V in a medium whose resistance varies as velocity. Prove that the particle returns to the point of projection with speed V_1 such that

   $$V + V_1 = \frac{g}{k}\log\left(\frac{g + kV_1}{g - kV_1}\right).$$

9. A particle is projected vertically upwards in a medium whose resistance varies as v^2. It has a kinetic energy k in its upwards path at a given point. Show that when it passes the same point in the way down loss of K.E. is $\frac{k^2}{k+k'}$ where k' is the limit to which its energy approaches in its downwards course.

10. A particle of mass m is thrown vertically up with speed u. If the air resistance at speed v is $mk\,v^2$, where k is a positive constant, show that the particle returns to its starting point with speed

$$u\sqrt{1+\frac{ku^2}{g}}.$$

11. A particle moves from rest at a distance a from a fixed point O under the action of a force to O equal to μ times the distance per unit mass. If the resistance of the medium in which it moves is k times the square of the velocity per unit mass show that square of the velocity when it is at a distance x from O is

$$\frac{\mu x}{k}-\frac{\mu a}{k}e^{2k(x-a)}+\frac{\mu}{2k^2}\left\{1-e^{-2k(x-a)}\right\}.$$

ANSWER

3. $h=\frac{V^2}{6g}\log\frac{u^2+V^2-uV}{(u+V)^2}+\frac{\pi V^2\sqrt{3}}{18\,g}+\frac{V^2\sqrt{3}}{3\,g}\tan^{-1}\frac{2u-V}{V\sqrt{3}}$

V is terminal velocity.

5. Motion of Projectile in a Resisting Medium

A particle is projected under gravity inclined at an angle α to the horizontal in a medium whose resistance is *mk* times the velocity. To discuss the motion.

Let the particle be projected from O with velocity u and angle α with the horizon. Let the particle be at some point P (x, y) after time t. The forces acting on the particle at P are weight mg acting vertically downwards and resisting force mkv along the tangent at P. Resolving horizontally and vertically the equations of motion are

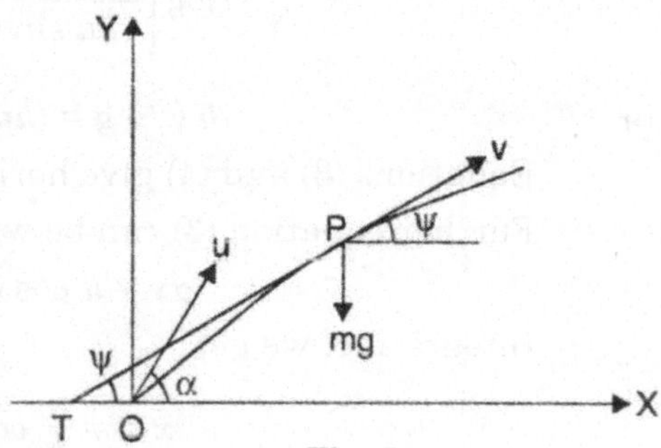

Fig. 3

$$m\frac{d^2x}{dt^2}=-mk\,v\cos\psi \qquad \left(\cos\psi=\frac{dx}{ds}\right)$$

or
$$\frac{d^2x}{dt^2}=-k\frac{ds}{dt}\cdot\frac{dx}{ds} \qquad \left(\sin\psi=\frac{dy}{ds}\right)$$

or
$$\frac{d^2x}{dt^2}=-k\frac{dx}{dt} \qquad ...(1)$$

$$m\frac{d^2y}{dt^2}=-mg-mk\,v\sin\psi=-m\left[g+k\frac{dy}{dt}\right]$$

or $$\frac{d^2y}{dt^2} = -\left(g + k\frac{dy}{dt}\right) \quad ...(2)$$

Equation (1) can be written as

$$\frac{\ddot{x}}{\dot{x}} = -k$$

Integrating both sides, we get

$$\log \dot{x} = -kt + C$$

Initially at O, $t = 0,\ \dot{x} = u\cos\alpha$

$\therefore$ $C = \log(u\cos\alpha)$

Hence $\log \dot{x} = -kt + \log(u\cos\alpha)$

or $$\dot{x} = u\cos\alpha\, e^{-kt} \quad ...(3)$$

Equation (2) can be written as

$$\frac{\ddot{y}}{k\dot{y} + g} = -1$$

Multiplying both sides by k and integrating, we get

$$\log(k\dot{y} + g) = -kt + C'$$

At $t = 0,\ \dot{y} = u\sin\alpha \Rightarrow C' = \log(ku\sin\alpha + g)$

Substituting C′, we have

$$\log\left(\frac{(k\dot{y} + g)}{(ku\sin\alpha + g)}\right) = -kt$$

or $$k\dot{y} + g = (ku\sin\alpha + g)\, e^{-kt} \quad ...(4)$$

Equations (3) and (4) give horizontal and vertical components of velocity at any time t.

Further equation (3) can be written as

$$dx = u\cos\alpha\, e^{-kt}\, dt$$

Integrating, we get

$$x = -\frac{u}{k}\cos\alpha\, e^{-kt} + C''$$

At $t = 0,\ x = 0 \Rightarrow C'' = \frac{u}{k}\cos\alpha$

$\therefore$ $$x = \frac{u}{k}\cos\alpha\,(1 - e^{-kt}) \quad ...(5)$$

Equation (4) can be written as

$$k\frac{dy}{dt} + g = (ku\sin\alpha + g)\, e^{-kt}$$

Integrating, we get

$$ky + gt = \left(\frac{ku\sin\alpha + g}{-k}\right)e^{-kt} + C'''$$

At $\quad t = 0, y = 0 \quad \Rightarrow \quad C''' = \frac{1}{k}(ku \sin \alpha + g)$

Substituting the value of C‴, we have

$$ky + gt = \frac{(ku \sin \alpha + g)}{k}(1 - e^{-kt}) \qquad ...(6)$$

Equations (5) and (6) gives the position of the projectile at any time t. The equation of path can be obtained by eliminating t from these equations.

From (5) $\quad 1 - e^{-kt} = \dfrac{kx}{u \cos \alpha}$

or
$$e^{-kt} = 1 - \frac{kx}{u \cos \alpha}$$

or
$$t = -\frac{1}{k} \log\left(1 - \frac{kx}{u \cos \alpha}\right)$$

Replacing t in equation (6), we get the equation of the path of the projectile.

i.e.,
$$ky - \frac{g}{k}\left[\log\left(1 - \frac{kx}{u \cos \alpha}\right)\right] = \frac{kx}{u \cos \alpha} \frac{(ku \sin \alpha + g)}{k}$$

or
$$y = \frac{g}{k^2} \log\left(1 - \frac{kx}{u \cos \alpha}\right) + \frac{x}{ku \cos \alpha}(ku \sin \alpha + g) \qquad ...(7)$$

This is the equation of the required path.

Range on the horizontal plane

If R is the range on the horizontal plane, the particle meets x axis at (R,0). Putting in equation (7)

$$0 = \frac{g}{k^2} \log\left(1 - \frac{kR}{u \cos \alpha}\right) + \frac{R}{ku \cos \alpha}(ku \sin \alpha + g) \qquad ...(8)$$

or
$$0 = \frac{g}{k^2}\left[\left(\frac{-kR}{u \cos \alpha}\right) - \frac{\left(\frac{kR}{u \cos \alpha}\right)^2}{2} - \frac{\left(\frac{kR}{u \cos \alpha}\right)^3}{3} - \ldots\ldots\right] + \frac{R}{ku \cos \alpha}(ku \sin \alpha + g)$$

or
$$ku \sin \alpha + g = \frac{ku \cos \alpha}{R} \cdot \frac{g}{k^2}\left[\frac{kR}{u \cos \alpha} + \frac{k^2R^2}{2u^2 \cos^2 \alpha} + \frac{k^3R^3}{3u^3 \cos^3 \alpha} + \ldots\ldots\right]$$

or
$$ku \sin \alpha + g = g + \frac{kRg}{2u \cos \alpha} + \frac{k^2R^2g}{3u^2 \cos^2 \alpha} + \ldots\ldots$$

If k is very small then neglecting k^3 and higher powers.

We have

$$ku \sin \alpha = \frac{kRg}{2u \cos \alpha} + \frac{k^2R^2g}{3u^2 \cos^2 \alpha}$$

or $$u \sin\alpha = \frac{Rg}{2u\cos\alpha} + \frac{kR^2 g}{3u^2\cos^2\alpha}$$

or $$R = \frac{2u^2\sin\alpha\cos\alpha}{g} - \frac{2kR^2}{3u\cos\alpha} \quad ...(9)$$

Taking first order approximation $R = \dfrac{2u^2\cos\alpha\sin\alpha}{g}$

Substituting in R.H.S. of (9), we get a second order approximation of R as

$$R = \frac{2u^2\sin\alpha\cos\alpha}{g} - \frac{8ku^3\sin^2\alpha\cos\alpha}{3g^2} \quad ...(10)$$

Time of flight. If T is the time of flight of the projectile it means that in $t = T$ vertical displacement is zero, *i.e.*, at $t = T$, $y = 0$ in equation 6.

$$0 + gT = \frac{(g + ku\sin\alpha)}{k}\{1 - e^{-kT}\}$$

or $$gT = \frac{(g + ku\sin\alpha)}{k}\left\{1 - \left(1 - kT + \frac{k^2T^2}{2!} - \frac{k^3T^3}{3!} + ...\right)\right\}$$

$$gT = gT + kTu\sin\alpha - \frac{1}{2}gkT^2 - \frac{1}{2}k^2T^2u\sin\alpha + \frac{1}{6}gk^2T^3 + ...$$

$$0 = kTu\sin\alpha - \frac{1}{2}gkT^2 - \frac{1}{2}k^2T^2u\sin\alpha + \frac{1}{6}gk^2T^3 + ...$$

Dividing by $\frac{1}{2}kT$ and neglecting higher powers of k.

$$0 = 2u\sin\alpha - gT - kTu\sin\alpha + \frac{1}{3}gkT^2 ....$$

or $$T = \frac{2u\sin\alpha}{g} + \frac{k}{g}\left(\frac{gT^2}{3} - Tu\sin\alpha\right)$$

Taking first order approximation $T = \dfrac{2u\sin\alpha}{g}$.

Substituting in the R.H.S. we get a second order approximation for T as

$$T = \frac{2u\sin\alpha}{g} + \frac{k}{g}\left\{\frac{g}{3}\frac{4u^2\sin^2\alpha}{g^2} - \frac{2u\sin\alpha}{g}u\sin\alpha\right\}$$

$$= \frac{2u\sin\alpha}{g} - \frac{2}{3}\frac{ku^2\sin^2\alpha}{g^2} \quad ...(11)$$

Greatest height attained by the projectile

Let h be the greatest height and t_1 be the time to attain it. At the greatest height vertical component of velocity is zero. Putting $\dot{y} = 0$ in equation (4), we get

$$e^{kt_1} = \frac{ku \sin \alpha + g}{g}$$

or $$t_1 = \frac{1}{k} \log\left(\frac{ku \sin \alpha}{g} + 1\right) \qquad \text{...(12)}$$

Putting this value of t_1 and $y = h$ in equation (6)

$$kh + \frac{g}{k} \log\left(1 + \frac{k}{g} u \sin \alpha\right) = \frac{g + ku \sin \alpha}{k} - \frac{g}{k}$$

or $$h_{\text{max}} = \frac{1}{k} u \sin \alpha - \frac{g}{k^2} \log\left(1 + \frac{k}{g} u \sin \alpha\right) \qquad \text{...(13)}$$

6. Motion of a Projectile when Resistance Varies as Square of the Velocity

Let the particle be projected from O. Suppose the particle be at P at time t with velocity v. Let u be the horizontal component of velocity at P.

$$\therefore \quad u = v \cos \psi \qquad \text{...(1)}$$

Equations of motion along the normal at P and horizontally are

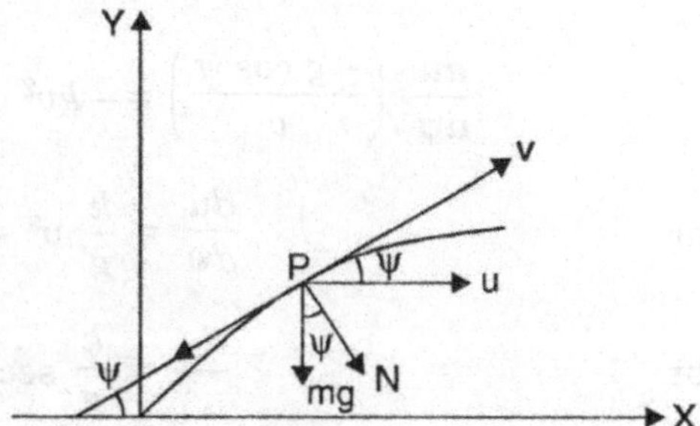

Fig. 4

$$\frac{mv^2}{\rho} = mg \cos \psi$$

or $$\frac{v^2}{\rho} = g \cos \psi \qquad \text{...(2)}$$

and $$m \frac{du}{dt} = -mkv^2 \cos \psi$$

or $$\frac{du}{dt} = -k v^2 \cos \psi \qquad \text{...(3)}$$

$$\rho = -\frac{ds}{d\psi}$$ (Negative sign is because ψ decreases as s increases)

Putting the value of ρ in equation (2)

$$v^2 \left(\frac{-d\psi}{ds}\right) = g \cos \psi$$

$$v \cdot \frac{ds}{dt}\left(\frac{d\psi}{ds}\right) = -g \cos \psi$$

or $$v \frac{d\psi}{dt} = -g \cos \psi \qquad \text{...(4)}$$

Equation (3) can be written as

$$\frac{du}{ds} \frac{ds}{dt} = -kv^2 \cos \psi$$

or $$\frac{du}{ds}\,v = -kv^2\cos\psi$$

or $$\frac{du}{ds} = -ku \qquad (\because\ u = v\cos\psi \text{ from (1)})$$

or $$\frac{du}{u} = -k\,ds$$

Integrating, $\log u = -ks + C$

At $s = 0,\quad u = u_0 \qquad \Rightarrow\ C = \log u_0$

Hence $\log u - \log u_0 = -ks$

or $$\log\frac{u}{u_0} = -ks$$

or $$u = u_0 e^{-ks} \qquad \text{...(5)}$$

Equation (3) can be written as

$$\frac{du}{d\psi}\frac{d\psi}{dt} = -kv^2\cos\psi \qquad \text{Using (4)}$$

$$\frac{du}{d\psi}\left(\frac{-g\cos\psi}{v}\right) = -kv^2\cos\psi$$

or $$\frac{du}{d\psi} = \frac{k}{g}v^3 = \frac{k}{g}u^3\sec^3\psi \qquad (v = u\sec\psi)$$

or $$\frac{du}{u^3} = \frac{k}{g}\sec^3\psi\,d\psi$$

or $$-2\frac{du}{u^3} = \frac{-2k}{g}\sec^3\psi\,d\psi$$

Integrating

$$\frac{1}{u^2} = \frac{-2k}{g}\left[\frac{\sec\psi\tan\psi}{2} + \frac{1}{2}\log(\sec\psi + \tan\psi)\right] + C'$$

Putting the value of u from (5)

$$\frac{e^{2ks}}{u_0^2} = \frac{-k}{g}[\sec\psi\tan\psi + \log(\sec\psi + \tan\psi)] + C'$$

This is the required intrinsic equation of path of the projectile.

7. A Bead Moves on a Smooth Wire in a Vertical Plane Under a Resistance kv^2; to Discuss the Motion

Let P be the position of the bead at any time t and the arcual distance travelled from initial position O be s.

Forces acting on the bead are

(i) Weight mg of the bead acting vertically downwards.

(ii) The resisting force $mk\,v^2$ acting along the tangent PT.

(iii) Normal reaction R acting along the normal at P.

The equations of motion along the tangent is

$$mv\frac{dv}{ds} = mg \sin \psi - mk\,v^2$$

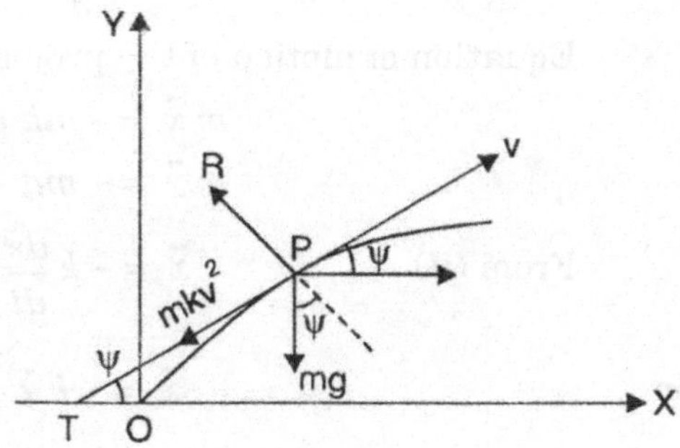

Fig. 5

or $$v\frac{dv}{ds} = g \sin \psi - kv^2$$

or $$\frac{1}{2}\frac{d}{ds}(v^2) + kv^2 = g \sin \psi$$

or $$\frac{d}{d\psi}(v^2)\frac{d\psi}{ds} + 2k\,v^2 = 2g \sin \psi$$

or $$\frac{d}{d\psi}(v^2) + 2k\rho v^2 = 2g\rho \sin \psi \quad \text{...(1)}$$

The equation of motion along the normal is

$$\frac{v^2}{\rho} = g \cos \psi - R \quad \text{...(2)}$$

If the equation of the curve is $s = f(\psi)$ then $\rho = f'(\psi)$

Equation (1) becomes

$$\frac{d}{d\psi}v^2 + 2k\,v^2 f'(\psi) = 2g \sin \psi\, f'(\psi)$$

This is a linear differential equation in v^2.

If particle moves along a circular bead $s = a\psi$

$$\frac{d}{d\psi}(v^2) + 2k\,v^2 a = 2g \sin \psi\, a \quad \text{...(3)}$$

Integrating factor $$= e^{\int 2ka\, d\psi} = e^{2ka\psi}$$

The solution of (3) is $$v^2 . e^{2ka\psi} = \int 2ag \sin \psi\, e^{2ka\psi}\, d\psi + C$$

$$= 2ag . \frac{1}{1+4a^2k^2} e^{2ka\psi}[2ka \sin \psi - \cos \psi] + C$$

or $$v^2 = \frac{2ag}{1+4a^2k^2}[2ak \sin \psi - \cos \psi] + Ce^{-2ak\psi}.$$

Example 1. *If the resistance vary as the velocity and the range on the horizontal plane through the point of projection is maximum, show that the angle* α *which the direction of projection makes with the vertical is given by* $\frac{\lambda(1+\lambda \cos \alpha)}{\lambda + \cos \alpha} = \log(1 + \lambda \sec \alpha)$ *where* λ *is the ratio of the velocity of projection to the terminal velocity.*

Sol. Equation of motion of a falling body under gravity is

$$m\frac{d^2x}{dt^2} = mg - mk\,v$$

If V is the terminal velocity then $mg - mk\,\text{V} = 0$

or $$\text{V} = \frac{g}{k} \qquad \text{...(1)}$$

Equation of motion of the projectile along the coordinate axis are

$$m\ddot{x} = -mk\,v\cos\psi \qquad \text{...(2)}$$

$$m\ddot{y} = -mg - mk\,v\sin\psi \qquad \text{...(3)}$$

From (*ii*) $$\ddot{x} = -k\frac{ds}{dt}\cdot\frac{dx}{ds}$$

or $$\ddot{x} = -k\dot{x} \quad \text{or} \quad \frac{\ddot{x}}{\dot{x}} = -k$$

Integrating, $\log\dot{x} = -kt + \text{C}$

At $t = 0$, $\dot{x} = u\sin\alpha$, $\Rightarrow$ $\text{C} = \log u\sin\alpha$

Substituting C we have

$$\log\frac{\dot{x}}{u\sin\alpha} = -kt$$

or $$\dot{x} = u\sin\alpha\,.\,e^{-kt} \qquad \text{...(4)}$$

Again integrating

$$x = -\frac{u\sin\alpha}{k}e^{-kt} + \text{C}_1$$

At $t = 0$, $x = 0$ $\Rightarrow$ $\text{C}_1 = \dfrac{u\sin\alpha}{k}$

This gives $$x = \frac{u\sin\alpha}{k}(1 - e^{-kt}) \qquad \text{...(5)}$$

From (3) $$\ddot{y} = -\left(g + k\frac{ds}{dt}\frac{dy}{ds}\right)$$

or $$\ddot{y} = -(g + k\dot{y})$$

or $$\frac{k\ddot{y}}{g + k\dot{y}} = -k$$

Integrating,

$$\log(g + k\dot{y}) = -kt + \text{C}_2$$

At $t = 0$, $\dot{y} = u\cos\alpha$ $\Rightarrow$ $\text{C}_2 = \log(g + ku\cos\alpha)$

Substituting C_2 $$\log\left(\frac{x + k\dot{y}}{g + ku\cos\alpha}\right) = -kt$$

or $$k\dot{y} = (g + ku\cos\alpha)e^{-kt} - g$$

Integrating again

$$ky = \frac{(g + ku\cos\alpha)e^{-kt}}{-k} - gt + C_3$$

At $t = 0, y = 0 \Rightarrow C_3 = \frac{g + ku\cos\alpha}{k}$

Substituting C_3, we have

$$ky = \frac{g + ku\cos\alpha}{k}[1 - e^{-kt}] - gt \quad ...(6)$$

Let the particle strike, the horizontal plane *i.e.*, at $t = t_1, y = 0$

$$gt_1 = \left(\frac{g + ku\cos\alpha}{k}\right)(1 - e^{-kt_1}) \quad ...(7)$$

If R be horizontal range, then at $t = t_1, x = R, y = 0$

∴ From equation (5)

$$R = \frac{u\sin\alpha}{k}(1 - e^{-kt_1}) \quad ...(8)$$

For maximum value of R,

$$\frac{dR}{d\alpha} = 0 \quad \text{and} \quad \frac{d^2R}{d\alpha^2} = (-)\text{ve}$$

From (8),
$$\frac{dR}{d\alpha} = \frac{u\cos\alpha}{k}(1 - e^{-kt_1}) + \frac{u\sin\alpha}{k}\left(ke^{-kt_1}\frac{dt_1}{d\alpha}\right) = 0$$

or $$\cos\alpha(1 - e^{-kt_1}) + k\sin\alpha\, e^{-kt_1}\frac{dt_1}{d\alpha} = 0 \quad ...(9)$$

$\frac{dt_1}{d\alpha}$ can be obtained from (7)

$$g\frac{dt_1}{d\alpha} = -u\sin\alpha(1 - e^{-kt_1}) + \left(\frac{g + ku\cos\alpha}{k}\right)e^{-kt_1}\, k\frac{dt_1}{d\alpha}$$

or $$\frac{dt_1}{d\alpha}[g - (g + ku\cos\alpha)e^{-kt_1}] = -u\sin\alpha(1 - e^{-kt_1})$$

or $$\frac{dt_1}{d\alpha} = \frac{u\sin\alpha(1 - e^{-kt_1})}{(g + ku\cos\alpha)e^{-kt_1} - g}$$

Substituting in equation (9)

$$\cos\alpha(1 - e^{-kt_1}) + k\sin\alpha\,.\,e^{-kt_1}\frac{[u\sin\alpha(1 - e^{-kt_1})]}{(g + ku\cos\alpha)e^{-kt_1} - g} = 0$$

or $$[(g + ku\cos\alpha)e^{-kt_1} - g][\cos\alpha(1 - e^{-kt_1})] + k\sin\alpha\, e^{-kt_1}[u\sin\alpha(1 - e^{-kt_1})] = 0$$

or $$[(g + ku\cos\alpha)e^{-kt_1} - g]\cos\alpha + ku\sin^2\alpha\, e^{-kt_1} = 0$$

or $$g\cos\alpha\, e^{-kt_1} + ku\cos^2\alpha\, e^{-kt_1} - g\cos\alpha + ku\sin^2\alpha\, e^{-kt_1} = 0$$

or $$g\cos\alpha\, e^{-kt_1} + e^{-kt_1}ku - g\cos\alpha = 0$$

or $$e^{-kt_1} = \frac{g \cos \alpha}{ku + g \cos \alpha} \quad \text{...(10)}$$

or $$t_1 = -\frac{1}{k} \log\left[\frac{g \cos \alpha}{ku + g \cos \alpha}\right] \quad \text{...(11)}$$

Putting the values of t_1 and e^{-kt_1}, in (7)

$$\frac{-g}{k} \log\left[\frac{g \cos \alpha}{ku + g \cos \alpha}\right] = \left(\frac{g + ku \cos \alpha}{k}\right)\left(1 - \frac{g \cos \alpha}{ku + g \cos \alpha}\right)$$

$$\frac{-g}{k} \log\left(\frac{g \cos \alpha}{ku + g \cos \alpha}\right) = \frac{(g + ku \cos \alpha) u}{ku + g \cos \alpha}$$

But $V = \frac{g}{k}$ from (1)

$$-V \log\left[\frac{Vk \cos \alpha}{ku + Vk \cos \alpha}\right] = \left[\frac{Vk + ku \cos \alpha}{ku + Vk \cos \alpha}\right] u$$

or $$V \log\left[\frac{u + V \cos \alpha}{V \cos \alpha}\right] = \left(\frac{V + u \cos \alpha}{u + V \cos \alpha}\right) u$$

Putting $$\lambda = \frac{u}{V} \quad \text{or} \quad u = V\lambda$$

$$V \log\left[\frac{V\lambda + V \cos \alpha}{V \cos \alpha}\right] = \left(\frac{V + V\lambda \cos \alpha}{V\lambda + V \cos \alpha}\right) V\lambda$$

or $$\log (1 + \lambda \sec \alpha) = \left(\frac{1 + \lambda \cos \alpha}{\lambda + \cos \alpha}\right) V.$$

Example 2. *A heavy bead of mass m slides on a smooth wire in the shape of a cycloid with axis vertical and vertex upwards, in a medium whose resistance is* $\frac{mv^2}{2c}$. *The distance of the starting point from the vertex is c. Show that the time of descent to the cusp is* $\sqrt{\frac{8a (4a - c)}{gc}}$, *where 2a is the length of the axis of the cycloid.*

Sol. Let the bead of mass m be at P at any time t

Equation of motion along the tangent is

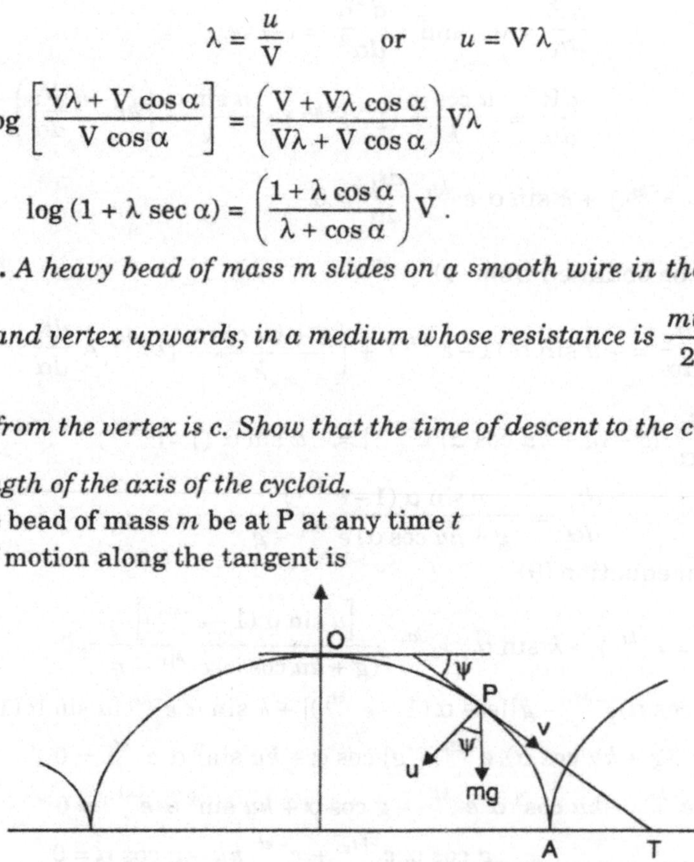

Fig. 6

$$mv\frac{dv}{ds} = mg\sin\psi - \frac{mv^2}{2c}$$

or $$v\frac{dv}{ds} = g\sin\psi - \frac{v^2}{2c} \quad ...(1)$$

Equation of the cycloid is $s = 4a\sin\psi$

Equation (1) can be written as

$$\therefore \quad \frac{1}{2}\frac{d}{ds}(v^2) + \frac{v^2}{2c} = g\frac{s}{4a}$$

or $$\frac{d}{ds}(v^2) + \frac{1}{c}(v^2) = \frac{g}{2a}s \quad ...(2)$$

Its integrating factor is $e^{\int \frac{1}{c}ds} = e^{s/c}$

Solution of equation (2) is

$$v^2 \cdot e^{s/c} = \frac{g}{2a}\int se^{s/c}\,ds + C \qquad \text{Integrating by parts}$$

$$= \frac{g}{2a}\left[sc\,e^{s/c} - \int c\,e^{s/c}\,ds\right] + C$$

$$= \frac{g}{2a}\left[sc\,e^{s/c} - c^2 e^{s/c}\right] + C$$

Initially at $s = c,\ v = 0 \quad \Rightarrow \quad C = 0$

$$\therefore \quad v^2 e^{s/c} = \frac{g}{2a}\left[sc\,e^{s/c} - c^2 e^{s/c}\right]$$

or $$v^2 = \frac{gc}{2a}[s - c] \qquad \text{or} \qquad v = \sqrt{\frac{gc}{2a}(s-c)}$$

or $$\frac{ds}{dt} = \sqrt{\frac{gc}{2a}(s-c)}$$

or $$dt = \sqrt{\frac{2a\,ds}{gc\,(s-c)}} \quad ...(3)$$

Time of descent from $s = c$ to $s = 4a$ (position of the cusp at A) is obtained by integrating (3)

$$t = \int_c^{4a}\sqrt{\frac{2a}{gc}}\,(s-c)^{-\frac{1}{2}}\,ds$$

$$= \sqrt{\frac{2a}{gc}}\left[\frac{\sqrt{s-c}}{\frac{1}{2}}\right]_c^{4a} = 2\sqrt{\frac{2a}{gc}}\sqrt{4a-c} = \sqrt{\frac{8a\,(4a-c)}{gc}}.$$

EXERCISE 2

1. A particle acted on by gravity is projected in a medium of which the resistance varies as the velocity. Show that its acceleration retains a fixed direction and diminishes without limit to zero.

2. If a particle be moving in a medium whose resistance varies as the velocity of the particle. Show that the equation of the trajectory can, by a proper choice of axes be put into the form $y + ax = b \log x$.

3. A particle, moving in a resisting medium is acted upon by a central force $\frac{\mu}{r^n}$. If the path be an equiangular spiral of angle α, whose pole is at the centre of force, show that the resistance is $\left(\frac{n-3}{3}\right) \frac{\mu \cos \alpha}{r^n}$.

4. A particle of unit mass is projected with velocity u at an inclination α above the horizon in a medium whose resistance is k times the velocity. Show that its direction will make an angle α with the horizon after a time $\frac{1}{k} \log \left\{1 + \frac{2ku \sin \alpha}{g}\right\}$.

5. A heavy particle is projected in a resisting medium. If v be the velocity at any time t and ϕ be the inclination to the vertical of the direction of motion and f be the retardation, prove that

$$\frac{1}{v}\frac{dv}{d\phi} + \cot \phi + \frac{f}{g \sin \phi} = 0.$$

6. If the resistance of the air be n times the weight of the particle and the particle be projected horizontally with velocity V, show that the velocity of the particle, when it is moving at an inclination ϕ to the horizontal is

$$V(1 - \sin \phi)^{\frac{n-1}{2}} . (1 + \sin \phi)^{\frac{-(n+1)}{2}}.$$

7. If the resistance per unit mass is $g\left(\frac{v}{V}\right)^2$, prove that $\frac{du}{ds} = \frac{-gu}{V^2}$, $\frac{d\psi}{ds} = \frac{g}{u^2} \cos^3 \psi$, where u is horizontal component of velocity and V is the terminal velocity.

8. A heavy particle describes a path given by $\cos \psi = f(\rho \cos \psi)$. show that the law of resistance is given by $Rv \frac{df}{dv} = -g\sqrt{1 - f^2} \frac{d}{dv}(fv)$, where $f = f\left(\frac{v^2}{g}\right)$.

9. A particle of unit mass is projected under gravity in a medium whose resistance is kv^n, where v is the speed. Show that $\frac{d^2y}{dx^2} = \frac{-g}{u^2}$, where u is the horizontal component of velocity v.

10. A bead moves on a smooth wire in the form of a circle in a vertical plane under a resistance kv^2. Find the velocity of the bead at any point.

8. Motion of the Particle When Mass Varies

There are many problems in Dynamics in which the moving particle changes its mass during the motion. For example, when a rocket is fired, it loses its mass continuously in the form of hot gases due to burning of the fuel, or when a body passes through a dusty cloud gathers dust particles on its surface and increasing the mass. In such cases the formula $P = mf$ is not applicable, as it is true only if mass is constant throughout the motion.

If mass varies, then the Newton's second law is

$$P = \frac{d}{dt}(mv) \qquad ...(1)$$

Let a particle gains a mass δm in time δt. Let this increment δm was moving with velocity V. Then in time δt the increment in the momentum of the particle is

$$= m\,(v + \delta v - v) + \delta m\,(v + \delta v - V) = m\,\delta v + \delta m\,(v + \delta v - V)$$

$$= P.\,\delta t \qquad \text{where P is force acting for time } \delta t$$

$$\therefore \qquad P = m\frac{\delta v}{\delta t} + v\frac{\delta m}{\delta t} + \frac{\delta m\,.\,\delta v}{\delta t} - V\frac{\delta m}{\delta t}$$

Taking limit $\delta t \to 0$ and Neglecting second order terms, we get

$$P = m\frac{dv}{dt} + v\frac{dm}{dt} - V\frac{dm}{dt}$$

or
$$m\frac{dv}{dt} + v\frac{dm}{dt} = P + V\frac{dm}{dt}$$

or
$$\mathbf{\frac{d}{dt}(mv) = P + V\frac{dm}{dt}} \qquad \text{...(2)}$$

If V = 0, then equation (2) reduces to equation (1)

i.e.,
$$P = \frac{d}{dt}(mv).$$

Example 1. *A particle of mass M is at rest and begins to move under the action of a constant force F in a fixed direction. It encounters the resistance of a steam of fine dust moving in opposite direction with velocity V, which deposits the matter on it at a constant rate c. Show that the mass will be m when it has travelled a distance* $\dfrac{k}{c^2}\left[m - M\left\{1 + \log\dfrac{m}{M}\right\}\right]$ *where* $k = F - cV$.

Sol. We know that Newton's second law in case of varying mass is

$$\frac{d}{dt}(mv) = P + V\frac{dm}{dt}$$

Here P = F and the dust is moving in opposite direction

$\therefore$ V is taken as – V.

$$\therefore \qquad \frac{d}{dt}(mv) = F - V\frac{dm}{dt} \qquad \text{...(1)}$$

$$\frac{dm}{dt} = c \qquad \text{(given)} \qquad \text{...(2)}$$

$$F - cV = k \qquad \text{(given)}$$

Equation (1) reduces to

$$\frac{d}{dt}(mv) = k$$

Integrating we get, $mv = kt + A$

Applying initial condition $t = 0$, V = 0, we have A = 0

$$\therefore \qquad mv = kt \qquad \text{...(3)}$$

Integrating equation (2)

$$m = ct + A'$$

at
$$t = 0,\ m = M \qquad \Rightarrow \quad A' = M$$

or
$$m = ct + M \qquad \text{(4)}$$

Putting the value of m in equation (3), we have

$$v = \frac{kt}{ct + M}$$

or

$$\frac{dx}{dt} = \frac{kt}{ct + M}$$

Integrating
$$x = \int \frac{kt}{ct + M}\, dt + A''$$

$$= \int \frac{k}{c} - \frac{Mk/c}{(ct + M)}\, dt + A'' = \frac{k}{c}\left[t - \frac{M}{c}\log(ct + M)\right] + A''$$

At $t = 0, x = 0,$ $\Rightarrow$ $A'' = \frac{kM}{c^2}\log M$

Putting the value of A″ the displacement

$$x = \frac{k}{c}\left[t - \frac{M}{c}\log(ct + M)\right] + \frac{Mk}{c^2}\log M \qquad \ldots(5)$$

But $t = \frac{m - M}{c}$ from equation (4)

Putting the value of t in equation (5)

$$x = \frac{k}{c}\left[\frac{m - M}{c} - \frac{M}{c}\log(m - M + M) + \frac{M}{c}\log M\right]$$

$$= \frac{k}{c^2}\left[m - M\left\{1 + \log\frac{m}{M}\right\}\right].$$

Example 2. *A sperical raindrop, falling freely receives at each instant an increase of volume equal to λ times its surface at that instant. Find the velocity at the end of time t, and the distance fallen through in that time.* (M.D.U. 2005)

Sol. Let the raindrop be of mass M and radius r at time t, and it is falling under gravity only. Then

$$\frac{d}{dt}[Mv] = Mg \qquad \ldots(1)$$

where $M = \frac{4}{3}\pi r^3\rho$ and $\frac{dM}{dt} = 4\pi r^2\lambda$

or
$$\frac{d}{dt}\left[\frac{4}{3}\pi r^3\rho\right] = \lambda 4\pi r^2 \quad \Rightarrow \quad 4\pi r^2\rho\frac{dr}{dt} = 4\pi\lambda r^2$$

or
$$\frac{dr}{dt} = \lambda$$

Integrating we get, $r = \lambda t + c$ ...(2)

Let a be the initial radius, then at $t = 0$, $r = a$, hence $c = a$

or
$$r = \lambda t + a \qquad \ldots(3)$$

From (1),
$$\frac{d}{dt}[4\pi r^3\rho v] = 4\pi r^3\rho g$$

or
$$\frac{d}{dt}[(a + \lambda t)^3 v] = (a + \lambda t)^3 g$$

Integrating $\quad (a + \lambda t)^3 v = \dfrac{(a+\lambda t)^4}{4\lambda} g + c'$

Initially at $t = 0, v = 0 \qquad \Rightarrow \quad c' = -g\dfrac{a^4}{4\lambda}$

or $$(a+\lambda t)^3 \frac{dx}{dt} = \frac{(a+\lambda t)^4}{4\lambda} g - \frac{a^4}{4\lambda} g$$

or $$\frac{dx}{dt} = \frac{g}{4\lambda}\left[a + \lambda t - \frac{a^4}{(a+\lambda t)^3}\right] \qquad ...(4)$$

Integrating we get, $\quad x = \dfrac{g}{4\lambda}\left[\dfrac{(a+\lambda t)^2}{2\lambda} + \dfrac{a^4}{2(a+\lambda t)^2 . \lambda}\right] + c''$

At $t = 0, x = 0 \quad \Rightarrow \quad c'' = \dfrac{-ga^2}{4\lambda^2}$

Hence, $$x = \frac{g}{8\lambda^2}\left[(a+\lambda t)^2 + \frac{a^4}{(a+\lambda t)^2} - 2a^2\right]$$

$$= \frac{g}{8\lambda^2}\left[(a+\lambda t) - \frac{a^2}{(a+\lambda t)^2}\right]^2$$

$$= \frac{g}{8\lambda^2}\left[\frac{2a\lambda t + \lambda^2 t^2}{a+\lambda t}\right]^2 = \frac{gt^2}{8}\left[\frac{2a+\lambda t}{a+\lambda t}\right]^2.$$

Example 3. *A chain of length l is coiled at the edge of a table. One end is fastened to a particle whose mass is equal to that of the whole chain and the other end is put over the edge. Show that, immediately after leaving the table, the particle is moving with velocity $\dfrac{1}{2}\sqrt{\dfrac{5}{6}gl}$.*

Sol. Let m be the mass of the chain.

Let portion of the chain uncoiled in time t be x.

$\therefore \quad$ Mass of length $x = \dfrac{m}{l}(x)$

Equation of motion is

$$\frac{d}{dt}\left[\frac{mx}{l}\frac{dx}{dt}\right] = \left(\frac{mx}{l}\right) g \qquad ...(1)$$

or $$\frac{d}{dt}\left[x\frac{dx}{dt}\right] = xg$$

or $$\frac{d}{dx}\left[x\frac{dx}{dt}\right]\frac{dx}{dt} = xg$$

or $$\left(x\frac{dx}{dt}\right)\frac{d}{dx}\left[x\frac{dx}{dt}\right] = x^2 g$$

or $$\frac{1}{2}\frac{d}{dx}\left(x\frac{dx}{dt}\right)^2 = x^2 g$$

Integrating we get, $$\frac{1}{2}\left(x\frac{dx}{dt}\right)^2 = \frac{x^3}{3}g + c$$

Initially the coil is at rest $\therefore$ $t = 0, x = 0, \frac{dx}{dt} = 0 \Rightarrow c = 0$

Hence, $$\frac{1}{2}\left(x\frac{dx}{dt}\right)^2 = \frac{x^3 g}{3}$$

or $$\left(\frac{dx}{dt}\right)^2 = \frac{2xg}{3}.$$

When the whole chain is uncoiled $x = l$ and $\frac{dx}{dt} = v$ then

$$v^2 = \frac{2}{3}lg$$

Now, just after uncoiling the chain, the particle of mass m attached with the chain is set into the motion with a jerk. If V be the velocity of the system after jerk, then we have

$$(m + m)V = mv \Rightarrow V = v/2$$

Just after leaving the table, the system gets a vertical impulse. If v' be the vertical velocity of the system then

$$(m + m)v' = mV \Rightarrow v' = \frac{V}{2} = \frac{v}{4}.$$

Thus, the particle has horizontal velocity $v/2$ and vertical velocity $v/4$.

$\therefore$ Resultant velocity

$$= \sqrt{\frac{v^2}{16} + \frac{v^2}{4}} = \frac{1}{4}v\sqrt{5} = \frac{1}{4} \times \sqrt{\frac{2}{3}gl}\sqrt{5} = \frac{1}{2}\sqrt{\frac{5}{6}gl}.$$

EXERCISE 3

1. A spherical raindrop of a liquid falling freely in vapour acquires mass by condensation at a constant rate c.
   Show that the velocity after falling from rest in time t is
   $\frac{1}{2}gt\left[\frac{2M + ct}{M + ct}\right]$ where M is the initial mass of the drop.
2. A solid cylinder of mass m, area of cross-section A moves parallel to its axis under a constant force F. It moves through a uniform cloud of fine dust of density ρ which is moving in opposite direction with constant velocity V. If all the dust that meets the cylinder clings to it, find the velocity and distance described in any time t, the cylinder being initially at rest.
3. A uniform chain is coiled up on a horizontal plane and one end passes over a small light pulley at a height a above the plane. Initially a length $b > a$, hangs freely on the other side. Find the motion.
4. A trailer full of sand is pulled by a constant force F. Sand leaks out at the rate of λ units of mass per second. Show that the velocity at the end of time t is $\frac{-F}{\lambda}\log\left(1 - \frac{\lambda t}{M}\right)$ and the distance

moved during that time is $\frac{FM}{\lambda^2}\left[\left(1-\frac{\lambda t}{M}\right)\log\left(1-\frac{\lambda t}{M}\right)+\frac{\lambda t}{M}\right]$ where M is the initial mass of trailer and sand.

5. A rocket whose total mass (fuel + shell) is m_0 ejects fuel at a constant rate cm_0 and a velocity V relative to the case. Show that the lowest rate of fuel consumption that will permit the rocket to rise at once is $c = g/v$. Assuming this design condition is met, find the greatest speed and height reached by the rocket.
6. A falling raindrop has its radius uniformly increased by access of moisture. If it have given to it a horizontal velocity show that it will than describes a hyperbola.
7. A mass in the form of a solid cylinder of radius c, acted upon by no forces, moves parallel to its axis through a uniform cloud of fine dust of density ρ at rest. If the dust which meets the cylinder adhere to it and if M and u be the mass and the velocity at the beginning of the motion, prove that the distance x traversed in time t is given by

$$(M + \pi\rho c^2 x)^2 = M^2 + 2\pi\rho\mu c^2 t M.$$

ANSWERS

5. $\frac{-g}{c}\left(1-\frac{M}{m_0}\right) - V\log\frac{M}{m_0}$, $\frac{V}{c}\left(1-\frac{M}{m_0}+\log\frac{M}{m_0}\right)+\frac{V^2}{2g}\left(\log\frac{M}{m_0}\right)^2$

(where M is mass of the shell).

12

Motion of a Particle on Smooth and Rough Plane Curves

MOTION ON A SMOOTH CURVE IN A VERTICAL PLANE

1. A heavy particle is made to move on a smooth curve in a vertical plane, to discuss its motion

Let the particle of mass m starts from a fixed point A with velocity u and moves along the curve. Let at some instant at time t particle be at P such that $\overset{\frown}{AP} = s$. Let A(a, b) and P(x, y) be two points on its path.

Forces acting on the particle at P are

(*i*) Weight mg acting vertically downwards.

(*ii*) Normal reaction R between the particle and curve along inward normal drawn at P.

Resolving the weight in tangential and normal direction.

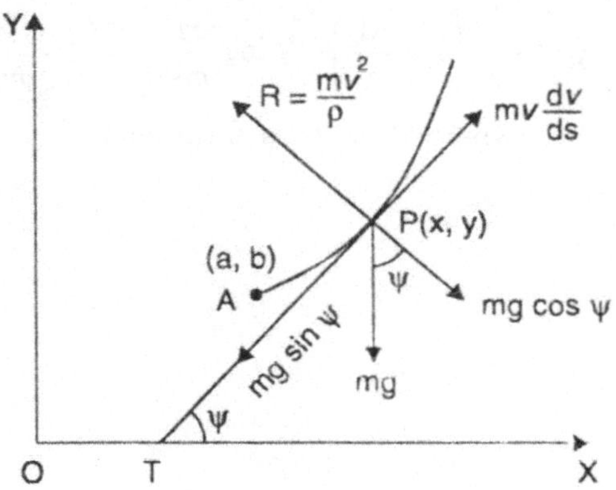

Fig. 1

Equations of motion are

$$mv\frac{dv}{ds} = -mg\sin\psi \quad ...(1)$$

and $$\frac{mv^2}{\rho} = R - mg\cos\psi \quad \left(\sin\psi = \frac{dy}{ds}\right) \quad ...(2)$$

From (1), we get

$$v\frac{dv}{ds} = -g\frac{dy}{ds}$$

or $$2v\,dv = -2g\,dy$$

Integrating $v^2 = -2gy + c$

At $t = 0, y = b, v = u \therefore c = u^2 + 2gb$

Hence, $v^2 = -2gy + u^2 + 2gb$

or $$v^2 = u^2 - 2g(y - b)$$

or $$\mathbf{v^2 = u^2 - 2gh}$$

where $h = y - b$ is the height of P above A.

However, if particle moves downward, then

$$v^2 = u^2 + 2gh,$$

where h is the depth descended by the particle from initial position.

2. Motion on the Outside of a Vertical Circle

A particle starts sliding down the outside of a smooth vertical circle from the highest point ; to discuss its motion. (KU 2004)

Let the particle of mass m starts with velocity zero from the highest point A of a vertical circle and slides down. Let at time t particle be at P and its velocity be v.

Forces acting on the particle are :

(*i*) Weight mg vertically downwards

(*ii*) Normal reaction R between curve and particle along outward normal.

Let $\angle AOP = \theta$, OA = OP = a (radii of the circle).

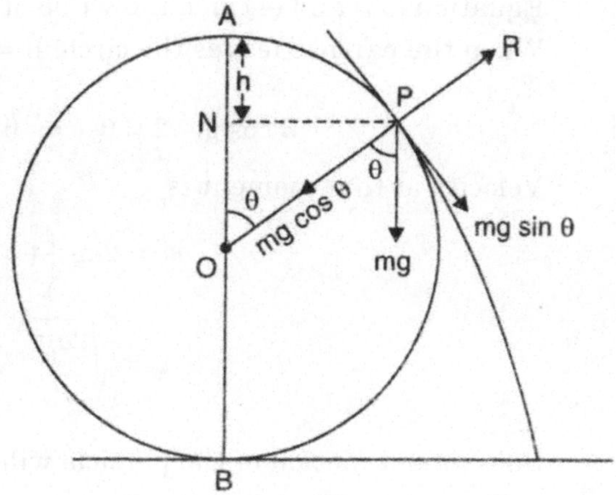

Fig. 2

Equations of motion along tangential and normal directions are

$$m\frac{d^2s}{dt^2} = mg\sin\theta \qquad ...(1)$$

$$\frac{mv^2}{a} = mg\cos\theta - R \qquad ...(2)$$

Let $\overset{\frown}{AP} = s = a\theta \quad \therefore \quad \dfrac{d^2s}{dt^2} = a\dfrac{d^2\theta}{dt^2}$

Putting in (1), we get

$$a\frac{d^2\theta}{dt^2} = g\sin\theta$$

Multiplying both sides by $2a\dfrac{d\theta}{dt}$

$$2a^2\frac{d\theta}{dt}\frac{d^2\theta}{dt^2} = 2ag\sin\theta\frac{d\theta}{dt}$$

or $$\frac{d}{dt}\left(a\frac{d\theta}{dt}\right)^2 = -\frac{d}{dt}(2ag\cos\theta)$$

or $$\left(a\frac{d\theta}{dt}\right)^2 = -2ag\cos\theta + c$$

Linear velocity $v = a\omega = a\dfrac{d\theta}{dt}$

$$v^2 = -2ag\cos\theta + c$$

At $\theta = 0, v = 0 \;\Rightarrow\; c = 2ag$

Hence, $v^2 = 2ag\,(1 - \cos\theta)$ ...(3)

From equation (2),

$$R = mg\cos\theta - \frac{mv^2}{a} \qquad \text{(using (3))}$$

$$= mg\cos\theta - \frac{m}{a}\{2ag\,(1 - \cos\theta)\}$$

$$R = mg\,(3\cos\theta - 2) \qquad ...(4)$$

Equations (3) and (4) give the value of velocity v and reaction R at any point P.

When the particle leaves the circle R = 0

or $$3\cos\theta - 2 = 0 \Rightarrow \theta = \left(\cos^{-1}\frac{2}{3}\right)$$

Velocity at that moment is

$$v^2 = 2ag\left(1 - \frac{2}{3}\right)$$

$$v = \sqrt{\frac{2ag}{3}}$$

Subsequent motion of the particle will be like a projectile thrown with velocity $\sqrt{\frac{2ag}{3}}$ and angle $\cos^{-1} 2/3$. Its path is

$$y = x\tan\theta + \frac{1}{2}\frac{gx^2}{v^2\cos^2\theta} \qquad ...(5)$$

i.e., $$y = \frac{x\sqrt{5}}{2} + \frac{27}{16a}x^2.$$

Example 1. *A particle moves along the outside of the arc of a vertical circle. If the particle starts from rest from a point whose angular distance from the highest point is α, show that it leaves the circle at angular distance* $\cos^{-1}\left(\frac{2}{3}\cos\alpha\right)$ *measured from the highest point of the circle.*

Sol. Let AB be the vertical diameter of the circle with centre O and particle starts from rest at C such that ∠AOC = α. Let the particle be at P at some instant t, Arc AP = s. Resolving the forces along the tangent and normal, the equations of motion are

$$m\frac{d^2s}{dt^2} = mg\sin\theta \qquad ...(1)$$

$$\frac{mv^2}{a} = mg\cos\theta - R \qquad ...(2)$$

Taking $s = a\theta$ $\quad$ a being the radius of circle.

Equation (1) becomes

$$a\frac{d^2\theta}{dt^2} = g\sin\theta$$

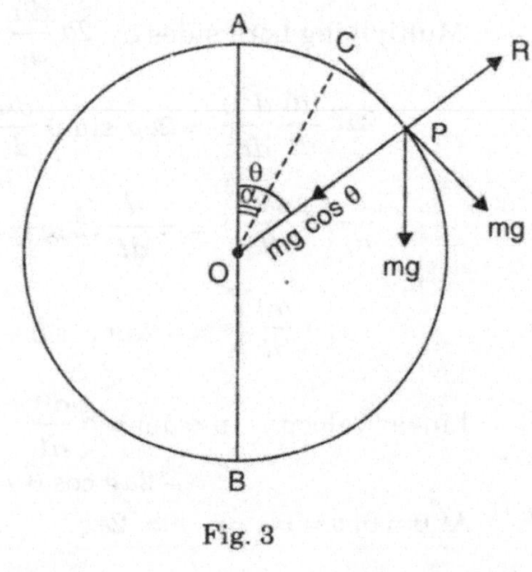

Fig. 3

Multiplying both sides by $2a\frac{d\theta}{dt}$ and integrating, we get

$$\left(a\frac{d\theta}{dt}\right)^2 = -2ag\cos\theta + c \quad ...(3)$$

Applying initial conditions *i.e.*, $\theta = \alpha$, $v = a\frac{d\theta}{dt} = 0$,

we have, $c = 2ag\cos\alpha$

Putting the value of c in (3), we get

$$v^2 = 2ag(\cos\alpha - \cos\theta) \quad ...(4)$$

Putting the value of v^2 in equation (2), we get

$$R = mg\cos\theta - \frac{m}{a}\cdot 2ag(\cos\alpha - \cos\theta)$$

$$R = mg\,[3\cos\theta - 2\cos\alpha] \quad ...(5)$$

When the particle leaves the circle, R = 0.

Hence, $3\cos\theta = 2\cos\alpha$

or $$\theta = \cos^{-1}\left(\frac{2}{3}\cos\alpha\right).$$

Example 2. *A particle is projected horizontally with a velocity $\sqrt{\frac{ag}{2}}$ from the highest point of the outside of a fixed smooth sphere of radius a. Show that it will leave the sphere at the point whose vertical distance below the point of projection is a/6.*

Sol. Particle starts from highest point A with velocity

$$\sqrt{\frac{ag}{2}} \quad i.e., \text{ at } \theta = 0, v = \sqrt{\frac{ag}{2}}$$

Equations of motion are

$$m\frac{d^2s}{dt^2} = mg\sin\theta \quad ...(1)$$

$$\frac{mv^2}{a} = mg\cos\theta - R \quad ...(2)$$

$\overset{\frown}{AP} = s$, $s = a\theta$ putting in (1)

$$\therefore \quad a\frac{d^2\theta}{dt^2} = g\sin\theta$$

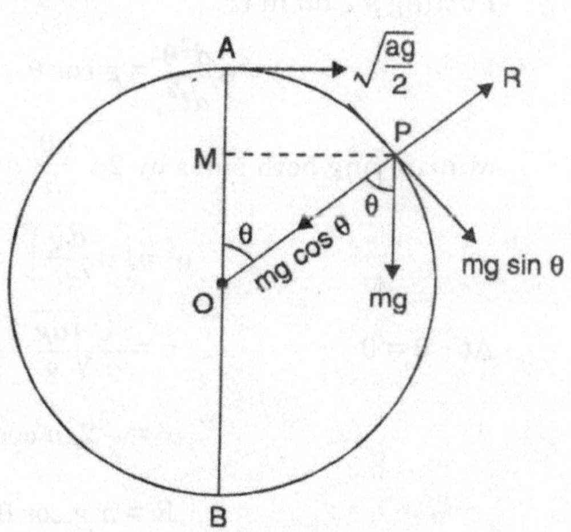

Fig. 4

Multiplying both sides by $2a\frac{d\theta}{dt}$ and integrating

$$v^2 = \left(a\frac{d\theta}{dt}\right)^2 = -2ag\cos\theta + c \quad \left(\because v = a\frac{d\theta}{dt}\right)$$

Applying initial condition $c = \frac{ag}{2} + 2ag = \frac{5ag}{2}$

$$\therefore \qquad v^2 = -2ag\cos\theta + \frac{5ag}{2} \qquad ...(3)$$

Substituting the value of v^2 in equation (2)

$$R = mg\cos\theta - \frac{m}{a}\left[-2ag\cos\theta + \frac{5ag}{2}\right] = mg\left[3\cos\theta - \frac{5}{2}\right]$$

When the particle leaves contact with the sphere R = 0.

$$\therefore \qquad \cos\theta = \frac{5}{6}$$

Depth of this point below A is AM

$$AM = a - a\cos\theta = a - \frac{5a}{6} = \frac{a}{6}.$$

Example 3. *A particle moves under gravity down from the highest point of a smooth vertical circle of radius a with velocity $\frac{1}{2}\sqrt{ag}$. Prove that it will strike the horizontal plane through the centre at a distance $\frac{(9\sqrt{39}+7\sqrt{7})}{64}$ a from the centre.*

Sol. Proceeding as in example 2. Equations of motion are

$$m\frac{d^2s}{dt^2} = mg\sin\theta \qquad ...(1)$$

$$\frac{mv^2}{a} = mg\cos\theta - R \qquad ...(2)$$

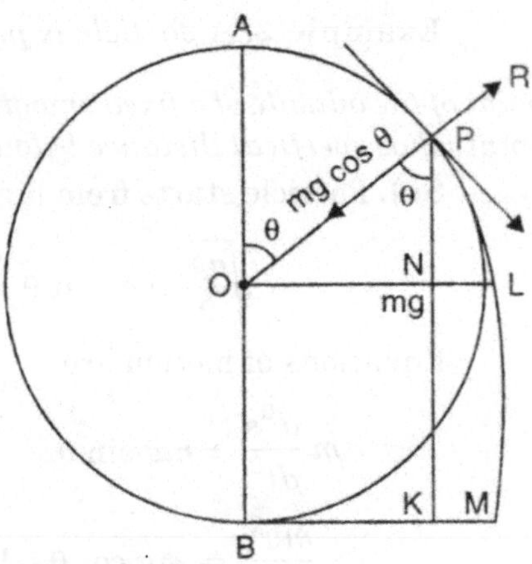

Fig. 5

Putting $s = a\theta$ in (1)

$$a\frac{d^2\theta}{dt^2} = g\cos\theta$$

Multiplying both sides by $2a\frac{d\theta}{dt}$ and integrating

$$v^2 = \left(a\frac{d\theta}{dt}\right)^2 = -2ga\cos\theta + c$$

At $\theta = 0$ $\qquad v = \frac{1}{2}\sqrt{\frac{ag}{2}} \Rightarrow c = \frac{9ag}{4}$

$$\therefore \qquad v^2 = -2ga\cos\theta + \frac{9ag}{4} \qquad ...(3)$$

$$R = mg\cos\theta - \frac{m}{a}\left\{-2ag\cos\theta + \frac{9ag}{4}\right\} = mg\left(3\cos\theta - \frac{9}{4}\right)$$

when the particle leaves the circle R = 0

$$\therefore \qquad \cos\theta = 3/4, \quad \tan\theta = \frac{\sqrt{7}}{3}$$

Velocity at the moment when the particle leaves the circle is

$$v^2 = -2ag\cos\theta + \frac{9ag}{4} = -2ag\cdot\frac{3}{4} + \frac{9ag}{4} = \frac{3ag}{4}$$

The equation of the projectile is

$$y = x\tan\theta - \frac{1}{2}\frac{gx^2}{\cos^2\theta} \qquad \text{...(4)}$$

If the particle strikes the horizontal plane through the centre at L then

$$PN = a\cos\theta = \frac{3a}{4}$$

Coordinate of L referred to P as origin are (NL, – PN)

Putting in equation (4), we get

$$-\frac{3a}{4} = -\frac{x\sqrt{7}}{3} - \frac{1}{2}g\cdot\frac{x^2}{\frac{3ag}{4}\cdot\frac{9}{16}}$$

$$\frac{3a}{4} = \frac{x\sqrt{7}}{3} + \frac{32x^2}{27a}$$

$$128x^2 + 36\sqrt{7}\,xa - 81a^2 = 0$$

(tan θ is taken (–)ve because the particle moves below the horizontal line in IVth quadrant)

$$x = -\frac{36\sqrt{7}a \pm \sqrt{9072a^2 + 41472a^2}}{256} = -\frac{9\sqrt{7}a + 9\sqrt{39}a}{64}$$

(–)ve sign is discarded, since x in 4th quadrant is (+)ve.

Thus, the horizontal distance from the centre is

$$OL = ON + x = a\sin\theta + x$$

$$= \frac{\sqrt{7}}{4} + \frac{-9\sqrt{7}a + 9\sqrt{39}a}{64} = \frac{(7\sqrt{7} + 9\sqrt{39})a}{64}.$$

EXERCISE 1

1. A heavy particle slides down a smooth vertical circle of radius $27a$ from rest at the highest point. Show that on leaving the circle it moves in a parabola of latus rectum $16a$.
2. A particle moves under gravity in a vertical circle sliding down the convex side of circular arc. If its initial velocity is that due to the starting point from a height h above the centre, show that it will fly off the circle when at a height $\frac{2}{3}h$ above the centre.
3. A heavy particle slides under gravity down the inside of a smooth vertical tube held in a vertical plane. It starts from the highest point with velocity $\sqrt{2ag}$, where a is the radius. Prove that when in the subsequent motion the vertical component of the acceleration is maximum, the pressure on the curve is equal to twice the weight of the particle.
4. A particle slides down from the highest point of a smooth vertical circle of radius a with velocity $\sqrt{\frac{1}{2}ag}$. Prove that it will strike the horizontal plane through the centre at a distance $\frac{9\sqrt{39} + 7\sqrt{7}}{64}a$ from the centre.

5. A particle slides down the arc of a smooth circle or radius a from rest from the highest point. Find where it will leave the circle and prove that it will strike the horizontal plane through the lowest point of the circle at a distance $\frac{5(\sqrt{5}+4\sqrt{2})a}{27}$ from the vertical diameter.

6. A particle is projected horizontally from the highest point of a vertical circle of radius r with tangential velocity $\sqrt{\frac{1}{2}gr(3\sqrt{2}-4)}$. Show that it will leave the circle when radius to the particle makes an angle of 45° with the upward vertical.

7. If a particle starts from rest at a depth $\frac{r}{2}$ below the highest point of a smooth vertical circle of radius r. Prove that it will leave the circle at a distance $\frac{1}{3}r$ above the centre.

8. The roadway of a bridge over a canal is in the form of a circular arc of radius $60\frac{1}{2}$ m. What is the greatest velocity with which a motorcycle can cross the bridge without leaving the ground at the highest point ?

3. Motion on Inside a Vertical Circle

To discuss of the motion of a particle projected from the lowest point of the circle with velocity *u* and moving along the inside of a smooth vertical circle.

(K.U. 2005)

Let a particle of mass m be projected from the lowest point A of the vertical circle with velocity u. Let the particle be at P at certain instant t and moving with velocity v. $\angle AOP = \theta$.

Forces on the particle are

(*i*) Weight mg acting vertically downwards.

(*ii*) Reaction R between circle and point P passing through the centre O towards the centre.

Resolving the forces along the tangent and normal, we have

$$m\frac{d^2s}{dt^2} = -mg\sin\theta \qquad ...(1)$$

$$\frac{mv^2}{a} = R - mg\cos\theta \qquad ...(2)$$

$$s = a\theta$$

where a is radius of the circle

$$a\frac{d^2\theta}{dt^2} = -g\sin\theta$$

Multiplying both sides with $2a\frac{d\theta}{dt}$ and integrating

$$\left(a\frac{d\theta}{dt}\right)^2 = 2ag\cos\theta + c \qquad \left(v = a\frac{d\theta}{dt}\right)$$

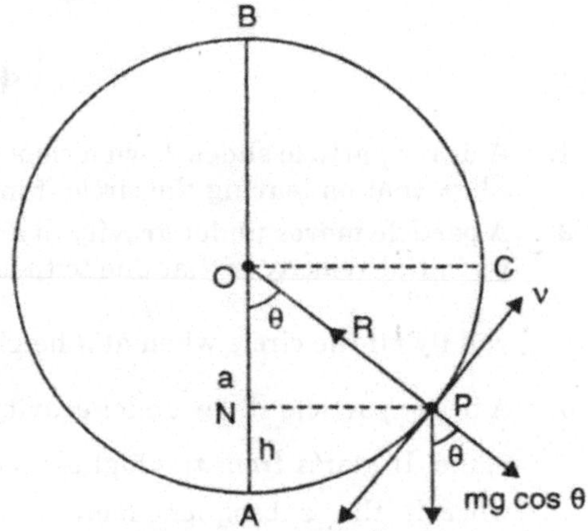

Fig. 6

At $\theta = 0, v = u \Rightarrow c = u^2 - 2ag$

Hence, $v^2 = u^2 - 2ag(1-\cos\theta)$ ...(3)

$$AN = a - a\cos\theta$$

If h is the height of P above A, then

$$h = a - a\cos\theta$$

$$\therefore \quad v^2 = u^2 - 2gh \qquad ...(4)$$

From (2), $R = \dfrac{mv^2}{a} + mg\cos\theta$

$$= \frac{m}{a}\{u^2 - 2gh\} + mg\left\{\frac{a-h}{a}\right\}$$

$$R = \frac{m}{a}\{u^2 + ag - 3gh\} \qquad ...(5)$$

Let the velocity vanishes at $h = h_1$ and R vanishes at $h = h_2$, then from (4) and (5)

$$h_1 = \frac{u^2}{2g} \quad \text{and} \quad h_2 = \frac{u^2 + ag}{3g}$$

Now three cases arise :

(*i*) Velocity vanishes earlier than the reaction *i.e.*,

$$h_1 < h_2 \Rightarrow \frac{u^2}{2g} < \frac{u^2 + ag}{3g} \Rightarrow u^2 < 2ag$$

If the particle is projected with velocity $\sqrt{2ag}$, it will rise upto horizontal diameter and since reaction does not vanish, the particle remain in contact with the circle and it retraces the path and oscillate.

(*ii*) If R vanishes at the highest point, then $h_2 = 2a$

$$\frac{u^2 + ag}{3g} = 2a \Rightarrow u^2 = 5ag$$

If the particle is projected with velocity $u \geq \sqrt{5ag}$ reaction and velocity does not vanish upto the highest point of the circle, the particle does not leave contact from the circle and trace complete circle.

(*iii*) If R vanishes before the velocity *i.e.*, $\sqrt{2ag} \leq u \leq \sqrt{5ag}$ the particle leave contact from the circle and moves further in a parabolic path.

Motion of a particle tied to one end of a light inextensible string whose other end is attached to a fixed point. It is projected with velocity u from the vertical position.

This motion is similar to the motion in vertical circle Art. 3. Reaction R will be replaced by tension T.

In case the particle moves in a thin smooth tube, the particle can not leave the tube, the cases are as follows :

(*i*) If $u^2 < 2ag$, the particle will oscillate about A in lower half of the tube.

(*ii*) If $u^2 > 5ag$, the particle makes complete revolution on the outer side of the circle.

(*iii*) If $4ag < u^2 < 5ag$, the particle completes revolution partly outside and partly on the inner side where $u^2 = 4ag$ or $u = 2\sqrt{ag}$ is the velocity required to reach the particle upto highest point of the tube.

(*iv*) If $u^2 < 4ag$, the particle rises above the horizontal diameter, returns and oscillates.

Example 1. *Find the velocity with which a particle must be projected along the interior of a smooth vertical hoop of radius a from the lowest point so that it may leave the hoop at an angular distance of 30° from the vertical. Show that it will strike the hoop again at an extremity of the horizontal diameter.*

Sol. Let u be the velocity of projection of the particle from lowest point A. When the particle travelled the angular distance θ, velocity and reaction are

$$v^2 = u^2 + 2ag\cos\theta - 2ag \quad ...(1)$$

$$R = \frac{m}{a}\{u^2 + 3ag\cos\theta - 2ag\} \quad ...(2)$$

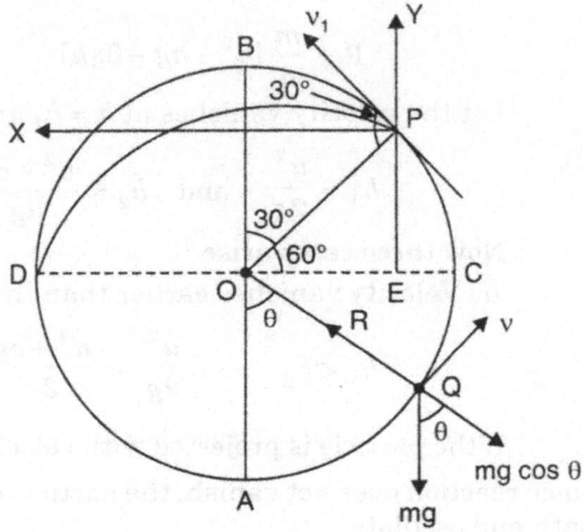

Fig. 7

Since particle leaves the circle when angular distance from the vertical is 30° *i.e.*, ∠BOP = 30°

$$\angle AOP = 180° - 30° = 150°$$

$\therefore$ R = 0 when θ = 150°

Hence, $u^2 + 3ag\cos 150° - 2ag = 0$

$$\therefore \quad u^2 = 2ag + \frac{3\sqrt{3}}{2}ag$$

$$= \frac{1}{2}(3\sqrt{3} + 4)\,ag$$

$\therefore$ Velocity of projection must be $\sqrt{\frac{1}{2}(3\sqrt{3}+4)\,ag}$

Particle starts moving in a parabolic path at P.

Let velocity at P be v_1

$$v_1^2 = u^2 + 2ag\cos\theta - 2ag = \frac{1}{2}[3\sqrt{3}+4]\,ag + 2ag\cos 150° - 2ag$$

$$v_1^2 = \frac{1}{2}\sqrt{3}\,ag \quad ...(3)$$

Taking P as origin and PX and PY as x and y-axes equation of the parabolic path is

$$\left(y = x\tan\theta - \frac{gx^2}{2v^2\cos^2\theta}\right)$$

$$y = x\tan 30° - \frac{1}{2}\frac{gx^2}{\frac{\sqrt{3}\,ag\cos^2 30°}{2}}$$

$$y = \frac{x}{\sqrt{3}} - \frac{4x^2}{3\sqrt{3}a}$$

or $$3\sqrt{3}ay = 3ax - 4x^2 \qquad ...(4)$$

$\because$ The particle meets the horizontal diameter CD at D.

$\therefore$ Co-ordinates of D must lie on the equation of the path.

Let co-ordinate of D be (x, y)

$$x = \text{ED} = \text{EO} + \text{OD} = a\cos 60 + a = \frac{3a}{2}$$

$$y = -\text{PE} = -a\sin 60 = \frac{-a\sqrt{3}}{2}$$

$\therefore$ $\left(\frac{3a}{2}, \frac{-a\sqrt{3}}{2}\right)$ lies on (4)

$$-3\sqrt{3}\,a\,.\frac{a\sqrt{3}}{2} = 3a\,.\frac{3a}{2} - 4\,.\frac{9a^2}{4}$$

$$\frac{-9a^2}{2} = \frac{9a^2}{2} - 9a^2 = \frac{-9a^2}{2}$$

Hence, the particle passes through D.

Example 2. *Show that the greatest angle through which a person can swing on the rope, which can support twice the person's weight is 120°.*

If the ropes are strong enough and he can swing through 180°, v is the speed at any point ; prove that the tension in the rope at that point is $\frac{3mv^2}{2l}$, *when m is the mass of the person and l the length of the rope.*

Sol. We know that velocity at a point where angular distance from the lowest point is θ is

$$v^2 = u^2 + 2gl\cos\theta - 2gl \qquad ...(1)$$

and $$\text{T} = \frac{m}{l}\{u^2 + 3gl\cos\theta - 2gl\} \qquad ...(2)$$

T is maximum at the lowest position where θ = 0,

$$\text{T} = 2mg \qquad \text{(given)}$$

$$\therefore \quad 2mg = \frac{m}{l}\{u^2 + 3gl\cos 0 - 2gl\}$$

$$\therefore \quad u^2 = gl$$

When $v = 0$, $$0 = u^2 + 2gl\cos\theta - 2gl$$

or $$\cos\theta = \frac{2gl - gl}{2gl} = \frac{1}{2} \Rightarrow \theta = 60°$$

$\therefore$ Man oscillates both sides at angle 60°.

$\therefore$ Man oscillates through 120°.

When he oscillates through 180°, the velocity is zero at $\theta = 90°$

$$\therefore \quad 0 = u^2 + 2gl \cos 90° - 2gl$$

Hence, $u^2 = 2gl$ ...(3)

$\therefore$ Tension at any point where velocity is v is given by

$$T = \frac{m}{l}\{u^2 + 3gl \cos\theta - 2gl\} \quad ...(4)$$

$$v^2 = u^2 + 2gl \cos\theta - 2gl$$

$$= 2gl + 2gl \cos\theta - 2gl = 2gl \cos\theta$$

$$\therefore \quad \cos\theta = \frac{v^2}{2gl} \quad ...(5)$$

Putting in (5)

$$T = \frac{m}{l}\left\{2gl + 3gl \cdot \frac{v^2}{2gl} - 2gl\right\}$$

$$\therefore \quad T = \frac{3mv^2}{2l}.$$

Example 3. *A heavy particle of mass m makes complete revolution in a smooth circular tube fixed in a vertical plane. Its greatest speed is n times the least speed. Prove that pressure on the tube when the particle is moving vertically is* $\frac{2mg(n^2+1)}{n^2-1}$ *and the speed is* $\sqrt{\frac{1}{2}(n^2+1)}$ *times its speed at the highest point.*

Sol. Let the least speed of the particle be u.

Greatest speed be nu

$$u^2 = (nu)^2 - 2g \cdot 2a$$

(a is radius of the tube)

$$\text{or} \quad (n^2 - 1)u^2 = 4ag \quad ...(1)$$

Velocity at the horizontal position when particle is moving vertically is

$$v^2 = (nu)^2 - 2ga$$

$$= \frac{1}{2}[2n^2u^2 - 4ag]$$

$$= \frac{1}{2}[2n^2u^2 - (n^2-1)u^2]$$

using (1)

$$= \frac{1}{2}[n^2 + 1]u^2$$

$$\therefore \quad v = u\sqrt{\frac{1}{2}(n^2+1)}$$

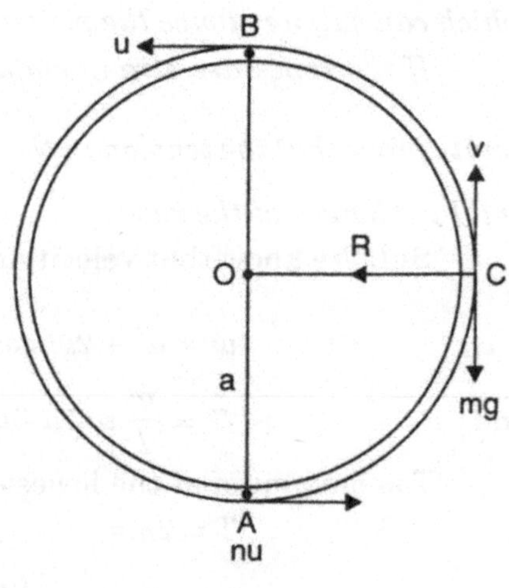

Fig. 8

Thus, velocity is $\sqrt{\frac{1}{2}(n^2+1)}$ times the least speed.

Reaction R at this position is

$$R = \frac{mv^2}{a} = \frac{m}{a}\frac{1}{2}(n^2+1)u^2$$

$$= \frac{m}{2a}(u^2+1) \cdot \frac{4ag}{n^2-1} = 2mg\,\frac{(n^2+1)}{n^2-1}.$$

EXERCISE 2

1. A heavy particle hangs from a point O, by string of length a. It is projected horizontally with a velocity $\sqrt{(2+\sqrt{3})\,ag}$. Show that the string becomes stack when it has described an angle $\cos^{-1}\left(-\frac{1}{\sqrt{3}}\right)$.
2. A particle is projected along the inner side of a smooth vertical circle of radius a from the lowest point with velocity u. Show that when the particle leaves the circle it describes a parabola whose latus rectum is $2(u^2 - 2ga)^3/(27g^3a^2)$.
3. A heavy particle of weight W attached to a fixed point by a light inextensible string describes a circle in a vertical plane. The least and highest tension in the string are mW and nW respectively show that $n = m + 6$.
4. A heavy particle attached to a fixed peg O by a string of length l. It is lifted up with the string horizontal and then let go. Prove that when the string makes an angle θ with the horizontal, the resultant acceleration is $g\sqrt{1+3\sin^2\theta}$.
5. A particle is projected along the inside of a smooth vertical circle of radius a from the lowest point. Show that the velocity of projection required in order that after leaving the circle, the particle may pass through the centre is $\sqrt{\frac{1}{2}ag\,(1+\sqrt{3})}$.
6. A particle is projected along the inner surface of a smooth vertical circle of radius r, its velocity at the lowest point being $\frac{1}{5}\sqrt{95gr}$. Show that it will leave the circle at angular distance $\cos^{-1}\frac{3}{5}$ from the highest point and that its velocity then is $\frac{1}{5}\sqrt{15gr}$.
7. A particle makes complete revolutions in a vertical circle. If ω_1 and ω_2 be the greatest and least angular velocities and R_1 and R_2 be the greatest and least reactions ; prove that when the particle projected from the lowest point of the circle makes an angle θ at the centre its angular velocity is $\sqrt{\omega_1^2\cos^2\theta/2 + \omega_2^2\sin^2\theta/2}$ and that the reaction is $R_1\cos^2\theta/2 + R_2\sin^2\theta/2$.
8. A heavy particle hanging vertically from a point by a light inextensible string of length l is started so as to make complete revolution in a vertical plane. Prove that sum of the tension at the ends of any diameter is constant.
9. A particle of mass m is tied to one end of a string and is describing a circle in a vertical plane. If the maximum speed is twice the minimum speed ; prove that when the string is horizontal its tension is $\frac{10}{3}mg$.

10. A tumbler of water is suspended by a string of length l and made to revolve along a vertical circle of radius equal to length of string. Show that the least velocity for water not to spill out of the tumbler at the highest point is $\sqrt{gl}$.

11. A particle hanging from a fixed point by a light inextensible string of length a is projected horizontally with velocity u. Show that during the circular motion, the tension of the string at any time is proportional to the depth of the particle at that moment below a certain horizontal line.

4. Motion on Other Smooth Curves

A particle slides down the arc of a smooth cycloid* whose axis is vertical and vertex downwards. Discuss its motion.

Let the cycloid (smooth) with vertex O be placed on the horizontal surface. Let the particle be at P at certain time t. Let $\overset{\frown}{OP} = s$.

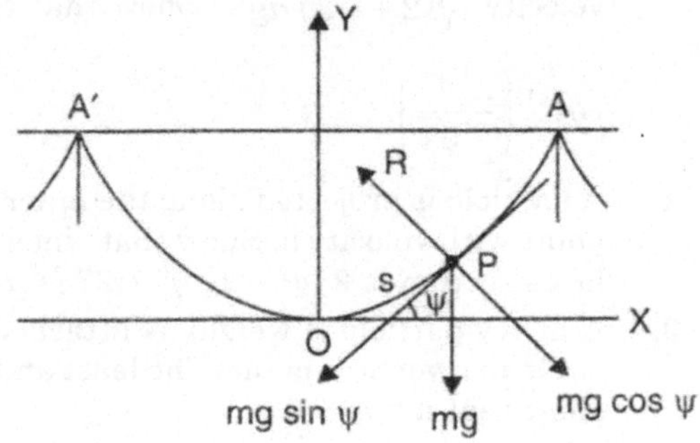

Fig. 9

Forces on the particle at P are :

(*i*) Weight mg acting vertically downwards.

(*ii*) Normal reaction R acting along the normal.

∴ Equations of motion in tangential and normal direction are

$$m\frac{d^2s}{dt^2} = -mg\sin\psi \quad \text{...(1)}$$

$$m\frac{v^2}{\rho} = R - mg\cos\psi \quad \text{...(2)}$$

where ρ is the radius of curvature.

Equation of cycloid $\quad s = 4a\sin\psi \quad$ (Intrinsic equation)

$$\rho = \frac{ds}{d\psi} = 4a\cos\psi$$

Putting in equation (1), we get

$$\frac{d^2s}{dt^2} = -\left(\frac{g}{4a}\right)s \quad \text{...(3)}$$

This is the equation of S.H.M. $\frac{d^2s}{dt^2} = -\mu s$ with time period

$$\therefore \quad T = \frac{2\pi}{\sqrt{u}} = \frac{2\pi}{\sqrt{g/4a}} = 4\pi\sqrt{\frac{a}{g}}$$

Thus, we conclude that from whatever point the particle starts the motion it is simple harmonic and period remains the same.

---

*Cycloid is locus of a point on the circumference of a circle which rolls along a fixed straight line (O is the vertex of the cycloid. AA' is base of the cycloid). $x = a(\theta + \sin\theta)$, $y = a(1 - \cos\theta)$ is Parametric equation of the cycloid. $s = 4a\sin\psi$ is intrinsic equation of the cycloid. $s^2 = 8ay$ is a relation in s and y. Cycloid has cusps at A and A', Arc length AA' = $8a$. a is the radius of the generating circle.

Further multiplying both sides of equation (3) by $2\dfrac{ds}{dt}$

$$2\frac{ds}{dt}\frac{d^2s}{dt^2} = -\frac{g}{4a}.s.2\frac{ds}{dt}$$

Integrating, we get

$$\left(\frac{ds}{dt}\right)^2 = -\frac{g}{4a}s^2 + C$$

If particle starts from the cusp at A with velocity 0,

we have, $$0 = -\frac{g}{4a}(4a)^2 + C \Rightarrow C = 4ag$$

Putting the value of C, we get

$$v^2 = -\frac{g}{4a}s^2 + 4ag$$

or $$v^2 = \frac{g}{4a}[16a^2 - s^2]$$

$$= \frac{g}{4a}[16a^2 - (4a\sin\psi)^2] = 4ag\cos^2\psi$$

$$v^2 = 4ag\cos^2\psi \quad \text{...(4)}$$

Putting the value of ρ and v^2 in equation (2)

$$R = \frac{mv^2}{\rho} + mg\cos\psi = \frac{m.4ag\cos^2\psi}{4a\cos\psi} + mg\cos\psi$$

$$R = 2\,mg\cos\psi \quad \text{...(5)}$$

Equations (4) and (5) give velocity and reaction at any point P of the cycloid.

Example 1. *A smooth parabolic tube is placed, vertex downwards in a vertical plane ; A particle of mass m slides down the tube from rest under gravity from a position of height h above the vertex. Prove that at any position the reaction of the tube is* $\dfrac{2mg(h+a)}{\rho}$ *where 4a is latus rectum and ρ is radius of curvature.*

Sol. Let the particle be at rest at Q on the parabola $x^2 = 4ay$. Height of Q above the vertex is QN = h (given). Let the particle be at P at certain time t and moving with velocity v. Let the co-ordinates of P be (x, y).

By the principle of work and energy

Change in K.E. = Work done by the forces

$$\frac{1}{2}mv^2 - 0 = mg\,(QL)$$

$$= mg\,(h - y)$$

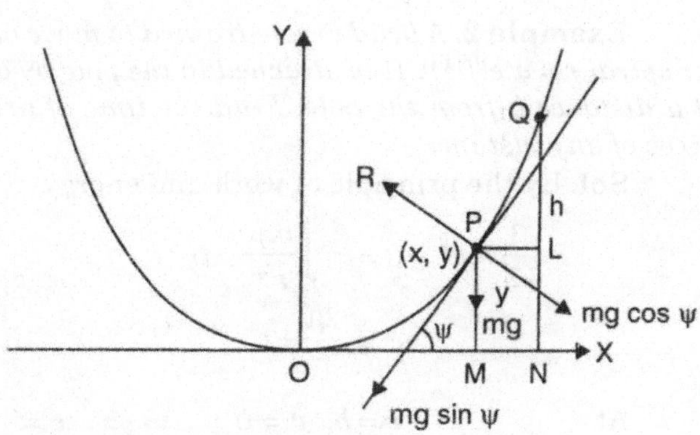

Fig. 10

$\therefore \quad v^2 = 2g\,(h - y)$ ...(1)

Equation of motion along the normal is

$$\frac{mv^2}{\rho} = R - mg\cos\psi$$

or

$$R = mg\cos\psi + \frac{mv^2}{\rho} \quad ...(2)$$

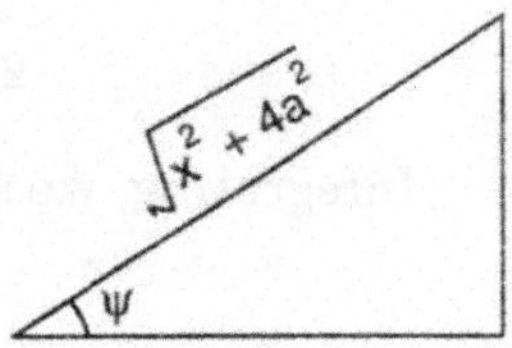

Fig. 11

$\because \quad x^2 = 4ay \Rightarrow \dfrac{dy}{dx} = \dfrac{x}{2a} = \tan\psi$

$\therefore \quad \cos\psi = \dfrac{2a}{\sqrt{x^2+4a^2}} = \dfrac{2a}{\sqrt{4ay+4a^2}}$

$\therefore \quad \cos\psi = \sqrt{\dfrac{a}{a+y}}$ ...(3)

$$\rho = \frac{\left\{1+\left(\dfrac{dy}{dx}\right)^2\right\}^{3/2}}{\dfrac{d^2y}{dx^2}} = \frac{\left(1+\dfrac{x^2}{4a^2}\right)^{3/2}}{\dfrac{1}{2a}} = \left(1+\frac{y}{a}\right)^{3/2}.\,2a$$

Putting the value of v^2, ρ and $\cos\psi$ in equation (2), we get

$$R = mg\sqrt{\frac{a}{a+y}} + m\,2g\,(h-y)/\rho$$

$$= \frac{2mg}{\rho}\left[\frac{1}{2}\rho\sqrt{\frac{a}{a+y}} + h - y\right]$$

$$= \frac{2\,mg}{\rho}\left[\frac{1}{2}2a\left(\frac{y+a}{a}\right)^{3/2}\sqrt{\frac{a}{a+y}} + h - y\right]$$

$$= \frac{2mg}{\rho}[a+y+h-y] = \frac{2\,mg\,(h+a)}{\rho}.$$

Example 2. *A bead is constrained to move on a smooth wire in the form of an equiangular spiral $r = a\,e^{\theta\cot\alpha}$. It is attached to the pole by a force = mμ/(distance)^2 and starts from rest at a distance b from the pole. Find the time of arriving at the pole. Also find reaction of the curve of any instance.*

Sol. By the principle of work and energy

$$\frac{1}{2}mv^2 - 0 = -\int\frac{m\mu}{r^2}\,.\,dr$$

$$v^2 = \frac{2\mu}{r} + c$$

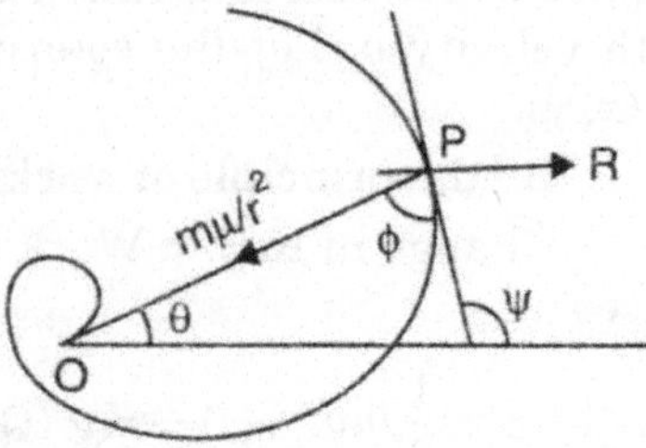

Fig. 12

At $\quad r = b,\ v = 0 \Rightarrow c = \dfrac{-2\mu}{b}$

$$\therefore \qquad v^2 = 2\mu\left(\frac{1}{r} - \frac{1}{b}\right)$$

or
$$v = \frac{ds}{dt} = -\sqrt{2\mu\left(\frac{1}{r} - \frac{1}{b}\right)}$$

or
$$\frac{ds}{dr} \cdot \frac{dr}{dt} = -\sqrt{2\mu\left(\frac{1}{r} - \frac{1}{b}\right)} \qquad \left(\frac{dr}{ds} = \cos\phi\right) \qquad \text{...(1)}$$

In case of equiangular spiral $\phi = \alpha$*

$$\therefore \qquad \frac{dr}{dt} = -\frac{dr}{ds}\sqrt{2\mu\left(\frac{1}{r} - \frac{1}{b}\right)}$$

$$\frac{dr}{dt} = -\cos\alpha\sqrt{2\mu\left(\frac{b-r}{br}\right)}$$

$$-\sec\alpha \int_b^0 \sqrt{\frac{r}{b-r}}\, dr = \sqrt{\frac{2u}{b}} \int dt$$

$$\therefore \qquad t = \sec\alpha\sqrt{\frac{b}{2\mu}} \int_0^b \sqrt{\frac{r}{b-r}}\, dr \qquad r = b\sin^2\theta,$$

$$dr = 2b\sin\theta\cos\theta\, d\theta$$

$$= \sec\alpha\sqrt{\frac{b}{2\mu}} \int_0^{\pi/2} \frac{\sin\theta}{\cos\theta} \cdot 2b\sin\theta\cos\theta\, d\theta$$

$$= 2b\sqrt{\frac{b}{2\mu}} \cdot \frac{1}{2} \cdot \frac{\pi}{2} \cdot \sec\alpha$$

$$t = \frac{\pi}{2}\sqrt{\frac{b^3}{2\mu}} \sec\alpha$$

Equation of motion in normal direction is

$$\frac{mv^2}{\rho} = \frac{m\mu}{r^2}\sin\phi - R$$

$$\therefore \qquad R = \frac{m\mu}{r^2}\sin\phi - \frac{mv^2}{\rho} \qquad \text{...(2)}$$

$$\because \qquad p = r\sin\phi = r\sin\alpha$$

$$1 = \frac{dr}{dp}\sin\alpha$$

$$\frac{dr}{dp} = \operatorname{cosec}\alpha \qquad \text{or} \qquad r\frac{dr}{dp} = r\operatorname{cosec}\alpha \qquad \left(\because \rho = r\frac{dr}{dp}\right)$$

$$\therefore \qquad \rho = r\operatorname{cosec}\alpha$$

---

* $r = a\,e^{\theta\cot\alpha} \Rightarrow \log r = \log a + \theta\cot\alpha$

$\frac{1}{r}\frac{dr}{d\theta} = \cot\alpha \Rightarrow \frac{rd\theta}{dr} = \tan\alpha$

$\Rightarrow \tan\phi = \tan\alpha \Rightarrow \phi = \alpha$

Putting the values of ρ, v^2 in equation (2)

$$R = \frac{m\mu}{r^2}\sin\alpha - \frac{m\,2\mu}{r\,\text{cosec}\,\alpha}\left(\frac{1}{r}+\frac{1}{b}\right)$$

$$= \frac{m\mu}{r}\sin\alpha\left[\frac{1}{r}-\frac{2}{r}+\frac{2}{b}\right] = \frac{m\mu}{r}\sin\alpha\left[\frac{2}{b}-\frac{1}{r}\right]$$

$$\therefore \quad R = \frac{m\mu}{r}\sin\alpha\left[\frac{2}{b}-\frac{1}{r}\right].$$

Example 3. *A particle starts from rest at the cusp of a smooth cycloid whose axis is vertical and vertex downwards. Prove that when it has fallen through half the distance measured along the arc to the vertex, two third of the time of descent will have elapsed.*

Sol. The equations of motion of the particle along the tangent and normal are

$$m\frac{d^2s}{dt^2} = -\,mg\sin\psi \qquad \text{...(1)}$$

$$\frac{mv^2}{\rho} = R - mg\cos\psi \qquad (\because \; s = 4a\sin\psi) \qquad \text{...(2)}$$

$$\therefore \quad \frac{d^2s}{dt^2} = \frac{-gs}{4a}$$

Multiplying both sides by $2\frac{ds}{dt}$ and integrating, we get

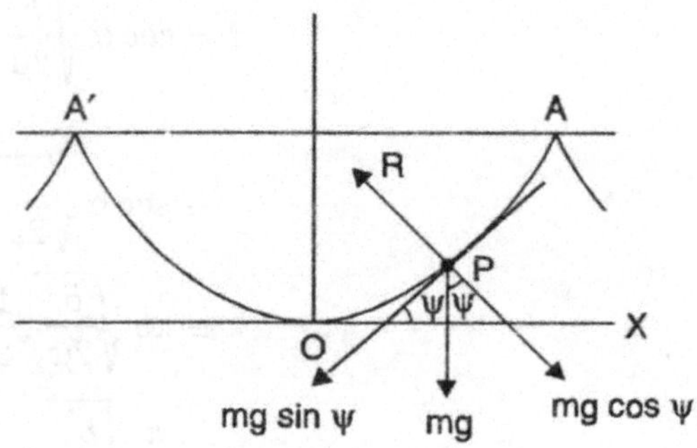

Fig. 13

$$\left(\frac{ds}{dt}\right)^2 = \frac{-g}{4a}s^2 + c$$

At the cusp A,

$$s = 4a, \quad \frac{ds}{dt} = 0$$

$$\therefore \quad c = 4ag$$

Hence, $$\left(\frac{ds}{dt}\right)^2 = \frac{-g}{4a}s^2 + 4ag = \frac{g}{4a}[16a^2 - s^2]$$

or $$\frac{ds}{dt} = -\sqrt{\frac{g}{4a}(16a^2 - s^2)}$$

With the increase of t, s decreases $\therefore$ $\frac{ds}{dt}$ is (–) ve.

Let t_1 be the time of descent from $s = 4a$ to $s = 2a$.

$$\therefore \quad t_1 = -\int_{4a}^{2a}\sqrt{\frac{4a}{g}}\frac{ds}{\sqrt{16a^2 - s^2}}$$

$$= 2\sqrt{\frac{a}{g}}\left[\cos^{-1}\frac{s}{4a}\right]_{4a}^{2a} = 2\sqrt{\frac{a}{g}}\left[\frac{\pi}{3}-0\right] = \frac{2\pi}{3}\sqrt{\frac{a}{g}}$$

Let t be the time from A to O *i.e.*, $s = 4a$ to $s = 0$

$$t = -\int_{4a}^{0} 2\sqrt{\frac{a}{g}} \frac{ds}{\sqrt{16a^2 - s^2}} = 2\sqrt{\frac{a}{g}}\left[\cos^{-1}\frac{s}{4a}\right]_{4a}^{0} = 2\sqrt{\frac{a}{g}}\frac{\pi}{2} = \pi\sqrt{\frac{a}{g}}$$

Hence, $t_1 = \frac{2}{3}t$.

Example 4. *From the lowest point of a smooth hollow cylinder whose cross-section is one half of the lemniscate of Bernaulli $r^2 = a^2 \cos 2\theta$ with axis vertical and node downwards, a particle is projected with velocity V along the inner surface in the plane of cross-section ; show that it will make complete oscillation if $3V^2 > 7ag$.*

Sol. Let the initial line be taken vertical and particle be projected from O with velocity V. Let at any time t the particle be at P and moving with velocity v.

Equations of motion along the tangent and normal are

$$mv\frac{dv}{ds} = -mg\cos\psi \qquad ...(1)$$

$$\frac{mv^2}{\rho} = R + mg\sin\psi \qquad ...(2)$$

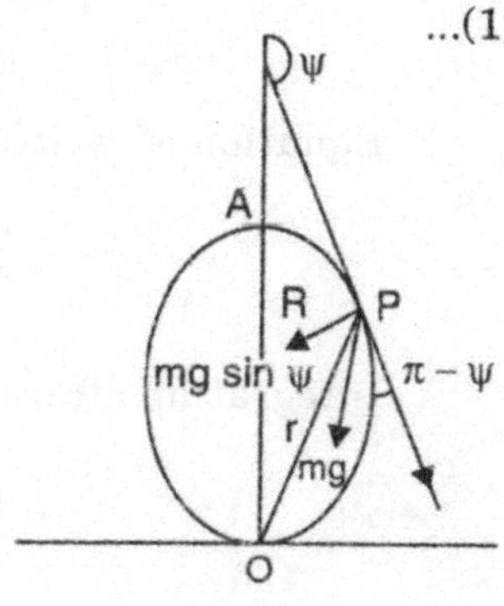

Fig. 14

$$\therefore \quad v\frac{dv}{ds} = -g\frac{dx}{ds} \qquad \left(\cos\psi = \frac{dx}{ds}\right)$$

Integrating $v^2 = -2gx + c$

Initially at O, $x = 0, v = V \Rightarrow c = v^2$

Hence, $v^2 = V^2 - 2gx$

$$v^2 = V^2 - 2gr\cos\theta \qquad ...(3) \qquad (x = r\cos\theta)$$

To find ρ

$$r^2 = a^2\cos 2\theta$$

Taking logarithm on both sides

$$2\log r = \log a^2 + \log\cos 2\theta$$

Differentiating,

$$\frac{2}{r}\frac{dr}{d\theta} = \frac{-2\sin 2\theta}{\cos 2\theta} \Rightarrow r\frac{d\theta}{dr} = -\cot 2\theta,$$

$$\tan\phi = \tan\left(\frac{\pi}{2} + 2\theta\right)$$

$$\therefore \quad \phi = \frac{\pi}{2} + 2\theta \quad \text{and} \quad \psi = \theta + \phi = \frac{\pi}{2} + 3\theta$$

$$p = r\sin\phi = r\cos 2\theta = r\cdot\frac{r^2}{a^2}$$

$$r^3 = a^2 p$$

$$3r^2\frac{dr}{dp} = a^2 \Rightarrow \rho = r\frac{dr}{dp} = \frac{a^2}{3r} \qquad ...(4)$$

From (2), (3) and (4)

$$R = \frac{mv^2}{\rho} - mg \sin \psi$$

$$R = m\,\frac{3r}{a^2}\{V^2 - 2gr \cos \theta\} - mg \sin\left(\frac{\pi}{2} + 3\theta\right)$$

$$= m\left[\frac{3r}{a^2}(V^2 - 2gr \cos \theta) - g \cos 3\theta\right]$$

At the highest point A, $r = a$, $\theta = 0$, $R \geq 0 \Rightarrow 3V^2 \geq 7ga$.

Example 5. *A particle slides down the arc of a smooth cycloid whose axis is vertical and vertex lowest, prove that the time occupied in falling down the first half of the vertical height is equal to the time of falling down the second half.* **(M.D.U. 2005)**

Sol. Refer to figure 9.

Equation of motion along the tangent is

$$m\frac{d^2s}{dt^2} = -\,mg \sin \psi \qquad ...(1)$$

Equation of cycloid is

$$s = 4a \sin \psi \qquad ...(2)$$

$$\therefore \quad \frac{d^2s}{dt^2} = -\frac{g}{4a}s$$

Integrating after multiplying by $\frac{2ds}{dt}$, we get

$$\left(\frac{ds}{dt}\right)^2 = -\frac{g}{4a}s^2 + c$$

At $\quad s = 4a, \frac{ds}{dt} = 0 \quad \therefore \quad c = \frac{g}{4a}(4a)^2$

$$\therefore \quad v^2 = \frac{g}{4a}(16a^2 - s^2)$$

$$\frac{ds}{dt} = -\sqrt{\frac{g}{4a}}\sqrt{16a^2 - s^2}$$

or

$$dt = -\sqrt{\frac{4a}{g}}\frac{ds}{\sqrt{(4a)^2 - s^2}}$$

Integrating, $\quad t = \sqrt{\frac{4a}{g}}\cos^{-1}\frac{s}{4a} + c_1$

At $t = 0$, $s = 4a$ $\quad c_1 = 0$

$$\therefore \quad t = 2\sqrt{\frac{a}{g}}\cos^{-1}\left(\frac{s}{4a}\right) \qquad \because \; s^2 = 8ay$$

$$= 2\sqrt{\frac{a}{g}}\cos^{-1}\left(\frac{2\sqrt{2ay}}{4a}\right)$$

If t_1 is time from cusp to half the vertical height

i.e., $\quad y = 2a$ to $y = a$

$$t_1 = 2\sqrt{\frac{a}{g}}\left[\cos^{-1}\frac{\sqrt{2ay}}{2a}\right]_{2a}^{a} = 2\sqrt{\frac{a}{g}}\left[\cos^{-1}\frac{1}{\sqrt{2}} - \cos^{-1}1\right]$$

$$= 2\frac{\pi}{4}\sqrt{\frac{a}{g}} = \frac{\pi}{2}\sqrt{\frac{a}{g}}$$

If time to cover second half is t_2, then

$$t_2 = 2\sqrt{\frac{a}{g}}\left[\cos^{-1}\sqrt{\frac{y}{2a}}\right]_a^0 = 2\sqrt{\frac{a}{g}}\left[\frac{\pi}{2} - \frac{\pi}{4}\right] = \frac{\pi}{2}\sqrt{\frac{a}{g}}$$

$\therefore$ $t_1 = t_2$.

EXERCISE 3

1. A particle starts sliding down the arc of a smooth cycloid from the cusp, show that the vertical velocity is maximum when it has described half the vertical height.
2. A heavy particle slides down a smooth cycloid starting from rest at the cusp, the axis being vertical and vertex downwards. Prove that the magnitude of acceleration at every point of the path is g and the pressure at the vertex is equal to twice the weight of the particle.
3. A particle is projected with velocity V from the cusp of a smooth cycloid vertex downwards ; show that the time of reaching the vertex is $2\sqrt{a/g}\,\tan^{-1}\left(\frac{\sqrt{4ag}}{V}\right)$.
4. A particle is moving on a smooth curve under gravity and its velocity varies as the arcual distance from the highest point. Prove that the curve is a cycloid. Also show that the particle will leave the curve when moving in a direction making an angle 45° with the vertical.
5. A particle falls from rest from the vertex of an inverted catenary. Prove that the particle will leave the curve when the path described is $\sqrt{3}$ times the vertical distance through which it had fallen.
6. A cycloid is placed with its axis vertical and vertex upwards and a particle is projected up the concave side of the curve with velocity $\sqrt{2gh}$. Prove that when the particle leaves the curve the parabola described after leaving is of latus rectum $h^2/2a$, where a is radius of generating circle of the cycloid.
7. A particle is projected from the lowest point an ellipse placed with major axis horizontal and minor axis vertical. Show that the particle will leave it at certain point on the ellipse if the velocity of projection lies between $\sqrt{2gb}$ and $\sqrt{g(a^2+4b^2)/b}$ where $2a$ and $2b$ are the major and minor axis of the ellipse.
8. Two particles are let drop from the cusp of a cycloid down the curve at an interval of time t. Prove that they will meet in time $2\pi\sqrt{\frac{a}{9}} + \frac{t}{2}$. *(K.U. 2005)*

5. Motion along Rough Plane Curves

A particle slides down a rough curve in a vertical plane under gravity, to discuss its motion.

When the particle moves along a rough curve the motion is similar as along smooth curves. The only difference is that there is an additional force μR acting along the tangent in the direction opposite to the motion of the particle. Equations of motion in this case are as follows :

$$mv\frac{dv}{ds} = mg\sin\psi - \mu R \quad ...(1)$$

$$\frac{mv^2}{\rho} = mg\cos\psi - R \quad ...(2)$$

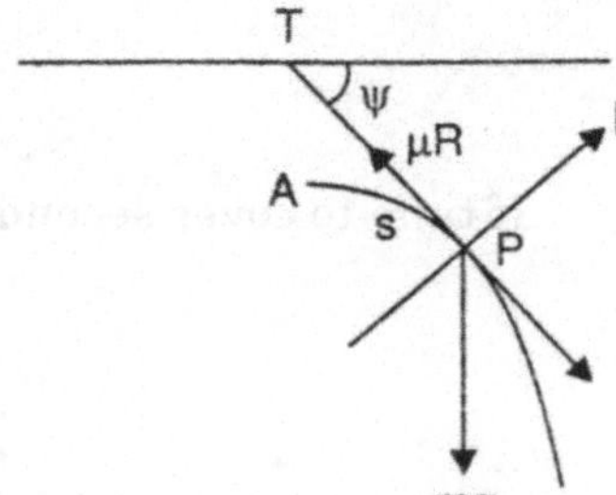

Fig. 15

Putting the value of R from (2) in (1)

$$mv\frac{dv}{ds} = mg\sin\psi - \mu\left[mg\cos\psi - \frac{mv^2}{\rho}\right]$$

or
$$\frac{vdv}{ds} = g\sin\psi - \mu g\cos\psi + \frac{\mu v^2}{\rho}$$

$$\frac{1}{2}\frac{d}{ds}(v^2) - \frac{\mu}{\rho}v^2 = g\,[\sin\psi - \mu\cos\psi]$$

or
$$\frac{d}{d\psi}(v^2)\frac{d\psi}{ds} - \frac{2\mu}{\rho}v^2 = 2g\,[\sin\psi - \mu\cos\psi] \qquad \left(\rho = \frac{ds}{d\psi}\right)$$

$$\therefore \quad \frac{d}{d\psi}(v^2) - 2\mu v^2 = 2g\rho\,[\sin\psi - \mu\cos\psi] \quad ...(3)$$

This is a linear differential equation of the type

$$\frac{dy}{dx} + Py = Q \quad \text{in } v^2 \text{ and } \psi$$

$\therefore$ Integrating factor is $e^{\int -2\mu\, d\psi} = e^{-2\mu\psi}$

Its solution is

$$v^2 . e^{-2\mu\psi} = 2g\int \rho(\sin\psi - \mu\cos\psi) . e^{-2\mu\psi}\, d\psi + c \quad ...(4)$$

where, c is a constant of integration and can be found by using initial conditions. If the equation of the curve is given, ρ can be obtained as a function of ψ.

$\therefore$ Equation (4) gives the velocity at some point at time t.

Example 1. *A particle, under no force, is projected with velocity V in a rough tube in the form of an equiangular spiral at a distance a from the pole and towards the pole. Show that it will arrive at the pole in time* $\frac{a}{V}\left(\frac{1}{\cos\alpha - \mu\sin\alpha}\right)$ *; α being the angle of the spiral and* $\mu < \cot\alpha$*, the coefficient of friction.*

Sol. Since the particle is moving under no forces, so the only forces acting upon it are normal reaction R and frictional force μR.

$\therefore$ Equations of motion along tangential and normal directions are

$$mv\frac{dv}{ds} = \mu R \quad ...(1)$$

$$\frac{mv^2}{\rho} = R \quad ...(2)$$

Eliminating R, we get

$$v\frac{dv}{ds} = \frac{\mu v^2}{\rho}$$

or $$\frac{dv}{ds} = \frac{\mu v}{\rho}$$

or $$\frac{dv}{d\psi}\frac{d\psi}{ds} = \frac{\mu v}{\rho} \qquad \rho = \frac{ds}{d\psi}$$

$\therefore$ $$\frac{dv}{d\psi} = \mu v$$

or $$\frac{dv}{v} = \mu\, d\psi$$

Integrating, we get

$$\log v = \mu\psi + \log c \quad \text{or} \quad v = ce^{\mu\psi} \qquad ...(3)$$

Equation of equiangular spiral is $r = ae^{\theta \cot \alpha}$

$$\log r = \log a + \theta \cot \alpha$$

$$\frac{1}{r}\frac{dr}{d\theta} = \cot \alpha \qquad \Rightarrow \qquad r\frac{d\theta}{dr} = \tan \alpha$$

or $$\tan \phi = \tan \alpha \qquad \Rightarrow \qquad \phi = \alpha$$

or $$\psi = \theta + \phi = \theta + \alpha$$

Putting the value of ψ in equation (3)

$$v = ce^{\mu(\theta+\alpha)} = ce^{\mu\alpha}\left[\frac{ae^{\theta \cot \alpha}}{a}\right]^{\mu \tan \alpha}$$

$$v = ce^{\mu\alpha}\left(\frac{r}{a}\right)^{\mu \tan \alpha} \qquad ...(4)$$

Initially at $\qquad r = a,\ v = -V$

$\therefore$ $$-V = ce^{\mu\alpha}$$

Hence, $$v = -V\left(\frac{r}{a}\right)^{\mu \tan \alpha} \qquad ...(5)$$

$$\frac{ds}{dt} = -V\left(\frac{r}{a}\right)^{\mu \tan \alpha}$$

or $$\frac{ds}{dr}.\frac{dr}{dt} = -V\left(\frac{r}{a}\right)^{\mu \tan \alpha} \qquad \left(\frac{dr}{ds} = \cos \phi = \cos \alpha\right)$$

$$\sec \alpha \frac{dr}{dt} = -V\left(\frac{r}{a}\right)^{\mu \tan \alpha}$$

or $$dt = -\frac{a^{\mu \tan \alpha}\ r^{-\mu \tan \alpha}}{V \cos \alpha}\, dr$$

Integrating both sides

$\therefore$ $$t = -\int_a^0 \frac{a^{\mu \tan \alpha}}{V \cos \alpha}.r^{-\mu \tan \alpha}\, dr$$

$$= -\frac{a^{\mu \tan \alpha}}{V \cos \alpha}\left[\frac{r^{-\mu \tan \alpha + 1}}{-\mu \tan \alpha + 1}\right]_a^0$$

$$= \frac{a^{\mu \tan \alpha}}{V \cos \alpha} \cdot \frac{a \cdot a^{-\mu \tan \alpha}}{1 - \mu \tan \alpha} = \frac{a}{V(\cos \alpha - \mu \sin \alpha)}.$$

Example 2. *A particle is projected horizontally with velocity V along the inside of a rough vertical circle from the lowest point. Prove that if it completes the circle it will return to the lowest point with a velocity given by*

$$v^2 = V^2 e^{-4\pi\mu} + \frac{2ga}{1 + 4\mu^2}(1 - 2\mu^2)(1 - e^{-4\mu\pi}).$$

Sol. Let the particle be projected from A, the lowest point of the circle with velocity V. Let it be at P at any time t moving up with velocity v. Let $\angle AOP = \theta$.

Equations of motion along tangent and normal are

$$mv\frac{dv}{ds} = -mg \sin \theta - \mu R \qquad ...(1)$$

$$\frac{mv^2}{a} = R - mg \cos \theta \qquad ...(2)$$

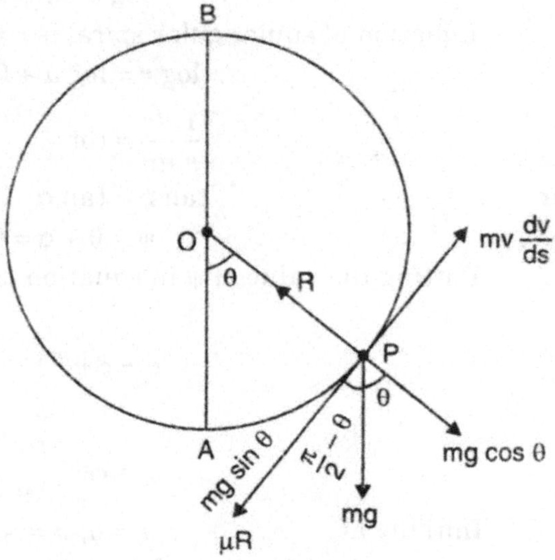

Fig. 16

Putting the value of R from (2) in (1)

$$mv\frac{dv}{ds} = -mg \sin \theta - \mu\left(mg \cos \theta + \frac{mv^2}{a}\right)$$

or $$v\frac{dv}{ds} + \frac{\mu v^2}{a} = -g(\sin \theta + \mu \cos \theta)$$

$$\frac{d}{ds}\left(\frac{1}{2}v^2\right) + \frac{\mu v^2}{a} = -g(\sin \theta + \mu \cos \theta)$$

$$\frac{d}{d\theta}\left(\frac{1}{2}v^2\right) \cdot \frac{d\theta}{ds} + \mu\frac{v^2}{a} = -g(\sin \theta + \mu \cos \theta) \qquad \left(s = a\theta, \frac{ds}{d\theta} = a\right)$$

$$\therefore \quad \frac{d}{d\theta}(v^2) + 2\mu v^2 = -2ag(\sin \theta + \mu \cos \theta) \qquad ...(3)$$

Equation (3) is a linear differential equation in v^2 and θ.

Its Integrating factor is

$$e^{\int 2\mu \, d\theta} = e^{2\mu\theta}$$

Solution of equation (3) is

$$v^2 . e^{2\mu\theta} = \int -2ga(\sin \theta + \mu \cos \theta) . e^{2\mu\theta} d\theta + C$$

$$= -2ga\left[\int e^{2\mu\theta} \sin \theta \, d\theta + \mu \int e^{2\mu\theta} \cos \theta \, d\theta\right] + C$$

$$= -2ga\left[\frac{1}{1+4\mu^2}e^{2\mu\theta}(2\mu\sin\theta-\cos\theta)+\frac{\mu e^{2\mu\theta}}{1+4\mu^2}(2\mu\cos\theta+\sin\theta)\right]+C$$

$$= -\frac{2ga\,e^{2\mu\theta}}{1+4\mu^2}[2\mu\sin\theta-\cos\theta+2\mu^2\cos\theta+\mu\sin\theta]+C$$

$$= \frac{-2ga\,e^{2\mu\theta}}{1+4\mu^2}[3\mu\sin\theta-(1-2\mu^2)\cos\theta]+C \quad ...(4)$$

At $\theta = 0, v = V \Rightarrow C = V^2 - \dfrac{2ga(1-2\mu^2)}{1+4\mu^2}$

Putting the value of C in equation (4), we get

$$v^2 e^{2\mu\theta} = \frac{-2ga\,e^{2\mu\theta}}{1+4\mu^2}[3\mu\sin\theta-(1-2\mu^2)\cos\theta] + V^2 - \frac{2ga(1-2\mu^2)}{1+4\mu^2} \quad ...(5)$$

This expression gives the velocity of the point when it is moving up.

When particle completes the circle and returns back to point A, it has described an angle $\theta = 2\pi$.

$\therefore$ Velocity on the return is given as

$$v^2 e^{4\mu\pi} = \frac{-2ga\,e^{4\mu\pi}}{1+4\mu^2}[0-(1-2\mu^2)] + V^2 - \frac{2ga}{1+4\mu^2}(1-2\mu^2)$$

or

$$v^2 = e^{-4\mu\pi}.V^2 + \frac{2ga}{1+4\mu^2}(1-2\mu^2)(1-e^{-4\mu\pi}).$$

Example 3. *A rough cycloid has its plane vertical and vertex downwards. A heavy particle slides down the curve from rest at a cusp and comes to rest again at the point on the other side of the vertex where the tangent is inclined at 45° to the vertical. Show that the coefficient of friction satisfies the equation*

$$3\mu\pi + \log(1+\mu) = 2\log 2. \quad \text{(KU 2004)}$$

Sol. Particle starts sliding from rest from A. Let the particle be at P at any time t.

The equations of motion along tangent and normal are

$$mv\frac{dv}{ds} = \mu R - mg\sin\psi \quad ...(1)$$

$$\frac{mv^2}{\rho} = R - mg\cos\psi \quad ...(2)$$

Eliminating R from (1) and (2)

$$v\frac{dv}{ds} - \frac{\mu v^2}{\rho} = g(\mu\cos\psi - \sin\psi) \quad (s = 4a\sin\psi)$$

$$\frac{d}{ds}\left(\frac{1}{2}v^2\right) - \frac{\mu v^2}{\rho} = g(\mu\cos\psi - \sin\psi)$$

$$\left(\rho = \frac{ds}{d\psi}\right)$$

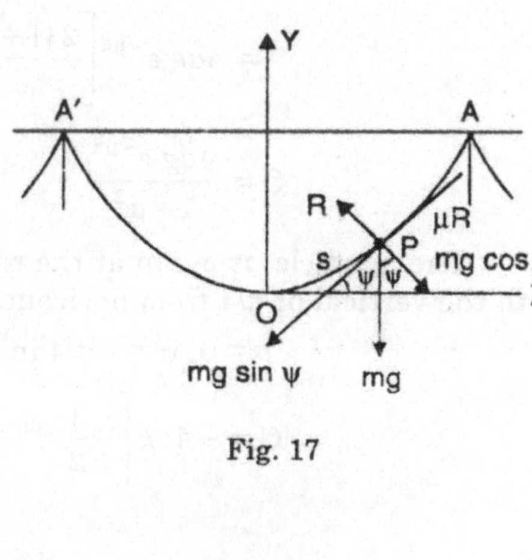

Fig. 17

$$\left(\frac{d}{d\psi}v^2\right)\cdot\frac{d\psi}{ds}-\frac{2\mu v^2}{\rho}=2g\,(\mu\cos\psi-\sin\psi)\qquad(\rho=4a\cos\psi)$$

$$\therefore\quad \frac{d}{d\psi}(v^2)-2\mu\,v^2=2g\,\rho\,(\mu\cos\psi-\sin\psi)\qquad\text{...(3)}$$

This is linear differential equation in v^2

$$\text{I.F.}=\int_e -2\mu\,d\psi=e^{-2\mu\psi}$$

Its solution is

$$v^2\,.\,e^{-2\mu\psi}=\int e^{-2\mu\psi}\,.\,2g\,.\,4a\cos\psi\,(\mu\cos\psi-\sin\psi)\,d\psi+\text{C}$$

$$v^2\,e^{-2\mu\psi}=\int 8ag\,.\,e^{-2\mu\psi}\,\{\mu\cos^2\psi-\sin\psi\cos\psi\}\,d\psi+\text{C}$$

$$=4ag\int e^{-2\mu\psi}\,\{\mu\,(1+\cos2\psi)-\sin2\psi\}\,d\psi+\text{C}$$

$$=4ag\left[\int e^{-2\mu\psi}\mu\,d\psi+\mu\int e^{-2\mu\psi}\cos2\psi\,d\psi-\int e^{-2\mu\psi}\sin w\psi\,d\psi\right]+\text{C}$$

$$=4ag\left[\frac{\mu\,.\,e^{-2\mu\psi}}{-2\mu}+\frac{e^{-2\mu\psi}}{4\,(1+\mu^2)}\mu\,\{-2\mu\cos2\psi+2\sin2\psi\}\right.$$

$$\left.-\frac{-e^{-2\mu\psi}}{4\,(1+\mu^2)}\{-2\mu\sin2\psi-2\cos2\psi\}\right]+\text{C}$$

$$=4ag\left[-\frac{1}{2}e^{-2\mu\psi}+\frac{e^{-2\mu\psi}}{4\,(1+\mu^2)}\{2(1-\mu^2)\cos2\psi+4\mu\sin2\psi\}\right]+\text{C}\qquad\text{...(4)}$$

When the particle is at the cusp A, $v=0$, $\psi=\pi/2$

$$\text{C}=-4ag\left[-\frac{1}{2}e^{-\mu\pi}+\frac{-e^{-\mu\pi}}{4\,(1+\mu^2)}\{-2\,(1-\mu^2)\}\right]$$

$$=4ag\,e^{-\mu\pi}\left[\frac{2\,(1+\mu^2)+2\,(1-\mu^2)}{4\,(1+\mu^2)}\right]$$

$$\text{C}=\frac{4ag\,e^{-\mu\pi}}{1+\mu^2}\qquad\text{...(5)}$$

The particle is again at the rest on the other side where tangent makes an angle $\pi/4$ with the vertical or $\pi/4$ from horizontal and (– ve).

$\therefore\quad v=0,\ \psi=-\pi/4$ in (4), we get

$$\text{C}=-4ag\left[-\frac{1}{2}e^{\mu\pi/2}+\frac{e^{\mu\pi/2}}{4(1+\mu^2)}(-4\mu)\right]=4\,ag\,e^{\mu\pi/2}\,\frac{1+\mu^2+2\mu}{2\,(1+\mu^2)}\qquad\text{...(6)}$$

Equating the values of C from (5) and (6)

$$\frac{4ag\,e^{-\mu\pi}}{1+\mu^2} = 4ag\,e^{\mu\pi/2}\,\frac{(1+\mu)^2}{2(1+\mu^2)}$$

$$2 = (1+\mu)^2 \,.\, e^{3\mu\pi/2}$$

Taking log of both sides

$$\log 2 = 2\log(1+\mu) + \frac{3\mu\pi}{2}$$

or $\qquad 3\mu\pi + 4\log(1+\mu) = 2\log 2.$

Example 4. *A small bead is projected with any velocity along a smooth circular wire under the action of a force varying inversely as the fifth power of the distance from a centre of force situated on the circumference. Prove that the pressure on the wire is constant.*

(K. U. 2007)

Sol. Let O be the point on the circumference of the circle taken as the centre of the force with diameter OA as initial line.

Let the particle be projected from A with velocity v. Let at any time t, the bead be at P.

Force along PO $= \frac{\mu}{r^5} = F$ say

From the equation of energy

$$\frac{1}{2}mv^2 - 0 = -\int \frac{\mu}{r^5}\,dr \qquad ...(1)$$

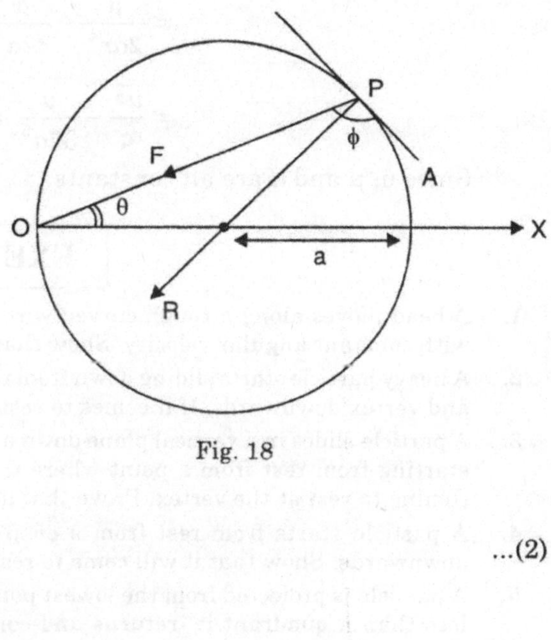

Fig. 18

$$\therefore \qquad v^2 = \frac{\mu}{2r^4} + C$$

At A, $r = 2a$, $v = u$

$$\therefore \qquad u^2 = \frac{\mu}{2\times 16a^4} + C$$

$$\therefore \qquad C = u^2 - \frac{\mu}{32a^4}$$

Putting the value of C in (1), we have

$$v^2 = \frac{\mu}{2r^4} - \frac{\mu}{32a^4} + u^2 \qquad ...(2)$$

Equation of motion along the normal is

$$\frac{v^2}{\rho} = R + F\sin\phi$$

$$\therefore \qquad R = \frac{v^2}{a} - \frac{\mu}{r^5} \sin \phi \qquad \text{...(3)} \quad (\because \text{ in circle, } \rho = a)$$

Equation of the circle in polar form is $r = 2a \cos \theta$

$$\frac{dr}{d\theta} = -2a \sin \theta$$

$$\therefore \qquad \frac{ds}{d\theta} = \sqrt{r^2 + \left(\frac{dr}{d\theta}\right)^2} = \sqrt{4a^2 \cos^2 \theta + 4a^2 \sin^2 \theta} = 2a$$

$$\because \qquad \sin \phi = r \frac{d\theta}{ds} = \frac{r}{\left(\dfrac{ds}{d\theta}\right)}$$

$$\therefore \qquad \sin \phi = \frac{r}{2a} \qquad \text{...(4)}$$

Putting (2) and (4) in (3), we get

$$R = \frac{1}{a}\left[\frac{\mu}{2r^4} - \frac{\mu}{32a^4} + u^2\right] - \frac{\mu}{r^5} \cdot \frac{r}{2a}$$

$$= \frac{\mu}{2ar^4} - \frac{\mu}{32a^5} + \frac{u^2}{a} - \frac{\mu}{2ar^4}$$

$$= \frac{u^2}{a} - \frac{\mu}{32a^5} = \text{constant}$$

Since u, μ and a are all constants.

EXERCISE 4

1. A bead moves along a rough curved wire which is such that it changes its direction of motion with constant angular velocity. Show that the wire is of the form of an equiangular spiral.
2. A heavy particle starts sliding down from rest from the cusp of a rough cycloid with base horizontal and vertex downwards. If it comes to rest at the vertex then show that $\mu^2 e^{\pi\mu} = 1$.
3. A particle slides in a vertical plane down a rough cycloid with axis vertical and vertex downwards, starting from rest from a point where the tangent makes an angle θ with the horizontal and coming to rest at the vertex. Prove that $\mu e^{\mu\theta} = \sin \theta - \mu \cos \theta$.
4. A particle starts from rest from a cusp of a rough cycloid whose axis is vertical and vertex downwards. Show that it will come to rest before reaching the vertex if $\mu e^{\mu\pi/2} > 1$.
5. A particle is projected from the lowest point of a rough sphere of radius a. After describing an arc less than a quadrant it returns and comes to rest at the lowest point. Show that the initial velocity must be $\sin \alpha \sqrt{2ga \dfrac{(1+\mu^2)}{1-2\mu^2}}$ where μ is the coefficient of friction and $a\alpha$ is the length of arc through which the particle moves.

6. A bead slides down a rough circular wire which is in a vertical plane, starting from rest at the end of a horizontal diameter. When it has described an angle θ about the centre its angular velocity is

$$\frac{2g}{a\,(1+4\,\mu^2)}\,[(1-2\mu^2)\sin\theta+3\mu\,(\cos\theta-e^{-2\mu\theta})]$$

where μ is coefficient of friction and a is the radius of the circle.

7. A particle is projected along the inner surface of a rough sphere and is acted on by no forces ; show that it will return to the point of projection at the end of time $\frac{a}{\mu V}\,(e^{2\mu\pi}-1)$, where a is the radius of the sphere, V is the velocity of projection and μ is the coefficient of friction.

SECTION IV

13

Central Orbits

Definitions

Central forces. If a force is always directed towards a fixed point it is called a central force and the fixed point is called as the centre of force.

Central orbit. The path, in which the particle is moving under the action of a central force, is called the central orbit.

1. Theorem

The central orbit is a plane curve.

Let the particle P be moving under the action of a central force mF directed towards the fixed point O. The equation of motion is

$$m\frac{d^2\vec{r}}{dt^2} = -m\vec{F}$$

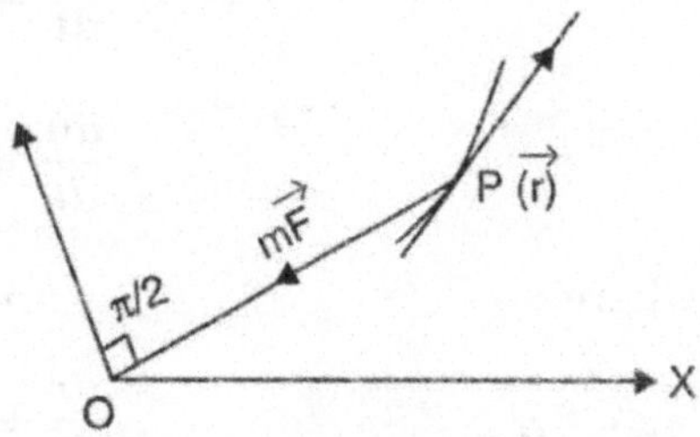

Fig. 1

Negative sign is taken since force acts along $\vec{PO}$

i.e., along the radius vector $-\vec{r}$

$$\therefore \quad \vec{r}\times\frac{d^2\vec{r}}{dt^2} = 0 \qquad (\vec{r}\times\vec{F} = 0)$$

or $$\vec{r}\times\frac{d^2\vec{r}}{dt^2} + \frac{d\vec{r}}{dt}\times\frac{d\vec{r}}{dt} = 0$$

or $$\frac{d}{dt}\left(\vec{r}\times\frac{d\vec{r}}{dt}\right) = 0$$

or $$\vec{r}\times\frac{d\vec{r}}{dt} = \vec{a} \text{ where } \vec{a} \text{ is a constant vector.}$$

or $$\vec{r}\,.\,\vec{r}\times\frac{d\vec{r}}{dt} = \vec{r}\,.\,\vec{a} \qquad \left(\because\ \vec{r}\,.\,\vec{r}\times\frac{d\vec{r}}{dt} = 0\right)$$

$$\therefore \quad \vec{r}\,.\,\vec{a} = 0 \qquad \text{for all values of } t.$$

$\therefore$ Radius vector $\vec{r} = \vec{OP}$ is always perpendicular to a fixed direction $\vec{a}$ its path must be a plane curve. Hence central orbit is a plane curve.

2. Differential Equation of a Central Orbit

A particle moves in a plane with an acceleration P towards a fixed point O in the plane ; to obtain the equation of the path. (K.U. 2005)

Let O be the origin OX be the initial line. Let at any time t particle be at $P(r, \theta)$ with acceleration P along PO.

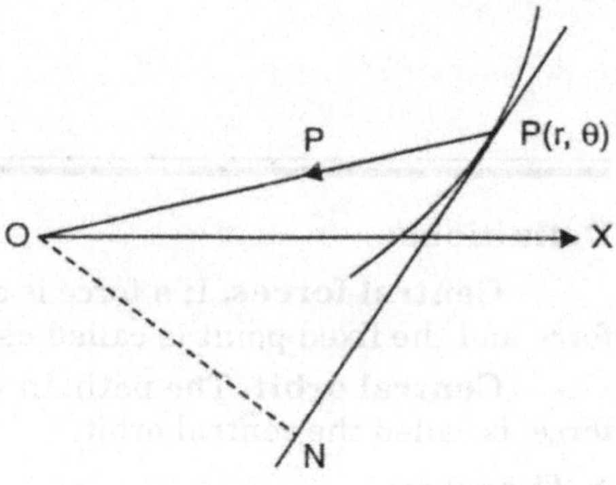

Fig. 2

$\therefore$ Equation of motion along radial direction is

$$\ddot{r} - r\dot{\theta}^2 = -P \quad ...(1)$$

Since there is no acceleration perpendicular to radius vector

$$\therefore \quad \frac{1}{r}\frac{d}{dt}\left(r^2 \frac{d\theta}{dt}\right) = 0 \quad ...(2)$$

Integrating (2), we get

$$\mathbf{r^2 \frac{d\theta}{dt} = constant = h} \text{ (say)} \quad ...(3)$$

$$\therefore \quad \frac{d\theta}{dt} = \frac{h}{r^2} = hu^2 \quad ...(4)$$

where $u = \frac{1}{r}$ or $r = \frac{1}{u}$

$$\therefore \quad \frac{dr}{dt} = -\frac{1}{u^2}\frac{du}{dt}$$

$$= -\frac{1}{u^2}\frac{du}{d\theta}\frac{d\theta}{dt} = -r^2\frac{d\theta}{dt}\frac{du}{d\theta}$$

$$\therefore \quad \frac{dr}{dt} = -h\frac{du}{d\theta} \quad \text{Using (3)}$$

$$\frac{d^2r}{dt^2} = \frac{d}{dt}\left(-h\frac{du}{d\theta}\right)$$

$$= -h\frac{d^2u}{d\theta^2}\cdot\frac{d\theta}{dt} = -h^2u^2\frac{d^2u}{d\theta^2} \quad \text{Using (4)}$$

Putting the values of $\frac{d^2r}{dt^2}$ and $\frac{d\theta}{dt}$ in (1), we get

$$-h^2u^2\frac{d^2u}{d\theta^2} - \frac{1}{u}h^2u^4 = -P$$

or

$$\boxed{\mathbf{\frac{d^2u}{d\theta^2} + u = \frac{P}{h^2u^2}}} \quad ...(5)$$

This is differential equation of the path in polar form.

If p be length of perpendicular from pole on the tangent vector.

$$\frac{1}{p^2} = \frac{1}{r^2} + \frac{1}{r^4}\left(\frac{dr}{d\theta}\right)^2 \qquad \textbf{From calculus}$$

$$r = \frac{1}{u}, \quad \frac{dr}{d\theta} = -\frac{1}{u^2}\frac{du}{d\theta}$$

$$\frac{1}{p^2} = u^2 + u^4 . \frac{1}{u^4}\left(\frac{du}{d\theta}\right)^2$$

or
$$\frac{1}{p^2} = u^2 + \left(\frac{du}{d\theta}\right)^2. \qquad ...(6)$$

Differentiating w.r.t. θ

$$-\frac{2}{p^3}\frac{dp}{d\theta} = 2u\frac{du}{d\theta} + 2\frac{du}{d\theta}\frac{d^2u}{d\theta^2}$$

or
$$-\frac{1}{p^3}\frac{dp}{d\theta} = \left(u + \frac{d^2u}{d\theta^2}\right)\frac{du}{d\theta}$$

Using equation (5)

$$-\frac{1}{p^3}\frac{dp}{d\theta} = \frac{\text{P}}{h^2u^2}\frac{du}{d\theta} \qquad \because \quad u = \frac{1}{r}$$

or
$$\frac{-h^2}{p^3}dp = \text{P}.\frac{1}{u^2}du \qquad du = -\frac{1}{r^2}dr$$

$$\therefore \quad \frac{-h^2}{p^3}dp = -\text{P}r^2 . \frac{1}{r^2}dr$$

or
$$\boxed{\frac{\mathbf{h^2}}{\mathbf{p^3}}\frac{\mathbf{dp}}{\mathbf{dr}} = \mathbf{P}} \qquad ...(7)$$

This is differential equation of the path in pedal form.

3. Rate of Description of Sectorial Area and Velocity at any Point of the Path

In every central orbit, the sectorial area traced out by the radius vector to the centre of force increases uniformly per unit of time and linear velocity varies inversely as the perpendicular from the centre upon the tangent to the path.

Let P(r, θ) be the position of the particle at time t and Q $(r + \delta r, \theta + \delta\theta)$ be the position at time $t + \delta t$ w.r.t. origin O and initial line OX.

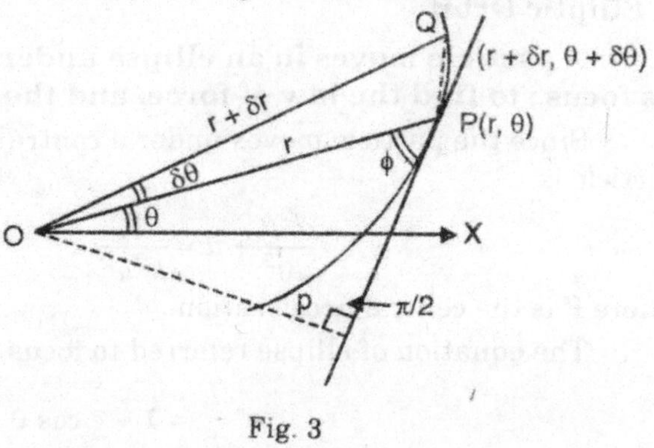

Fig. 3

Thus, OP = r, OQ = $r + \delta r$ and ∠POQ = $\delta\theta$

$\therefore$ Area of $\Delta POQ = \frac{1}{2} r. (r + \delta r) \sin \delta\theta$ $\left(\Delta = \frac{1}{2} bc \sin A \right)$

When Q is very close to P the sector POQ is nearly equal to ΔPOQ.
Hence, the rate of description of sectorial area

$$= \lim_{\delta t \to 0} \frac{1}{2} \frac{r(r + \delta r) \sin \delta\theta}{\delta t}$$

$\because$ $\delta\theta$ is small

$\therefore$ $\sin \delta\theta \simeq \delta\theta$

$$= \lim_{\delta t \to 0} \frac{1}{2} \frac{(r)(r + \delta r)\, \delta\theta}{\delta t} \qquad ...(1)$$

Rate of change of sectorial area

$$= \frac{1}{2} r^2 \frac{d\theta}{dt} = \frac{h}{2} = \text{constant} \qquad \left(\because r^2 \frac{d\theta}{dt} = h \right)$$

$\therefore$ Rate of change of sectorial area is constant. (Equation (3) of article 6.2)

Further $r^2 \frac{d\theta}{dt} = h$

$$\therefore \quad r^2 \frac{d\theta}{ds} \cdot \frac{ds}{dt} = h \qquad \left(r \frac{d\theta}{ds} = \sin \phi \right)$$

$$r \sin \phi \,.\, v = h$$ From calculus

$$pv = h \qquad (\because \; p = r \sin \phi)$$

or $\boxed{\mathbf{v = h/p}}$ ...(2)

$\therefore$ Linear velocity is inversely proportional to the perpendicular from the origin upon the tangent to the path at P.

Also, $$v^2 = \frac{h^2}{p^2} = h^2 \left[\frac{1}{r^2} + \frac{1}{r^4} \left(\frac{dr}{d\theta} \right)^2 \right]$$

$$\boxed{\mathbf{v^2 = h^2 \left[u^2 + \left(\frac{du}{d\theta} \right)^2 \right]}} \qquad ...(3)$$

Equation (3) gives the linear velocity at any point of the path of a central orbit.

4. Elliptic Orbit

A particle moves in an ellipse under a force which is always directed towards its focus; to find the law of force, and the velocity at any point of the path.

Since the particle moves under a central force therefore the equation of the path of the particle is

$$\frac{d^2u}{d\theta^2} + u = \frac{P}{h^2u^2} \qquad ...(1)$$

where P is the central acceleration.

The equation of ellipse referred to focus as a pole is

$$\frac{l}{r} = 1 + e \cos \theta$$ $e < 1$, l is semi latus rectum.

or $$u = \frac{1+e\cos\theta}{l} \qquad u = \frac{1}{r}$$

or $$\frac{du}{d\theta} = -\frac{e\sin\theta}{l},\ \frac{d^2u}{d\theta^2} = -\frac{e\cos\theta}{l}$$

Putting in equation (1)

$$\frac{-e\cos\theta}{l} + \frac{1+e\cos\theta}{l} = \frac{P}{h^2u^2}$$

or $$P = \frac{h^2u^2}{l} \quad ...(2)$$

or $$P = \mu\frac{1}{r^2} \quad \text{where} \quad \mu = \frac{h^2}{l}$$

or $$\mathbf{h} = \sqrt{\mu l} = \sqrt{\mu \times \textbf{Semi latus rectum}}$$

and $$P \propto \frac{1}{r^2}$$

Hence, the acceleration is inversely proportional to square of the distance from the focus.

Also, $$v^2 = h^2\left[u^2 + \left(\frac{du}{d\theta}\right)^2\right] = h^2\left[\left(\frac{1+e\cos\theta}{l}\right)^2 + \left(\frac{e\sin\theta}{l}\right)^2\right]$$

$$= \mu l\left[\frac{1+2e\cos\theta+e^2}{l^2}\right] = \mu\left[\frac{2(1+e\cos\theta)}{l} - \frac{(1-e^2)^*}{l}\right]$$

$$\mathbf{v}^2 = \mu\left[\frac{2}{\mathbf{r}} - \frac{1}{\mathbf{a}}\right] \quad ...(3)$$

where $2a$ is major axis of the ellipse.

Thus, velocity at any point of the path depends on the distance from the focus and independent of the direction of motion.

From equation (3) it follows that the square of the velocity V of projection from any point whose distance from the focus is r_0, must be less than $\frac{2\mu}{r_0}$ and that the semi major axis a of the ellipse is given by $V^2 = \mu\left(\frac{2}{r_0} - \frac{1}{a}\right)$.

We know that rate of describing the sectorial area per unit time is $\frac{1}{2}h$. Let T be the time taken by the particle in completing the elliptic orbit

---

*Semi latus rectum of the ellipse $l = \frac{b^2}{a} = \frac{a^2(1-e^2)}{a}$

$$\therefore \quad \frac{1}{a} = \frac{1-e^2}{l}.$$

$$\text{Total area described} = T \times \frac{1}{2}h = \pi ab$$

$$\therefore \qquad T = \frac{2\pi ab}{h} = \frac{2\pi a\,.\,b}{\sqrt{\mu l}} = \frac{2\pi a\, b}{\sqrt{\mu\,.\,\frac{b^2}{a}}} = \frac{2\pi a^{3/2}}{\sqrt{\mu}}.$$

$$\therefore \quad \textbf{Time period} = \frac{2\pi a^{3/2}}{\sqrt{\mu}}.$$

Hyperbolic orbit. If a particle moves in a hyperbola under a central force towards focus, proceeding in the similar manner as in case of elliptic orbit, we get

$$v^2 = \mu\left[\frac{2(1+e\cos\theta)}{l} - \frac{(1-e^2)}{l}\right] \qquad \text{Equation (3) of article 4}$$

can be written as

$$v^2 = \mu\left[\frac{2(1+e\cos\theta)}{l} + \frac{e^2-1}{l}\right] \qquad \because \ e > 1$$

$$\mathbf{v^2} = \mu\left[\frac{\mathbf{2}}{\mathbf{r}} + \frac{\mathbf{1}}{\mathbf{a}}\right] \qquad \text{where } l = \frac{b^2}{a} = a(e^2-1)$$

and $$P = \frac{\mu}{r^2} = \mu u^2.$$

Parabolic orbit $e = 1$

$$P = \frac{\mu}{r^2} \quad \text{and} \quad \mathbf{v^2} = \frac{2\mu}{\mathbf{r}}.$$

Example 1. *A particle describes the curve $r^n = a^n \cos n\theta$ under a force towards the pole. Find the law of force.* (KU 2004)

Sol. Equation of the path is

$$r^n = a^n \cos n\theta \qquad \text{...(1)}$$

Taking $r = \frac{1}{u}$

$$u^n a^n \cos n\theta = 1$$

Taking log on both sides

$$n \log u + n \log a + \log \cos n\theta = 0$$

Differentiating w.r.t. θ

$$n\frac{1}{u}\frac{du}{d\theta} = n \tan n\theta$$

or $$\frac{du}{d\theta} = u \tan n\theta$$

or $$\frac{d^2u}{d\theta^2} = un \sec^2 n\theta + \frac{du}{d\theta}\tan n\theta$$

$$= un \sec^2 n\theta + u \tan^2 n\theta$$

Putting in the equation of path $\dfrac{d^2u}{d\theta^2}+u=\dfrac{P}{h^2u^2}$, we get

$$un\sec^2 n\theta + u\tan^2 n\theta + u = \frac{P}{h^2u^2}$$

or $$(n+1)\,u\sec^2 n\theta = \frac{P}{h^2u^2}$$

or $$P = h^2u^3(n+1)\sec^2 n\theta$$

$$= \frac{h^2(n+1)}{r^3\cos^2 n\theta} = \frac{h^2(n+1)a^{2n}}{r^3 . r^{2n}}$$

or $$P \propto \frac{1}{r^{2n+3}}$$

$\therefore$ The force varies inversely as $(2n+3)$rd power of the distance from the pole.

Example 2. *A particle of unit mass describes an equiangular spiral of angle α, under a force which is always in the direction perpendicular to the straight line joining the particle to the pole of the spiral. Show that the force is $\mu r^{2\sec^2\alpha-3}$ and rate of description of sectorial area about the pole is $\frac{1}{2}\sqrt{\mu\sin\alpha\cos\alpha}\; r^{\sec^2\alpha}$.*

Sol. The equation of the equiangular spiral is

$$r = a\,e^{\theta\cot\alpha}$$

$$\frac{dr}{dt} = \dot{r} = a\cot\alpha\, e^{\theta\cot\alpha}.\dot{\theta}$$

$$= r\cot\alpha\,\dot{\theta} \qquad \text{...(1)}$$

$\because$ Force is acting perpendicular to radius vector

$\therefore$ $$\ddot{r} - r\dot{\theta}^2 = 0$$

or $$\ddot{r} - r.\left(\frac{\dot{r}}{r\cot\alpha}\right)^2 = 0 \qquad \text{From (1) } \dot{\theta} = \frac{\dot{r}}{r\cot\alpha}$$

or $$\frac{\ddot{r}}{\dot{r}} = \frac{\dot{r}}{r}\tan^2\alpha$$

Integrating both sides

$$\log\dot{r} = (\log r)\tan^2\alpha + \log c$$

or $$\log\frac{\dot{r}}{c} = \log r^{\tan^2\alpha}$$

or $$\dot{r} = c\,r^{\tan^2\alpha} \qquad \text{...(2)}$$

Also from equation (1)

$$r\dot{\theta} = \dot{r}\tan\alpha$$

or $$r^2\frac{d\theta}{dt} = r\tan\alpha\frac{dr}{dt}$$

$$= r \tan \alpha . c r^{\tan^2 \alpha} \qquad \text{Using (2)}$$

$$= c \tan \alpha \, r^{\sec^2 \alpha} \qquad \text{...(3)}$$

If P be the force on the unit mass at right angle to radius vector, then

$$P = \frac{1}{r}\frac{d}{dt}\left(r^2 \frac{d\theta}{dt}\right) = \frac{1}{r}\frac{d}{dt}\left(c \tan \alpha \, r^{\sec^2 \alpha}\right) \qquad \text{Using (3)}$$

$$= \frac{1}{r} c \tan \alpha . \sec^2 \alpha . r^{\sec^2 \alpha - 1} . \frac{dr}{dt}$$

$$= c \tan \alpha \sec^2 \alpha . r^{\sec^2 \alpha - 2} . \frac{dr}{dt}$$

$$= c \tan \alpha \sec^2 \alpha \, r^{\sec^2 \alpha - 2} \, c \, r^{\tan^2 \alpha}$$

$$= c^2 \tan \alpha . \sec^2 \alpha . r^{\sec^2 \alpha - 2} . r^{\sec^2 \alpha - 1}$$

$$= c^2 \tan \alpha \sec^2 \alpha \, r^{2\sec^2 \alpha - 3} \qquad \text{(using 2)}$$

$$= \mu . r^{2\sec^2 \alpha - 3} \qquad \text{...(4)}$$

where $\mu = c^2 \tan \alpha \sec^2 \alpha$

or $\qquad c = \sqrt{\mu \cot \alpha \cos^2 \alpha}$

Further rate of describing of sectorial area

$$= \frac{1}{2} r^2 \frac{d\theta}{dt} = \frac{1}{2} c \tan \alpha \, r^{\sec^2 \alpha} \qquad \text{(From 3)}$$

$$= \frac{1}{2} \sqrt{\mu \cot \alpha \cos^2 \alpha} \, \tan \alpha . r^{\sec^2 \alpha}$$

$$= \frac{1}{2} \sqrt{\mu \sin \alpha \cos \alpha} \, . r^{\sec^2 \alpha}. \qquad \text{Hence proved.}$$

Example 3. *A particle describes an ellipse under a force* μ/r^2 *and has a velocity V at a distance R from the centre of force. Show that the time period is* $\frac{2\pi}{\sqrt{\mu}}\left[\frac{2}{R} - \frac{V^2}{\mu}\right]^{-3/2}$.

Sol. We know that when the particle moves in an elliptic orbit,

$$v^2 = \mu\left[\frac{2}{r} - \frac{1}{a}\right]$$

at $\qquad r = R \quad v = V$ given

$$\therefore \qquad V^2 = \mu\left[\frac{2}{R} - \frac{1}{a}\right]$$

or $$\frac{1}{a} = \frac{2}{R} - \frac{V^2}{\mu}$$

or $$a = \left[\frac{2}{R} - \frac{V^2}{\mu}\right]^{-1}$$

But time period $T = \dfrac{2\pi a^{3/2}}{\sqrt{\mu}} = \dfrac{2\pi}{\sqrt{\mu}}\left[\dfrac{2}{R} - \dfrac{V^2}{\mu}\right]^{-3/2}$.

Example 4. *In a central orbit, the velocity at any point is inversely proportional to the distance of the point from the centre of the force. Show that the path is an equiangular spiral.*

Sol. From equation (3) of article 2

$$v = h/p$$

where p is the length of perpendicular from origin upon the tangent.

Also, $v = \dfrac{\mu}{r}$ (given)

$$\therefore \quad \frac{h}{p} = \frac{\mu}{r}$$

or $p = \left(\dfrac{h}{\mu}\right) r$ or $p = ar$ where $a = \dfrac{h}{\mu}$

This is the pedal equation of equiangular spiral.

Example 5. *A particle describes the ellipse whose pedal equation is* $\dfrac{a^2b^2}{p^2} = a^2 + b^2 - r^2$ *with pole at the centre. Find the law of force.*

Sol. The equation of the path is

$$\frac{a^2b^2}{p^2} = a^2 + b^2 - r^2 \qquad ...(1)$$

Differentiating w.r.t. r

$$-\frac{2a^2b^2}{p^3}\frac{dp}{dr} = -2r$$

or $$\frac{1}{p^3}\frac{dp}{dr} = \frac{r}{a^2b^2} \qquad ...(2)$$

The differential equation of the central orbit in pedal form is $\dfrac{h^2}{p^3}\dfrac{dp}{dr} = P$

or $$h^2 \frac{r}{a^2b^2} = P \qquad \text{(Using 2)}$$

or $$P \propto r$$

Hence the force is directly proportional to the distance from the centre.

EXERCISE 1

1. A particle describes the following curves under a force P towards the pole. Find the law of force
   (*i*) Equiangular spiral $r = a\, e^{\theta \cot \alpha}$ (*K.U. 2005*)
   (*ii*) $r^2 = a^2 \cos 2\theta$ (*iii*) $r = 2a \cos \theta$
   (*iv*) $r^n \cos n\theta = a^n$ (*v*) $r = a \sin n\theta$
   (*vi*) $r = a(1 + \cos \theta)$. (*vii*) $r \cosh n\theta = a$.

2. A particle describes the curve $au = \dfrac{\cosh\theta - 2}{\cosh\theta + 1}$ under a force P to the pole. Find the law of force.

3. A particle describes the curve $au = \tanh \theta/\sqrt{2}$ under a force P to the pole. Find the law of force.

4. A particle describes the following curve under a force P to the pole. Find the law of force

   (i) $p^2 = ar$ (ii) $r^2 = 2ap$ (iii) $\dfrac{b^2}{p^2} = \dfrac{2a}{r} + 1$.

5. If the central force varies as the distance from a fixed point, show that the orbit is a conic, centre being the pole.

6. If the central force varies as the cube of the distance from a fixed point, then find the orbit.

7. A particle moves in a conic $\dfrac{l}{r} = 1 + e\cos\theta$ under a force P towards the pole. Find the law of force and the velocity of the particle at any time.

8. A particle describes the path $r^4 = a^4 \cos 4\theta$ under a force which is always directed towards the pole. Find the law of force and the velocity at any point.

9. A particle describes a circle, pole on its circumference, under a force P to the pole. Find the law of force. (*K.U. 2007*)

10. If the central force varies as $\dfrac{1}{r^5}$, show that the path is a circle.

11. A particle describes a curve whose equation is $\dfrac{a}{r} = \theta^2 + b$ under a force to the pole. Find the law of force.

12. If the central force varies as distance from a fixed point, find the equation of the orbit.

13. The velocity of any point at a central orbit is $\dfrac{1}{n}$ th of what it would be for a circular orbit at the same distance. Show that the central force varies as $\dfrac{1}{r^{2n^2+1}}$ and that the equation of the orbit is

$$r^{n^2-1} = a^{n^2-1}\cos(n^2-1)\theta.$$

ANSWERS

1. (i) $P \propto \dfrac{1}{r^3}$ (ii) $P \propto \dfrac{1}{r^7}$ (iii) $P \propto \dfrac{1}{r^5}$ (iv) $P \propto r^{2n-3}$

   (v) $P \propto \dfrac{1-n^2}{r^3} + \dfrac{2a^2n^2}{r^5}$ (vi) $P \propto \dfrac{1}{r^4}$ (vii) $P \propto \dfrac{1}{r^4}$

2. $P \propto \dfrac{1}{r^4}$ 3. $P \propto \dfrac{1}{r^5}$

4. (i) $P \propto \dfrac{1}{r^2}$ (ii) $P \propto \dfrac{1}{r^5}$ (iii) $P \propto \dfrac{1}{r^2}$ 6. $\dfrac{2h^2}{p^2} = c - \mu r^4$.

7. $P \propto \dfrac{1}{r^2}$ 8. $P \propto \dfrac{1}{r^{11}}$ 9. $P \propto \dfrac{1}{r^5}$ 11. $P \propto \dfrac{1}{r^2}$

12. Conic, $\dfrac{h^2}{p^2} = c - \mu r^2$.

Definitions

Apse. An apse is a point in a central orbit at which the radius vector drawn from the centre of force to the orbit has a minimum and maximum value.

Apsidal distance. The distance of an apse from the centre of the force is called the Apsidal distance.

Apsidal angle. The angle between two apsidal distances is called apsidal angle.

5. Theorem

At an apse the motion of the particle is perpendicular to the radius vector.

Proof. Apse is the point on the orbit at which the distance r from the centre is maximum or minimum. Since $u = \frac{1}{r}$.

Therefore at an apse u is minimum or maximum.

$\therefore$ For the maxima or minima of u, $\frac{du}{d\theta} = 0$ and the first differential coefficient of u that does not vanish is of even order.

Also we know that

$$\frac{1}{p^2} = u^2 + \left(\frac{du}{d\theta}\right)^2 \qquad \text{(Equation VI of article 2)}$$

$$\because \quad \frac{du}{d\theta} = 0 \quad \therefore \quad \frac{1}{p^2} = u^2 = \frac{1}{r^2}$$

Hence $p = r$ or $r \sin\phi = r \Rightarrow \sin\phi = 1$ or $\phi = \frac{\pi}{2}$

$\therefore$ At an apse the radius vector is perpendicular to the tangent from the centre.

Hence Proved.

6. Theorem

When the central acceleration is a function of distance only and is always the same at the same distance, every apse line divides the orbit into two equal and similar portions and thus there can be only two apsidal distances.

Proof. Let ABC be a portion of orbit having three consecutive apses A, B and C. Let O be the centre of the force. Let V be the velocity of the particle at B when it is moving towards C it traces the path BQC. If velocity at B is reversed it will describe the path BPA.

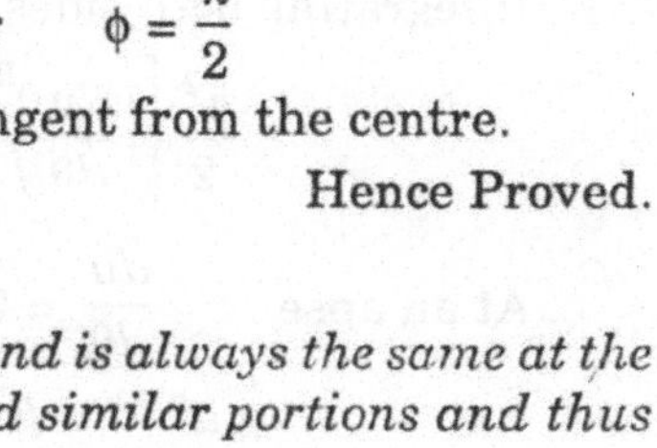

Fig. 4

We know that $\ddot{r} - r\dot{\theta}^2 = -P$ and $r^2 \frac{d\theta}{dt} = h.$

These two equation gives that velocity depends only on distance from O and not on the direction. Therefore r and angle θ at any time t for the particle moving towards c (*i.e.*, OQ and $\angle$BOQ) are equal OP and $\angle$BOP for the second particle, moving towards A. Hence the curve BQC and BPA are similar. Hence since A and C are the points where the radius vector is perpendicular to the tangent, we have OA = OC.

Further if there is some other apse D next to C we have OB = OD and so on.

Thus, there are only two different apsidal distances.

7. Theorem

If central acceleration varies as some integral power of the distance, say μu^n, there are at the most two apsidal distances. (KU 2004)

Proof. Let $P = \mu u^n$

We know that differential equation of central orbit is

$$\frac{d^2u}{d\theta^2} + u = \frac{P}{h^2u^2} = \frac{\mu u^n}{h^2u^2} \quad ...(1)$$

or
$$\frac{d^2u}{d\theta^2} + u = \frac{\mu u^{n-2}}{h^2}$$

or
$$h^2\left[\frac{d^2u}{d\theta^2} + u\right] = \mu u^{n-2}$$

or
$$\frac{h^2}{2}.2\frac{du}{d\theta}\left[\frac{d^2u}{d\theta^2} + u\right] = \mu u^{n-2}\frac{du}{d\theta}$$

$$\frac{h^2}{2}\left[\frac{d}{d\theta}\left\{\left(\frac{du}{d\theta}\right)^2 + u^2\right\}\right] = \frac{\mu}{n-1}\frac{d}{d\theta}\{u^{n-1}\}$$

Integrating both sides, we get

$$\frac{h^2}{2}\left[\left(\frac{du}{d\theta}\right)^2 + u^2\right] = \frac{\mu u^{n-1}}{n-1} + c \quad ...(2)$$

At an apse $\frac{du}{d\theta} = 0$, Equation (2) reduces to

$$\frac{\mu u^{n-1}}{n-1} - \frac{h^2u^2}{2} + c = 0$$

or
$$u^{n-1} - \frac{(n-1)h^2u^2}{2\mu} + c' = 0$$

where $c' = \frac{(n-1)c}{\mu}$

Whatever be the value of n and c' the equation in u has only two variations of sign. Therefore by Descartes' rule of sign, it can not have more than two positive roots. Hence at the most two apsidal distances are possible.

Example 1. *A particle is projected from an apse at a distance a with velocity from infinity under the action of a central acceleration $\frac{\mu}{r^{2n+3}}$. Prove that equation of the path is $r^n = a^n \cos n\theta$.*

Sol. Let v be the velocity at a distance x from the centre, then the acceleration

$$v\frac{dv}{dx} = -\frac{\mu}{x^{2n+3}}$$

(–) ve sign is taken because x decreases as time increases.

If V is the velocity of projection at an apse at a distance a then

$$\int_o^V v\,dv = -\mu \int_\infty^a \frac{dx}{x^{2n+3}}$$

$$\frac{V^2}{2} = \frac{\mu}{2n+2}\left[\frac{1}{x^{2n+2}}\right]_\infty^a$$

or
$$V^2 = \frac{\mu}{n+1}\left[\frac{1}{a^{2n+2}} - 0\right] \qquad ...(1)$$

Also differential equation of the path is

$$\frac{d^2u}{d\theta^2} + u = \frac{P}{h^2u^2} \qquad ...(2)$$

Here $\qquad P = \dfrac{\mu}{r^{2n+3}} = \mu\, u^{2n+3}$

$\therefore$ Equation (2) becomes

$$\frac{d^2u}{d\theta^2} + u = \frac{\mu}{h^2} u^{2n+1}$$

Multiplying both sides by $2\dfrac{du}{d\theta}$

$$h^2\left[\frac{d^2u}{d\theta} + u\right]2\frac{du}{d\theta} = \mu\, u^{2n+1}\left(2\frac{du}{d\theta}\right)$$

or
$$h^2 \frac{d}{d\theta}\left[\left(\frac{du}{d\theta}\right)^2 + u^2\right] = 2\mu \frac{d}{d\theta}\left[\frac{u^{2n+2}}{2n+2}\right]$$

Integrating

$$v^2 = h^2\left[\left(\frac{du}{d\theta}\right)^2 + u^2\right] = 2\mu \frac{u^{2n+2}}{2n+2} + c \qquad ...(3)$$

At $r = a$ or $u = \dfrac{1}{a}$, $\dfrac{du}{d\theta} = 0$ and $v = V$

$$V^2 = 0 + \frac{h^2}{a^2} = \frac{\mu}{(n+1)a^{2n+2}} + c \qquad ...(4)$$

Using (1) in equation (4), we get

$$V^2 = V^2 + c$$

$\therefore \qquad c = 0$

Putting $c = 0$ in (4), we get

$$h^2 = \frac{\mu}{(n+1)a^{2n}}$$

Putting h^2 and c in equation (3), we get

$$\frac{\mu}{(n+1)a^{2n}}\left[\left(\frac{du}{d\theta}\right)^2 + u^2\right] = \frac{\mu u^{2n+2}}{n+1}$$

or $$\left(\frac{du}{d\theta}\right)^2 + u^2 = u^{2n+2} \cdot a^{2n} \quad \text{....(5)}$$

$\because \quad u = \frac{1}{r} \quad \therefore \quad \frac{du}{d\theta} = -\frac{1}{r^2}\frac{dr}{d\theta}$

$\therefore$ Equation (5) becomes

$$\frac{1}{r^4}\left(\frac{dr}{d\theta}\right)^2 + \frac{1}{r^2} = \frac{a^{2n}}{r^{2n+2}}$$

or $$\left(\frac{dr}{d\theta}\right)^2 = \left[\frac{a^{2n}}{r^{2n}} - 1\right] r^2$$

or $$\frac{dr}{d\theta} = \frac{\pm\sqrt{a^{2n} - r^{2n}}}{r^{n-1}}$$

or $$\frac{-r^{n-1}\,dr}{\sqrt{(a^n)^2 - (r^n)^2}} = d\theta.$$

Integrating by substituting $r^n = z$, $nr^{n-1}dr = dz$

$$\int \frac{-dz}{\sqrt{(a^n)^2 - z^2}} = n\theta + c'$$

$$\cos^{-1}\left(\frac{r^n}{a^n}\right) = n\theta + c'$$

At $r = a$, $\theta = 0 \Rightarrow c' = 0$

$\therefore$ $$\cos^{-1}\left(\frac{r^n}{a^n}\right) = n\theta$$

or $r^n = a^n \cos n\theta$ is the required equation of the path.

Example 2. *A particle is moving with central acceleration $\mu(r^5 - c^4 r)$, being projected from an apse at a distance c with a velocity $\sqrt{\frac{2\mu}{3}} \cdot c^3$. Show that its path is the curve $x^4 + y^4 = c^4$.*

Sol. The differential equation of the path is

$$\frac{d^2u}{d\theta^2} + u = \frac{P}{h^2u^2}$$

$$P = \mu(r^5 - c^4 r)$$

$$= \mu\left(\frac{1}{u^5} - \frac{c^4}{u}\right) \quad \text{...(1)}$$

$$\therefore \qquad \frac{d^2u}{d\theta^2} + u = \frac{\mu}{h^2}\left[\frac{1}{u^7} - \frac{c^4}{u^3}\right]$$

Multiplying by $2\frac{du}{d\theta}$ both sides

$$\text{or} \qquad h^2\left[\frac{d^2u}{d\theta^2} + u\right] . \ 2\frac{du}{d\theta} = \mu\,[u^{-7} - c^4 u^{-3}]\ 2\frac{du}{d\theta}$$

Integrating

$$v^2 = h^2\left[\left(\frac{du}{d\theta}\right)^2 + u^2\right] = 2\mu\left[-\frac{1}{6u^6} + \frac{c^4}{2u^2}\right] + c_1 \qquad ...(2)$$

At $r = c$ *i.e.*, $u = \frac{1}{c}$, $v = c^3\sqrt{\frac{2\mu}{3}}$ and $\frac{du}{d\theta} = 0$

$$c^6\left(\frac{2\mu}{3}\right) = h^2\left[0 + \frac{1}{c^2}\right] = 2\mu\left[-\frac{c^6}{6} + \frac{c^6}{2}\right] + c_1$$

$$\therefore \qquad h^2 = \frac{2}{3}\mu c^8 \text{ and } c_1 = 0$$

Putting in equation (2), we get

$$\frac{2}{3}\mu c^8\left[\left(\frac{du}{d\theta}\right)^2 + u^2\right] = 2\mu\left[-\frac{1}{6u^6} + \frac{c^4}{2u^2}\right]$$

$$c^8\left[\left(\frac{du}{d\theta}\right)^2 + u^2\right] = \frac{-1}{2u^6} + \frac{3c^4}{2u^2}$$

$$c^8\left(\frac{du}{d\theta}\right)^2 = \frac{-1}{2u^6} + \frac{3c^4}{2u^2} - c^8u^2 = \frac{1}{u^6}\left[-\frac{1}{2} + \frac{3}{2}c^4u^4 - c^8u^8\right]$$

$$= \frac{1}{u^6}\left[-\frac{1}{2} - \left(c^4u^4 - \frac{3}{4}\right)^2 + \frac{9}{16}\right]$$

$$c^8\left(\frac{du}{d\theta}\right)^2 = \frac{1}{u^6}\left[\left(\frac{1}{4}\right)^2 - \left(c^4u^4 - \frac{3}{4}\right)^2\right]$$

$$\therefore \qquad c^4\frac{du}{d\theta} = \pm\frac{1}{4u^3}\sqrt{1 - (4c^4u^4 - 3)^2}$$

$$\frac{-4u^3c^4\,du}{\sqrt{1 - (4c^4u^4 - 3)^2}} = d\theta \qquad \text{Taking (–) ve sign}$$

$$-\frac{dz}{\sqrt{1 - z^2}} = +4d\theta \qquad \text{Putting } 4c^4u^4 - 3 = z,\ 16c^4u^3\,du = dz$$

$$\cos^{-1} z = 4\theta + c_2$$

Integrating both sides

$$\cos^{-1}(4c^4u^4 - 3) = 4\theta + c_2 \qquad ...(3)$$

Initially at $u = \dfrac{1}{c}$, $\theta = 0 \Rightarrow c_2 = 0$

$\therefore$ $$4c^4u^4 - 3 = \cos 4\theta$$

or $$\frac{4c^4}{r^4} - 3 = \cos 4\theta$$

$$4c^4 - 3r^4 = r^4\,[2\cos^2 2\theta - 1]$$

$$= r^4\,[2(2\cos^2\theta - 1)^2 - 1] = r^4\,[8\cos^4\theta - 8\cos^2\theta + 1]$$

Changing it to cartesian coordinates $x = r\cos\theta$, $y = r\sin\theta$, $x^2 + y^2 = r^2$, we get

$$4c^4 = 3(x^2 + y^2)^2 + [8x^4 - 8(x^2 + y^2)\,x^2 + (x^2 + y^2)^2]$$

$$4c^4 = 3(x^4 + y^4 + 2x^2y^2) + [8x^4 - 8x^4 - 8x^2y^2 + x^4 + y^4 + 2x^2y^2]$$

$$4c^4 = 4x^4 + 4y^4$$

or $x^4 + y^4 = c^4$. Hence Proved.

Example 3. *A particle moves with a central acceleration [$\mu/(distance)^5$] and projected from an apse at a distance a with a velocity equal to n times that which would be acquired in falling from infinity, show that the other apsidal distance is $a/\sqrt{n^2 - 1}$. If n = 1, the particle be projected in any direction, show that the path is a circle passing through the centre of force.*

Sol. If v be the velocity at a distance x from the centre, then

$$v\frac{dv}{dx} = \frac{-\mu}{x^5}$$

Let V be the velocity acquired in falling from infinity to a distance a from the centre, then

$$\int_0^V v\,dv = -\int_\infty^a \mu x^{-5}\,dx$$

$$\frac{V^2}{2} = \left[\frac{\mu}{4x^4}\right]_\infty^a = \frac{\mu}{4a^4}$$

or $$V = \sqrt{\frac{\mu}{2a^4}} \qquad ...(1)$$

Differential equation of the path is

$$\frac{d^2u}{d\theta^2} + u = \frac{P}{h^2u^2} \quad \text{where } P = \frac{\mu}{r^5} = \mu u^5 \qquad \text{(given)}$$

$$h^2\left[\frac{d^2u}{d\theta^2} + u\right] = \mu u^3 \qquad ...(2)$$

Multiplying both sides by $2\dfrac{du}{d\theta}$ and integrating, we get

$$v^2 = h^2\left[\left(\frac{du}{d\theta}\right)^2 + u^2\right] = 2\mu\frac{u^4}{4} + c \qquad ...(3)$$

At $r = a$ or $u = \dfrac{1}{a}$, $\dfrac{du}{d\theta} = 0$ and $v = n\sqrt{\dfrac{\mu}{2a^4}}$

$$\frac{n^2\mu}{2a^4} = h^2\left[\frac{1}{a^2}\right] = \frac{\mu}{2a^4} + c$$

$$\therefore \qquad c = \frac{\mu\,[n^2 - 1]}{2a^4} \text{ and } h^2 = \frac{n^2\mu}{2a^2}$$

Putting in equation (3)

$$\frac{n^2\mu}{2a^2}\left[\left(\frac{du}{d\theta}\right)^2 + u^2\right] = \frac{\mu\,(n^2-1)}{2a^4} + \frac{\mu u^4}{2}$$

or
$$\left(\frac{du}{d\theta}\right)^2 + u^2 = \frac{1}{a^2n^2}\,[a^4u^4 + (n^2-1)]$$

$$\left(\frac{du}{d\theta}\right)^2 = \frac{1}{a^2n^2}\,[a^4u^4 - a^2n^2u^2 + (n^2-1)] \qquad \text{...(4)}$$

At an apse $\dfrac{du}{d\theta} = 0$

$$\therefore \qquad a^4u^4 - a^2n^2u^2 + (n^2-1) = 0 \qquad \because \; u = \frac{1}{r}$$

or
$$(n^2-1)\,r^4 - a^2n^2r^2 + a^4 = 0$$

The equation is quadratic in r^2 and has two roots r_1^2 and r_2^2

$$\therefore \qquad r_1^2 r_2^2 = \frac{a^4}{n^2-1} \quad \text{or} \quad r_1 r_2 = \frac{a^2}{\sqrt{n^2-1}}$$

But one apsidal distance is $r_1 = a$ and hence the other apsidal distance $r_2 = \dfrac{a}{\sqrt{n^2-1}}$.

(*ii*) If $n = 1$ equation (4) reduces to

$$\left(\frac{du}{d\theta}\right)^2 = \frac{1}{a^2}\,[a^4u^4 - a^2u^2] = a^2u^4 - u^2$$

$$\frac{du}{d\theta} = u\sqrt{a^2u^2 - 1}$$

or
$$\frac{du}{u\sqrt{a^2u^2-1}} = d\theta \qquad \text{Integrating}$$

$$\sec^{-1}(au) = \theta + c_1$$

At $r = a$ or $u = \dfrac{1}{a}$, $\theta = 0$ $\quad\therefore\quad c_1 = 0$

Hence, $au = \sec\theta$ or $r = a\cos\theta$

which is polar equation of a circle passing through the pole.

Example 4. *A particle subject to a central acceleration $\frac{\mu}{r^3} + f$ is projected from an apse at a distance a with a velocity $\sqrt{\mu}/a$. Prove that at any subsequent time t, $r = a - \frac{1}{2}ft^2$.*

Sol. The equation of the path

$$h^2\left[\frac{d^2u}{dt^2} + u\right] = \frac{P}{u^2} = \frac{\mu u^3 + f}{u^2}$$

Multiplying both sides by $2\frac{du}{d\theta}$ and integrating

$$v^2 = h^2\left[\left(\frac{du}{d\theta}\right)^2 + u^2\right] = \mu u^2 + \frac{2fu^{-1}}{-1} + c \qquad ...(1)$$

At $r = a, u = \frac{1}{a}, \quad v^2 = \frac{\mu}{a^2}, \frac{du}{d\theta} = 0$

$$\frac{\mu}{a^2} = \frac{h^2}{a^2} = \frac{\mu}{a^2} - 2fa + c$$

$\therefore$ $h^2 = \mu$ and $c = 2fa$

Putting in (1)

$$\mu\left[\left(\frac{du}{d\theta}\right)^2 + u^2\right] = \mu u^2 - \frac{2f}{u} + 2fa$$

$$\mu\left[\left(\frac{du}{dt}\cdot\frac{dt}{d\theta}\right)^2 + u^2\right] = \mu u^2 - \frac{2f}{u} + 2fa \qquad ...(2)$$

But $h = r^2\frac{d\theta}{dt} \Rightarrow \sqrt{\mu} = \frac{1}{u^2}\frac{d\theta}{dt}$

$\therefore$ $\frac{dt}{d\theta} = \frac{1}{\sqrt{\mu}u^2}$

Putting in (2)

$$\mu\left[\left(\frac{du}{dt}\right)^2\cdot\frac{1}{\mu u^4} + u^2\right] = \mu u^2 - \frac{2f}{u} + 2fa$$

$$\frac{1}{u^4}\left(\frac{du}{dt}\right)^2 = 2fa - \frac{2f}{u} \qquad u = \frac{1}{r}, \frac{du}{dt} = -\frac{1}{r^2}\frac{dr}{dt}$$

$$r^4\cdot\frac{1}{r^4}\left(\frac{dr}{dt}\right)^2 = 2fa - 2fr$$

or $$\frac{dr}{dt} = \sqrt{2f}\sqrt{a-r}$$

$$\frac{dr}{\sqrt{a-r}} = \sqrt{2f}\,dt$$

Integrating, $-2\sqrt{a-r} = \sqrt{2f}\,t + c$

At $t = 0 \quad r = a \Rightarrow c = 0$

$2(a - r) = ft^2$ or $r = a - \frac{1}{2}ft^2$. Hence Proved.

Example 5. *A particle of mass m is attached to a fixed point by an elastic string of natural length a and co-efficient of elasticity being nmg. It is projected from an apse at a distance a with velocity* $\sqrt{2pgh}$. *Show that the other apsidal distance is given by the equation* $nr^2(r-a) - 2pha(r+a) = 0$.

Sol. The tension in the string is given by Hooke's law

$$T = \frac{\lambda(r-a)}{a} = \frac{nmg(r-a)}{a}$$

$\therefore$ Central acceleration $= \dfrac{ng}{a}(u^{-1} - a)$

Putting in the equation of path

$$h^2\left[\frac{d^2u}{d\theta^2} + u\right] = \frac{P}{u^2} = \frac{ng}{au^2}(u^{-1} - a)$$

Multiplying by $2\dfrac{du}{d\theta}$ and integrating

$$v^2 = h^2\left[\left(\frac{du}{d\theta}\right)^2 + u^2\right] = \frac{ng}{a}[-u^{-2} + 2au^{-1}] + c \qquad \text{...(1)}$$

At an apse $r = a,\ u = \dfrac{1}{a},\ \dfrac{du}{d\theta} = 0,\ v^2 = 2pgh$

$$2pgh = \frac{h^2}{a^2} = \frac{ng}{a}[-a^2 + 2a^2] + c \qquad \text{which gives}$$

$$h^2 = 2pgha^2,\ c = 2pgh - nga$$

Putting h^2 and c in (1), we get

$$2pgha^2\left[\left(\frac{du}{d\theta}\right)^2 + u^2\right] = \frac{ng}{a}\left[-\frac{1}{u^2} + \frac{2a}{u}\right] + 2pgh - nga$$

To find an apse we put $\dfrac{du}{d\theta} = 0,\ u = \dfrac{1}{r}$

$$\frac{2pgha^2}{r^2} = \frac{ng}{a}[-r^2 + 2ar] + 2pgh - nga$$

$$2pha^3 = -nr^4 + 2ar^3n + 2phar^2 - nga^2r^2$$

$$nr^2(r^2 - 2ar + a^2) + 2pha(a^2 - r^2) = 0$$

$\because$
$$nr^2(r-a)^2 + 2pha(a-r)(a+r) = 0$$

$r = a$ is one apsidal distance. Therefore removing one factor $(r-a)$, we get

$$nr^2(r-a) - 2pha(r+a) = 0$$

This equation will give another apsidal distance.

Example 6. *A particle moving under a constant force from a centre is projected at a distance a from the centre in a direction perpendicular to the radius vector with velocity acquired in falling to the point of projection from the centre. Show that its path is* $\left(\dfrac{a}{r}\right)^3 = \cos^2\dfrac{3\theta}{2}$.

(K.U. 2007)

Sol. Let the constant force be $m\mu$

$\therefore$ the acceleration $= \mu$ (constant)

Velocity acquired in falling a distance 'a' from the centre to the point of projection = $\sqrt{2\mu a}$ the equation of the central orbit is

$$\frac{d^2u}{d\theta^2} + u = \frac{P}{h^2u^2}$$

Here $\qquad P = -\mu$

$$\therefore \qquad \frac{d^2u}{d\theta^2} + u = -\frac{\mu}{h^2u^2}$$

or $$h^2\left[\frac{2du}{d\theta}\frac{d^2u}{d\theta^2} + 2u\frac{du}{d\theta}\right] = -\frac{\mu}{u^2}\cdot\frac{2du}{d\theta}$$

Integrating both sides, we get

$$\left(\frac{du}{d\theta}\right)^2 + u^2 = \frac{2\mu}{u} + c$$

or $$v^2 = h^2\left[\left(\frac{du}{d\theta}\right)^2 + u^2\right] = \frac{2\mu}{u} + c \qquad \text{...(1)}$$

At an apse $\qquad u = \frac{1}{a}, \frac{du}{d\theta} = 0, u = \sqrt{2\mu a}$

Putting in (1), we get

$$2\mu a = \frac{h^2}{a^2} = 2\mu a + c \Rightarrow c = 0, h^2 = 2\mu a^3$$

Putting c and h^2 in (1), we get

$$2\mu a^3\left[\left(\frac{du}{d\theta}\right)^2 + u^2\right] = \frac{2\mu}{u}$$

or $$\left(\frac{du}{d\theta}\right)^2 + u^2 = \frac{2\mu}{u^3}$$

or $$\left(\frac{du}{d\theta}\right)^2 = \frac{1}{ua^3} - u^2 = \frac{1 - a^3u^3}{a^3u}$$

or $$\frac{du}{d\theta} = -\frac{\sqrt{1 - a^3u^3}}{a^{3/2}\sqrt{u}}$$

or $$-\frac{a^{3/2}\sqrt{u}\,du}{\sqrt{1 - (a^{3/2}u^{3/2})^2}} = d\theta$$

Substituting $\quad a^{3/2}\, u^{3/2} = z$

or $$\frac{3}{2} a^{3/2}\, u^{1/2}\, du = dz, \text{ we get}$$

$$-\frac{2}{3}\frac{dz}{\sqrt{1-z^2}} = d\theta$$

Integrating, we get

$$\frac{2}{3}\cos^{-1} z = \theta + c$$

$$\cos^{-1} z = \frac{3}{2}\theta + \frac{3}{2}c$$

or $$\cos^{-1} a^{3/2}\, u^{3/2} = \frac{3}{2}\theta + \frac{3}{2}c$$

At $u = \frac{1}{a}$, $\quad \theta = 0 \;\Rightarrow\; \cos^{-1} 1 = 0 + \frac{3}{2}c \;\Rightarrow\; c = 0$

$\therefore$ $$\cos^{-1} a^{3/2}\, u^{3/2} = \frac{3}{2}\theta$$

or $$\left(\frac{a}{r}\right)^{3/2} = \cos\frac{3}{2}\theta$$

or $$\left(\frac{a}{r}\right)^{3} = \cos^2\left(\frac{3}{2}\theta\right)$$

This is the required equation of the path.

EXERCISE 2

1. **A particle moves with a central acceleration which varies as cube of the distance. If it be projected from an apse at a distance a from the origin with a velocity which is $\sqrt{2}$ times the velocity for a circle of radius a, show that equation to the path is $\frac{r\cos\theta}{\sqrt{2}} = a$.**

2. **A particle moves with central acceleration $= \frac{\mu}{(\text{distance})^3}$ and is projected from an apse at a distance a with velocity V. Show that the path is $r\cosh\left[\frac{\sqrt{\mu - a^2V^2}}{aV}\theta\right] = a$ or $r\cos\left[\frac{\sqrt{a^2V^3 - \mu}}{aV}\theta\right] = a$ according as V $\lesseqgtr$ the velocity from infinity.**

3. A particle moves with a central acceleration $\mu\left(r + \frac{a^4}{r^3}\right)$ being projected from an apse at a distance a with a velocity $2\sqrt{\mu a}$. Show that it describes the curves $r^2\,[2 + \cos\sqrt{3}\theta] = 3a^2$.

4. A particle, subject to a force producing an acceleration $\mu\left(\frac{r+2a}{r^5}\right)$ towards origin from the point $(a, 0)$ with a velocity equal to velocity from infinity at an angle $\cot^{-1} z$ with the initial line. Show that the equation to the path is $r = a\,(1 + 2\sin\theta)$.

5. A particle moves under a force $m\mu\,\{3au^4 - 2(a^2 - b^2)\,u^5\}$, $a > b$ and is projected from an apse at a distance $a + b$ with velocity $\sqrt{\mu}\,(a - b)$. Show that equation of the path is $r = a + b\cos\theta$.

6. A particle moves with a central acceleration $\mu\,\{2(a^2 + b^2)\,u^5 - 3a^2b^2\,u^7\}$ is projected from a distance a with velocity $\sqrt{\mu}/a$ in a direction at right angle to the initial distance. Show that the path is the curve $r^2 = a^2\cos^2\theta + b^2\sin^2\theta$.

7. A particle is moving under the force $\mu\,(3u^3 + 2a^2\,u^5)$ and is projected at a distance a with a velocity $\sqrt{\frac{5\mu}{a^2}}$ in a direction making an angle $\tan^{-1}\frac{1}{2}$ with the radius. Show that equation to the path is $r = a\tan\theta$.

8. A particle moves with a central acceleration $\frac{\mu}{(\text{distance})^3}$. Find the path and distinguish the cases.

9. A particle is projected from an apse at a distance a with a velocity from infinity, the acceleration being $\mu\,u^7$. Show that the equation of the path is $r^2 = a^2\cos 2\theta$.

10. A particle acted on by a central force $\frac{\mu}{r^3}$ is projected with velocity $\frac{\sqrt{\mu}}{a}$ at angle $\frac{\pi}{4}$ with its initial distance a from the centre of force. Prove that its orbit is $r = a\bar{e}^{\theta}$.

11. If the law of force be $\mu\left(u^4 - \frac{10}{9}au^5\right)$ and the particle projected from an apse at a distance $5a$ with velocity equal to $\frac{5}{7}$ of that in a circle at the same distance its orbit is $r = a(3 + 2\cos\theta)$.

12. If the central acceleration is μu^5, prove that the velocities at the two apsidal distances satisfy the relation $V_1^2 + V_2^2 = \frac{2h^4}{\mu}$.

In the previous articles we have discussed central orbits under general laws of forces.

Now we shall discuss the motion when the central acceleration follows the Newtonian law of attraction.

Under this law the mutual attraction between every two particles of masses m_1 and m_2 placed at a distance r apart is $\frac{\gamma m_1 m_2}{r^2}$ where γ is a constant known as constant of gravitation. The value of γ is 1.05×10^{-9} approximately, when masses are in pounds and distance is measured in feet. It is 6.669×10^{-8} when masses are in grammes and distance in centimetres.

The above law is found to hold good in case of all planets in the solar system.

8. Kepler's Laws (M.D.U. 2005)

The astronomer Kepler, discovered three laws connecting the motions of various planets about the sun.

(*i*) Each planet describes an ellipse having the sun in one of its foci.

(*ii*) The areas described by the radii drawn from the plant to the sun are, in the same orbit, is proportional to the time of describing them.

(*iii*) The squares of the periodic times of planets are proportional to the cubes of the major axis of their orbits.

From the first law, we find that the orbit is an ellipse and then the law of force must be inverse square distance *i.e.,* the acceleration of each planet varies as the square of its distance from the sun. Kepler's second law implies that the rate of description of the sectorial area is constant. We have seen in article 3 that this is true only when the acceleration of the planet is directed towards the centre.

Also we know from article 4 that periodic time $T = \frac{2\pi a^{3/2}}{\sqrt{\mu}}$. Hence, the square of the time period $T^2 = \frac{\pi^2}{2\mu}(2a)^3$ or $T^2 \propto (2a)^3$ *i.e.,* square of the periodic time varies as the cube of the major axes. Hence from Kepler's third law, we conclude that μ is same for all planets *i.e.,* absolute acceleration μ [acceleration at unit distance $r = 1$ from the sun] is same for all planets.

The first and second laws were enunciated by Kepler in 1609 in his book on the motion of the planet Mars. The third law was given ten years later in a book on the Harmonics of the world. The explanation of these laws was given by Newton in his 'Principle' published in 1687.

Third law is true only when the sun is fixed and mass of the planet is neglected in comparison with that of the sun.

9. Necessary Modification of Kepler's Third Law

Let P and S be the masses of the planet and the sun respectively. Let γ be the constant of gravitation. Then force of attraction between them is $\frac{\gamma P.S}{r^2}$, where r is the distance between the planet and the sun at some instant.

Fig. 5

The planet's acceleration towards the sun is $\frac{\gamma PS}{r^2 . P}$ and the sun's acceleration towards the planet is $\frac{\gamma PS}{r^2 S}$.

Thus, acceleration of the planet relative to the sun is

$$\frac{\gamma S}{r^2} + \frac{\gamma P}{r^2} = \frac{\gamma (P+S)}{r^2} = \frac{\mu}{r^2}, \text{ where } \mu = \gamma (P+S)$$

Thus, exact time period of one complete revolution is

$$T = \frac{2\pi a^{3/2}}{\sqrt{\mu}} = \frac{2\pi a^{3/2}}{\sqrt{\gamma (S+P)}} \quad ...(1)$$

This is the modification required in the third law of Kepler.

Let T_1 be the periodic time of other planet of mass P_1 and let $2a_1$ be the major axis of its orbit then

$$T_1 = \frac{2\pi a_1^{3/2}}{\sqrt{\gamma (S+P_1)}} \quad ...(2)$$

From (1) and (2)

$$\frac{T^2}{T_1^2} = \frac{S+P_1}{S+P} \cdot \frac{a^3}{a_1^3}$$

or

$$\frac{(S+P) T^2}{(S+P_1) T_1^2} = \frac{a^3}{a_1^3} \quad ...(3)$$

If d be the mean distance (*i.e.*, distance between the centres) of the planet from the sun, then we have

$$T = \frac{2\pi d^{3/2}}{\sqrt{\gamma (S+P)}} \quad ...(4)$$

If d_1 be the mean distance of a satellite (of mass p) from the planet (of mass P) then the time period of the satellite

$$t = \frac{2\pi d_1^{3/2}}{\sqrt{\gamma (P+p)}} \quad ...(5)$$

From (4) and (5), we get

$$\frac{T^2}{t^2} = \frac{(P+p)}{(S+P)} \frac{d^3}{d_1^3}$$

or

$$\frac{(S+P) T^2}{(P+p) t^2} = \frac{d^3}{d_1^3} \quad ...(6)$$

10. Motion under Inverse Square Law

If a particle is moving under central acceleration $\frac{\mu}{(\text{distance})^2}$, then discuss its motion.

Since the acceleration $P = \frac{\mu}{r^2}$. (KU 2004)

Pedal equation of the path is

$$\frac{h^2}{p^3}\frac{dp}{dr} = \text{P} = \frac{\mu}{r^2}$$

or $$h^2\frac{dp}{p^3} = \frac{\mu}{r^2}dr$$

Integrating, we have

$$\frac{-h^2}{2p^2} = \frac{-\mu}{r} + c$$

or $$\frac{h^2}{p^2} = \frac{2\mu}{r} + c_1 \qquad \text{But } \frac{h}{v} = p \text{ (Equation (2) article 3)}$$

$$\therefore \quad \frac{h^2}{p^2} = v^2 = \frac{2\mu}{r} + c_1 \qquad ...(1)$$

But pedal equations of ellipse, parabola and hyperbola are

$$\frac{b^2}{p^2} = \frac{2a}{r} - 1,\ p^2 = ar \quad \text{and} \quad \frac{b^2}{p^2} = \frac{2a}{r} + 1 \text{ respectively} \qquad ...(2)$$

respectively where $2a$ and $2b$ are major and minor axes of ellipse, $4a$ is latus rectum of parabola and $2a$, $2b$ are transverse and conjugate axes of hyperbola. Now three cases arise.

Elliptic orbit. Comparing equation (1) with equation of ellipse in (2)

$$\frac{h^2}{b^2} = \frac{2\mu}{2a} = \frac{c_1}{-a} \quad \therefore \quad h^2 = \frac{\mu b^2}{a},\ c_1 = \frac{-\mu}{a}$$

$$\therefore \quad \mathbf{v^2 = \mu\left[\frac{2}{r} - \frac{1}{a}\right]} \qquad ...(3)$$

$\therefore \quad v^2 < \frac{2\mu}{r}$ for elliptic path.

Case I. Parabolic path

Comparing (1) with equation of parabola

$$p^2 = ar$$

or $$\frac{1}{p^2} = \frac{\frac{1}{a}}{r}$$

Comparing with (1)

$$\frac{h^2}{1} = \frac{2\mu}{\frac{1}{a}} = \frac{c_1}{0} \qquad \Rightarrow \quad h^2 = 2\mu a,\ c_1 = 0$$

$$\therefore \quad v^2 = \frac{h^2}{p^2} = \frac{2\mu a}{ar} \qquad ...(4)$$

$\therefore \quad \mathbf{v^2 = \frac{2\mu}{r}}$ **for parabolic path.**

Case II. Hyperbolic path

Comparing (1) with the equation of hyperbola is (2), we get

$$\frac{h^2}{b^2} = \frac{\mu}{a} = \frac{c_1}{1} \quad \text{or} \quad h^2 = \frac{\mu b^2}{a} \quad c_1 = \frac{\mu}{a}$$

$$\therefore \quad v^2 = \frac{h^2}{p^2} = \frac{2\mu}{r} + \frac{\mu}{a} = \mu\left[\frac{2}{r} + \frac{1}{a}\right] \qquad ...(5)$$

$$\mathbf{v^2 = \mu\left[\frac{2}{r} + \frac{1}{a}\right]}$$

$\therefore \quad v^2 > \frac{2\mu}{r}$ for the hyperbolic path.

Periodic time of elliptic orbit

$$T = \frac{\text{Area of ellipse}}{\text{Rate of description of sectorial area}} = \frac{\pi ab}{\frac{h}{2}} = \frac{2\pi ab}{h} = \frac{2\pi ab}{\sqrt{\frac{\mu b^2}{a}}}$$

$$\mathbf{T = \frac{2\pi a^{3/2}}{\sqrt{\mu}}}. \qquad ...(6)$$

11. Perihelion and Aphelion

The point on an elliptic orbit nearest to the centre of force (Focus or Sun in cases of planetary motion) is called the perihelion and the farthest point is aphelion. In the following figure A is perihelion and A′ is aphelion.

Perihelion and aphelion coincide with the apses of the orbit.

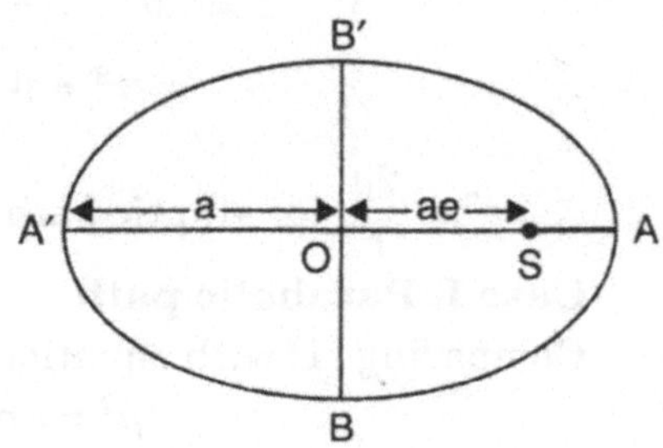

Fig. 6

If $SA = r_1$ and $SA' = r_2$

$$r_1 = OA - OS = a - ae$$

$$\mathbf{r_1 = a(1 - e)}$$

$$\mathbf{r_2 = a(1 + e)}$$

Circular orbit. In case of circular orbit

$$r = a = p$$

and
$$u = \frac{1}{r} = \frac{1}{a} \text{ (constant)}$$

$\therefore$ Differential equation of the path is

$$\frac{d^2u}{d\theta^2} + u = \frac{P}{h^2u^2} \text{ reduces to } P = h^2 u^3$$

But
$$h = pv = av$$

$$\therefore \quad P = a^2 v^2 . \frac{1}{a^3} \quad \text{or} \quad v^2 = aP.$$

12. Velocity from Infinity

If a particle starts from rest at infinity and moves under the central acceleration $P = \mu/r^2$, then

$$v\frac{dv}{dr} = \frac{-\mu}{r^2}$$

or $$\int_0^V v\,dv = -\mu\int_\infty^R \frac{dr}{r^2}$$

$$V^2 = \frac{2\mu}{R} \qquad ...(1)$$

This is the velocity which the particle will acquire in falling from infinity to a distance R from the centre.

If V_1 is the velocity of describing a circle of radius R, then

$$V_1^2 = aP = RP = R.\frac{\mu}{R^2} = \frac{\mu}{R} = \frac{1}{2}\left(\frac{2\mu}{R}\right)$$

$$\therefore \quad V_1^2 = \frac{1}{2}V^2$$

or $$V_1 = \frac{1}{\sqrt{2}}V$$

$\therefore$ Velocity for describing of a circle of radius R

$= \frac{1}{\sqrt{2}}$ velocity from infinity to the distance R from the centre of force.

Example 1. *The greatest and least velocities of a certain planet in its orbit round the sun are 30 km / sec and 29.2 km/sec. Find the ecentricity of the orbit.*

Sol. The velocity at any point on an elliptic path is

$$v^2 = \mu\left[\frac{2}{r} - \frac{1}{a}\right]$$

The velocity is greatest when r is least, and it is least when r is greatest.

Minimum value of r is at perihelion $= a(1-e)$

Maximum value of r is at aphelion $= a(1+e)$

$$\therefore \quad (30)^2 = \mu\left[\frac{2}{a(1-e)} - \frac{1}{a}\right]$$

$$(29.2)^2 = \mu\left[\frac{2}{a(1+e)} - \frac{1}{a}\right]$$

Dividing $$\frac{30\times 30}{29.2\times 29.2} = \frac{\frac{2}{1-e}-1}{\frac{2}{1+e}-1} = \frac{(1+e)^2}{(1-e)^2}$$

$$\therefore \quad \frac{1+e}{1-e} = \frac{30}{29.2} \quad \text{or} \quad e = \frac{1}{74}.$$

Example 2. *A particle moves with a central acceleration $\mu/(distance)^2$. It is projected with velocity V at a distance R. Show that its path is a rectangular hyperbola if angle of projection is* $\sin^{-1}\left[\frac{\mu}{VR\sqrt{V^2 - 2\mu/R}}\right]$.

Sol. In the case of hyperbola the velocity at any point at a distance r from the centre of force is $v^2 = \mu\left[\frac{2}{r} + \frac{1}{a}\right]$

at $r = \text{R}$, $v = \text{V}$

$$\therefore \quad \text{V}^2 = \mu\left[\frac{2}{\text{R}} + \frac{1}{a}\right]$$

or
$$\text{V}^2 - \frac{2\mu}{\text{R}} = \frac{\mu}{a} \qquad ...(1)$$

Let α be the angle of projection and p_0 be the initial value of p.

$$\therefore \quad p_0 = \text{R} \sin \alpha \qquad (\because \ p = r \sin \phi)$$

Also, $h = p_0\text{V} = \text{VR} \sin \alpha$ ...(2)

We also know that $h = \sqrt{\mu l} = \sqrt{\frac{\mu b^2}{a}}$

In case of rectangular hyperbola $a = b$

$$\therefore \quad h = \sqrt{\mu a} \qquad ...(3)$$

Comparing (2) and (3), we get

$$\sqrt{\mu a} = \text{R} \sin \alpha \,.\, \text{V}$$

$$\mu a = \text{R}^2 \sin^2 \alpha \,.\, \text{V}^2$$

or
$$\frac{\mu^2}{\left(\frac{\mu}{a}\right)} = \text{R}^2 \sin^2 \alpha \,.\, \text{V}^2$$

or
$$\frac{\mu^2}{\text{V}^2 - \frac{2\mu}{\text{R}}} = \text{R}^2 \sin^2 \alpha \,.\, \text{V}^2$$

$$\therefore \quad \sin \alpha = \frac{\mu}{\text{VR}\sqrt{\text{V}^2 - 2\mu/\text{R}}}$$

or
$$\alpha = \sin^{-1}\left[\frac{\mu}{\text{VR}\sqrt{\text{V}^2 - 2\mu/\text{R}}}\right].$$

Example 3. *Prove that the time taken by Earth to travel over half its orbit remote from the sun, separated by minor axis is two days more than half the year. The ecentricity of the orbit = 1/60.*

Sol. Time of describing the complete orbit is one year.

So the rate of description of area

$$= \frac{\text{Area of ellipse}}{\text{One year}} = \frac{\pi ab}{1} \qquad ...(1)$$

Half of the orbit remote from the sun is BA′ B′ in which case the sectorial area described is SB A′ B′ S. Let T be the time taken in describing it.

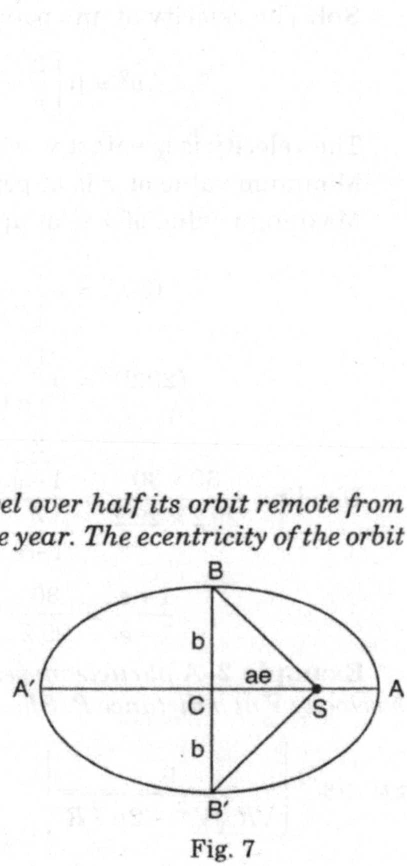

Fig. 7

$\therefore$ Rate of description of area

$$= \text{Area } \frac{SBA'B'S}{T} \qquad ...(2)$$

$\because$ Rate of description of sectorial area is same.

$$\therefore \quad \frac{\pi ab}{1 \text{ year}} = \text{Area } \frac{SBA'B'S}{T} = \frac{\text{Ar. } \Delta SBB' + \frac{1}{2} \text{ area of ellipse}}{T}$$

$$\frac{\pi ab}{1 \text{ year}} = \frac{\frac{1}{2} 2b \,.\, ae + \frac{1}{2} \pi ab}{T}$$

$$\therefore \quad T = \left(\frac{1}{2} + \frac{e}{\pi}\right) \text{ year} \qquad e = \frac{1}{60}$$

$$T = \left(\frac{1}{2} + \frac{1}{60\pi}\right) \text{ year} = \frac{1}{2} \text{ year} + \frac{365}{60 \times 3.1416} \text{ days}$$

$$= \frac{1}{2} \text{ year} + 2 \text{ days nearly.}$$

Example 4. *Two inelastic particles of masses 3m and m are describing the same ellipse of ecentricity e in opposite directions under a force to the focus. They collide and coalesce at an extremity of the minor axis. Prove that ecentricity of the new orbit is* $\frac{1}{4}\sqrt{9 + 7e^2}$.

Sol. Let the particles collide at B. Let velocity of each particle be V each. Since the path is elliptic, we have

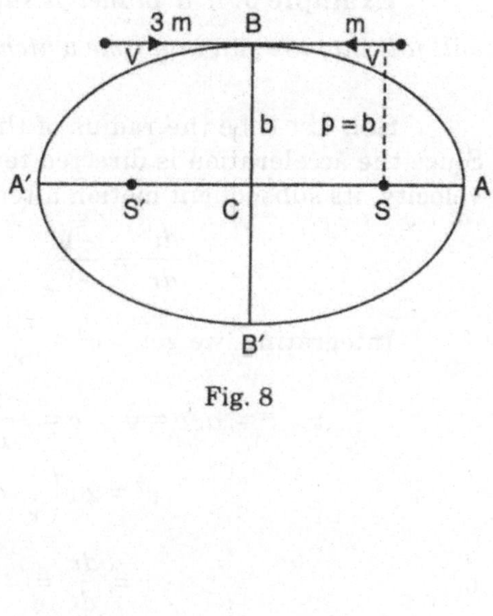

Fig. 8

$$V^2 = \mu \left[\frac{2}{SB} - \frac{1}{a}\right]$$

$$SB + S'B = 2a, \; SB = SB'$$

$$\Rightarrow \quad SB = a$$

$$V^2 = \mu \left[\frac{2}{a} - \frac{1}{a}\right] = \frac{\mu}{a} \qquad ...(1)$$

Let V_1 be the velocity of the combined body after into which the particles coalesce after impact.

$\therefore$ By conservation of momentum

$$3mV + m(-V) = 4mV_1$$

$$V_1 = \frac{V}{2}$$

or

$$V_1^2 = \frac{V^2}{4} = \frac{\mu}{4a} \qquad \text{using (1)} \qquad ...(2)$$

Let $2a'$ be the major axis of new orbit

$$\therefore \quad V_1^2 = \mu \left[\frac{2}{SB} - \frac{1}{a'}\right]$$

Putting V_1^2 from equation (2)

$$\frac{\mu}{4a} = \mu\left[\frac{2}{a} - \frac{1}{a'}\right] \Rightarrow \frac{1}{a'} = \frac{2}{a} - \frac{1}{4a}$$

or $$a' = \frac{4}{7}a \qquad ...(3)$$

We know that $Vp = h = \sqrt{\mu l}$

$$V_1 b = \sqrt{\mu \frac{b'^2}{a'}}$$ b' is semi minor axis of new orbit and $V = V_1$

$$V_1^2 b^2 = \frac{\mu b'^2}{a'}$$

or $$V_1^2 a^2 (1 - e^2) = \frac{\mu a'^2 (1 - e'^2)}{a'}$$ e' is ecentricity of new orbit

Putting the value of V_1^2, we get

$$\frac{\mu}{4a} a^2 (1 - e^2) = \mu a' (1 - e'^2) \qquad \text{Using (3)}$$

$$\frac{1}{4} a (1 - e^2) = \frac{4}{7} a (1 - e'^2)$$

or $$1 - e'^2 = \frac{7}{16}(1 - e^2). \quad \text{or} \quad e' = \frac{1}{4}\sqrt{9 + 7e^2}.$$

Example 5. *If a planet is suddenly stopped in its orbit supposed circular, show that it will fall into the sun in a time which is $\sqrt{2}/8$ times the period of the planet's revolution.*

(K.U. 2007)

Sol. Let a be the radius of the circular path. The sun S is at the centre of the circle. Since the acceleration is directed towards the sun, and planet P is suddenly reduced to zero velocity, its subsequent motion after release will be towards the sun.

$$\therefore \quad v\frac{dv}{dr} = \frac{-\mu}{r^2}$$

Integrating, we get $v^2 = \frac{2\mu}{r} + c$

$$\therefore \text{ at } r = a, v = 0 \quad c = \frac{-2\mu}{a}$$

$$\therefore \quad v^2 = 2\mu\left(\frac{1}{r} - \frac{1}{a}\right)$$

or $$v = \frac{dr}{dt} = -\sqrt{2\mu\left(\frac{a - r}{ar}\right)}$$

Negative sign is taken as r decreases with the increase of t.

Let T_1 be the time taken by the planet to reach the sun

$$\int_0^{T_1} -\sqrt{\frac{2\mu}{a}}\, dt = \int_a^0 \sqrt{\frac{r}{a - r}}\, dr$$

$$r = a\sin^2\theta$$
$$dr = 2a\sin\theta\cos\theta\, d\theta$$

$$-\sqrt{\frac{2\mu}{a}}\,T_1 = -\int_0^{\pi/2}\sqrt{\frac{a\sin^2\theta}{a\cos^2\theta}}\,.\,2a\sin\theta\cos\theta\, d\theta$$

$$\sqrt{\frac{2\mu}{a}}\,T_1 = 2a\int_0^{\pi/2}\sin^2\theta\, d\theta = 2a\,.\,\frac{1}{2}\,.\frac{\pi}{2}$$

$$T_1 = \frac{2\pi a^{3/2}}{4\sqrt{2\mu}} = \frac{1}{4\sqrt{2}}\left(\frac{2\pi a^{3/2}}{\sqrt{\mu}}\right) = \frac{\sqrt{2}}{8}T$$

where T is time period of planet's revolution.

EXERCISE 3

1. **A particle is projected from the earth's surface with velocity v. Show that if dimunition of gravity is taken into account and resistance of air is reglected, the path is an ellipse of major axis $\frac{2ga^2}{(2ga - v^2)}$, where a is radius of the earth.**

2. **A particle describes an ellipse under a force μ/r^2 towards the focus. If it was projected with velocity V from a point distance r from the centre of force, show that its periodic time is $\frac{2\pi}{\sqrt{\mu}}\left[\frac{2}{r} - \frac{V^2}{\mu}\right]^{-3/2}$.**

3. **If a particle describes an ellipse as a central orbit about the focus, the velocity at the end of the minor axis is geometric mean between the velocities at the ends of any diameter.**

4. **If v_1 and v_2 are linear velocities of a planet when it is respectively nearest to and farthest from the sun, prove that $(1 - e)v_1 = (1 + e)v_2$.**

5. **A body is describing an ellipse of ecentricity $\frac{1}{5}$ under the action of a force tending to a focus. When the body is at the nearer apse, the centre of force is transferred to the other focus. Prove that ecentricity of the new orbit is $\frac{4}{5}$.**

6. **Two inelastic particle of mass 3m and m are describing the same ellipse of ecentricity $\frac{1}{7}$ in opposite direction under a force to the focus. They collide and coalesce at an extremity of minor axis. Prove that ecentricity of new orbit is $\frac{2}{\sqrt{7}}$.**

7. **Show that the velocity of a planet at any point of its orbit is the same as it would have been if it had fallen to the point from rest at a distance from the sun equal to the length of the major axis.**

8. **Prove that if the velocity of a particle when it is at a distance r from the focus is v in a direction making an angle ϕ with the radius vector, then $e^2\mu^2 = (v^2r - \mu)^2\sin^2\phi + \mu^2\cos^2\phi$.**

9. **A particle describes an ellipse under a force to the focus S. When the particle is at one extremity of the minor axis, its kinetic energy is doubled without any change in the direction of motion. Prove that the particle proceeds to describe a parabola.**

10. A comet is moving in a parabola about the sun at focus. When it is at the end of its latus rectum its velocity suddenly changed in the ratio $n : 1$ $(n < 1)$. Show that the comet will describe an ellipse whose ecentricity is $\sqrt{1-2n^2+2n^4}$ and major axis $\frac{l}{1-n^2}$, where l is semi latus rectum of parabolic path.

11. A planet is describing an ellipse about the sun as focus. Show that its velocity away from the sun is greatest when the radius vector to the planet is at right angles to the major axis of its path and that it then is $\frac{2\pi ae}{T\sqrt{1-e^2}}$, where T is periodic time.

12. A body describing an ellipse of ecentricity e under the action of a central force. When it is at the nearer apse the centre of force is transferred to the other focus. Prove that ecentricity of the new orbit is $e(3+e)/(1-e)$.

13. If the velocity of the earth at any point of its orbit assumed circular, were increased by about one half, prove that it would describe a parabola about the sun as focus. Show also that if a body were projected from the earth with a velocity exceeding 11.2 km/sec ; it will not return to the earth.

14. Prove that Keplar's laws for Planetary motion is equivalent to Newton's law of Gravitation.

(K.U. 2005)

14

Motion of a Particle in Three Dimension

1. Motion of a Particle in Three Dimension

A rigid body is defined as a system of particles for which the distances between the particles remains unchanged. Thus if each particle of such a body is located from reference axis attached to and rotating with the body, there would be no change in any position vector as measured from these axis. In this chapter all of the material is based on the assumption of rigidity. A rigid body has its position determined if we know the three coordinates of any given point G and the angles which any two lines GA and GB fixed in a body make with the axis of coordinates.

Types of motion. In solving problems of rigid body kinematics it is extremely important to identify the type of motion involved.

(*i*) **Translation.** In this motion all lines in the body remain parallel to their original position.

(*ii*) **Fixed-axis rotation.** When a body rotates about a fixed line, all points in the body such as A and B, rotate in circles whose planes are perpendicular to the rotation axis.

(*iii*) **General plane motion.** Motion in which all points in a rigid body move in parallel planes. The rolling cylinder on a plane is an example of such motion.

(*iv*) **Fixed point rotation.** When a rigid body is free to pivot about any axis, through a single fixed point, the motion is known as fixed point rotation.

(*v*) **General space motion.** A regid body which does not have any such restrictions mentional above is said to move with general space motion.

2. Velocity and Acceleration of a Particle along a Curve

Consider a particle moving along a curve C, not necessarily a plane curve.

Let the particle be at P at time t and at Q at time $t + \delta t$. Let position vectors of P and Q be $\vec{r}$ and $\vec{r} + \vec{\delta r}$ respectively. $\vec{OP} = \vec{r}$ and $\vec{OQ} = \vec{r} + \vec{\delta r}$

$$\therefore \quad \vec{PQ} = \vec{\delta r}$$

$$\text{The velocity at P} = \lim_{\delta t \to 0} \frac{\delta r}{\delta t}$$

$$\therefore \quad \vec{V} = \frac{d\vec{r}}{dt}$$

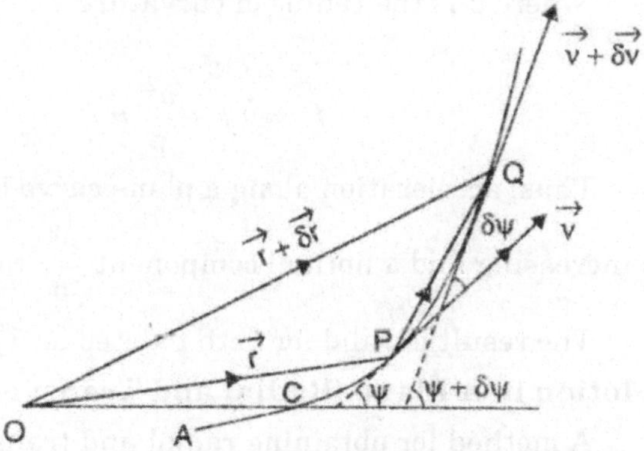

Fig. 1

If arc $\widehat{PQ} = \delta s$ then $\frac{d\vec{r}}{dt} = \frac{d\vec{r}}{ds}\frac{ds}{dt}$

where $\frac{d\vec{r}}{ds} = \lim_{Q \to P} \frac{\delta \vec{r}}{\delta s}$ and is in the direction of tangent at P.

Let $\vec{t}$ denote the unit tangent vector along the tangent at P then

$$\vec{v} = \frac{d\vec{r}}{dt} = \frac{d\vec{r}}{ds}\frac{ds}{dt} = v\vec{t} \qquad ...(1)$$

$$v = \frac{ds}{dt} = \dot{s} \text{ is magnitude of the velocity vector } \vec{v}$$

Acceleration at P is denoted by $\vec{f}$ is rate of change of velocity vector

$$\therefore \qquad \vec{f} = \frac{d}{dt}(\vec{v}) = \frac{d}{dt}(v\vec{t}) = \frac{dv}{dt}\vec{t} + v\frac{d}{dt}\vec{t} = \dot{v}\vec{t} + v\frac{d\vec{t}}{dt}$$

$$= \dot{v}\vec{t} + v\frac{d\vec{t}}{dt} = \dot{v}\vec{t} + v^2\frac{d\vec{t}}{ds} \qquad ...(2)$$

$\frac{d\vec{t}}{ds}$ denotes the rate of change tangent vector and acts in the direction perpendicular to $\vec{t}$ at

P. Let $\vec{n}$ denote the unit vector in such a direction. $\left|\frac{d\vec{t}}{ds}\right|$ is curvature at P and is denoted

by κ.

$$\therefore \qquad \vec{f} = \dot{v}\vec{t} + kv^2\vec{n} \qquad ...(3)$$

where $\vec{n}$ is called the principal normal at P and is measured inwards towards the centre of curvature at P $\because \kappa = \frac{1}{\rho}$

$\kappa = \frac{1}{\rho}$ where ρ is the radius of curvature

$$\therefore \qquad \vec{f} = \dot{v}\vec{t} + \frac{v^2}{\rho}\vec{n}.$$

Thus, acceleration along a plane curve has a tangential component $\dot{v}$ in the direction of s increasing and a normal component $\frac{v^2}{\rho}$ towards the centre of curvature.

The result is valid for both twisted and plane curves.

3. Motion in a Plane (Radial and Transverse Components)

A method for obtaining radial and transverse velocity and acceleration is explained in chapter *Radial and Transverse velocities* article 3.

$$\vec{v} = \dot{r}\,\hat{r} + r\dot{\theta}\,\hat{n}$$

$\dot{r}$ and $r\dot{\theta}$ are radial and transverse velocities acting along $\hat{r}$ and normal directions $\hat{n}$ respectively.

$$\vec{f} = (\ddot{r} - r\dot{\theta}^2)\,\hat{r} + (2\dot{r}\dot{\theta} + r\ddot{\theta})\,\hat{n}$$

$\ddot{r} - r\dot{\theta}^2$ is radial component of acceleration and $2\dot{r}\dot{\theta} + r\ddot{\theta}$ is transverse component of accelerations.

4. Relative Velocity and Acceleration

Let P and Q be the points having position vectors $\vec{r}_p$ and $\vec{r}_q$ with respect to a fixed origin O at time t. Let the coordinates of P and Q be (x_p, y_p, z_p) and (x_q, y_q, z_q) respectively.

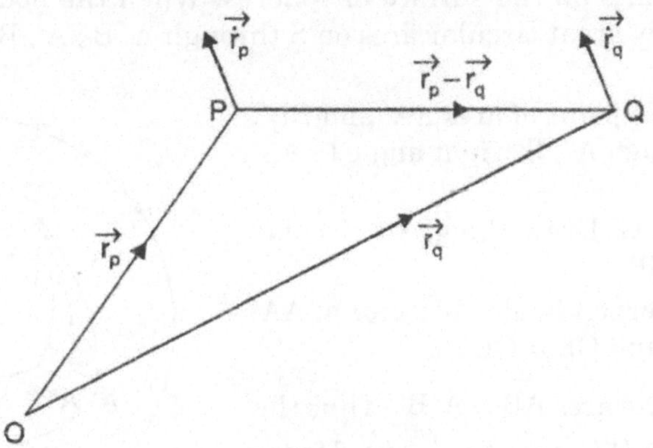

Fig. 2

then $\quad \vec{r}_p = x_p\,\hat{i} + y_p\,\hat{j} + z_p\,\hat{k}$ and $\vec{r}_q = x_q\,\hat{i} + y_q\,\hat{j} + z_q\,\hat{k}$

Velocity and acceleration of P and Q are given as

$$\dot{\vec{r}}_p = \dot{x}_p\,\hat{i} + \dot{y}_p\,\hat{j} + \dot{z}_p\,\hat{k}$$

$$\dot{\vec{r}}_q = \dot{x}_q\,\hat{i} + \dot{y}_q\,\hat{j} + \dot{z}_q\,\hat{k}$$

$$\ddot{\vec{r}}_p = \ddot{x}_p\,\hat{i} + \ddot{y}_p\,\hat{j} + \ddot{z}_p\,\hat{k}$$

$$\ddot{\vec{r}}_q = \ddot{x}_q\,\hat{i} + \ddot{y}_q\,\hat{j} + \ddot{z}_q\,\hat{k}$$

Relative velocity and acceleration of P with respect to Q is

$$\dot{\vec{r}}_p - \dot{\vec{r}}_q = (\dot{x}_p - \dot{x}_q)\,\hat{i} + (\dot{y}_p - \dot{y}_q)\,\hat{j} + (\dot{z}_p - \dot{z}_q)\,\hat{k}$$

$$\ddot{\vec{r}}_p - \ddot{\vec{r}}_q = (\ddot{x}_p - \ddot{x}_q)\,\hat{i} + (\ddot{y}_p - \ddot{y}_q)\,\hat{j} + (\ddot{z}_p - \ddot{z}_q)\,\hat{k}.$$

5. Rotation of a Rigid Body About a Fixed Point

Let a rigid body is moving quite generally in space. Let P_1, P_2, P_3 be three non-collinear points in the body. Let (x_i, y_i, z_i), $i = 1, 2, 3$ be the co-ordinates of these points. Thus to fix the position of these points 9 co-ordinates are required. But since the body is rigid, distances $P_1 P_2$, P_2P_3, P_3P_1 are constant through out the motion. Thus we have three equations of the type

$(x_1 - x_2)^2 + (y_1 - y_2)^2 + (z_1 - z_2)^2 = \text{constant}.$

Hence, only six independent quantities are required. Thus the motion has six degree of freedom.

Suppose the body rotate about a fixed point O. Its degree of freedom reduces to three. Two direction cosine are required to fix the direction of the axis of rotation through O and an angular rotation about this axis completely specifies the motion of such a body. Since the shape and size of the body are not involved in the motion we may consider a sphere S fixed in the body and having its centre at O.

Let A and B be two points on the surface of spheres. When the body rotates A and B moves to positions A′, B′. Draw great circular arcs on S through A, B ; A′, B′ and also through A, A′ ; B, B′.

Let A″, B″ be the middle point of arcs AA′ and BB′. Draw great circular arcs through A″, B″ right angle to AA′ and BB′.

Let these arcs meet at C. Draw the great circular arcs CA, CA′, CA″, CB, CB′, CB″.

∵ CA″ and CB″ are perpendicular bisector of AA′ and BB′, therefore CA = CA′ and CB = CB′.

Since the body is rigid the arcs $\overset{\frown}{AB} = \overset{\frown}{A'B'}$. Thus the spherical triangle ABC and A′B′C are congruent. Hence the portion of the sphere in triangle ABC is now within A′B′C. During the rotation C has remain fixed and hence the rotation has taken place about OC. Let this rotation be $\delta\theta$ in time δt, where $\delta\theta$ is the angle between the planes AOC, A′OC and between the planes BOC and B′OC.

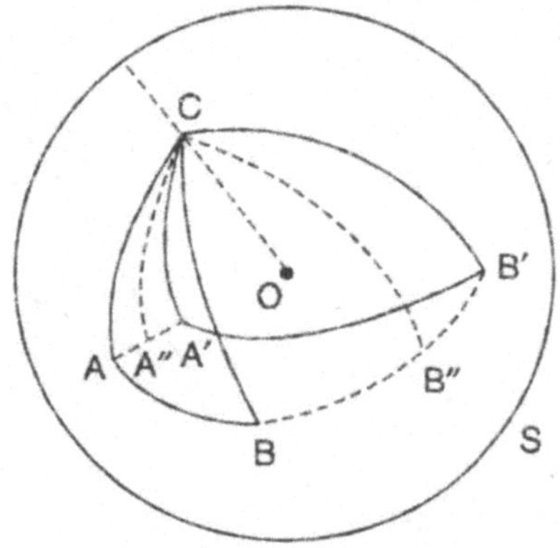

Fig. 3

∴ Mean angular velocity in time δt is $\delta\theta/\delta t$ and instantaneous angular velocity at time t about OC is

$$\omega = \lim_{\delta t \to 0} \frac{\delta\theta}{\delta t} = \frac{d\theta}{dt}.$$

Thus, any rotation about a point O is equivalent to a rotation about a definite axis, through O. This axis about which body rotates is called the instantaneous axis of rotation.

6. Vector Angular Velocity

Let a rigid body rotate about a fixed point O with angular velocity ω about instantaneous axis OC at time t. Let P be a point on the rigid body such that $\overrightarrow{OP} = \vec{r}$. Draw PM ⊥ OC.

Let $\angle POM = \theta$.

∴ $PM = r \sin\theta$

Linear speed P of the particle is $\omega . r \sin\theta$.

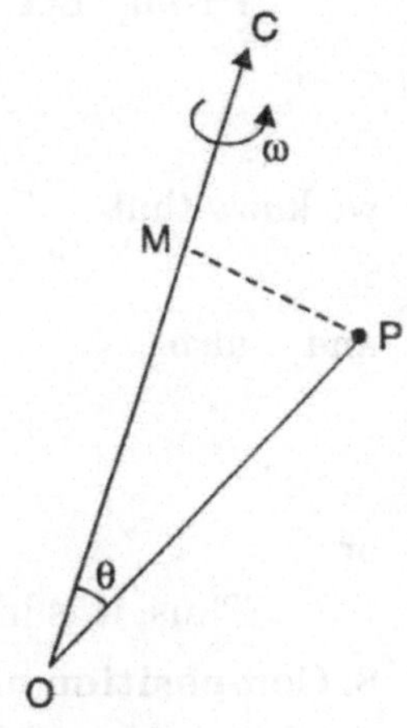

Fig. 4

The velocity $\vec{V}$ of P has the direction of $\hat{a} \times \vec{r}$. Where $\hat{a}$ is the unit vector along $\vec{OC}$ such that a right hand screw advancing along it in the sense of ω.

If $\hat{t}$ is the unit vector along $\hat{a} \times \vec{r}$ then

$$\vec{V} = (\omega r \sin\theta)\,\hat{t} = \vec{\omega} \times \vec{r} \text{ where } \vec{\omega} = \omega \hat{a}$$

$\vec{\omega}$ is the instantaneous vector angular velocity of rigid body about O.

$$\text{If } \vec{\omega} = \omega_1 \hat{i} + \omega_2 \hat{j} + \omega_3 \hat{k}$$

$$\vec{V} = \vec{\omega} \times \vec{r} = \begin{vmatrix} i & j & k \\ \omega_1 & \omega_2 & \omega_3 \\ x & y & z \end{vmatrix}$$

$$\vec{V} = (\omega_2 z - \omega_3 y)\hat{i} + (\omega_3 x - \omega_1 z)\hat{j} + (\omega_1 y - \omega_2 x)\hat{k}$$

$$\text{Curl } \overline{V} = \begin{vmatrix} i & j & k \\ \dfrac{\partial}{dx} & \dfrac{\partial}{dy} & \dfrac{\partial}{dz} \\ \omega_2 z - \omega_3 y & \omega_3 x - \omega_1 z & \omega_1 y - \omega_2 x \end{vmatrix}$$

$$= \hat{i}(\omega_1 + \omega_1) + \hat{j}(\omega_2 + \omega_2) + \hat{k}(\omega_3 + \omega_3) = 2(\omega_1 \hat{i} + \omega_2 \hat{j} + \omega_3 \hat{k}) = 2\vec{\omega}$$

$$\therefore \quad \vec{\omega} = \frac{1}{2} \text{curl } \vec{V}.$$

7. General Motion of a Rigid Body

When no point of a rigid body is fixed, we may consider a point O which is displaced to some point O′ followed by an appropriate turning about a certain axis through O′. Thus in general the motion is described as a translation and rotation.

When we take the limit O′ → O the motion is specified by a linear velocity $\vec{V}_o$ and angular velocity ω about O.

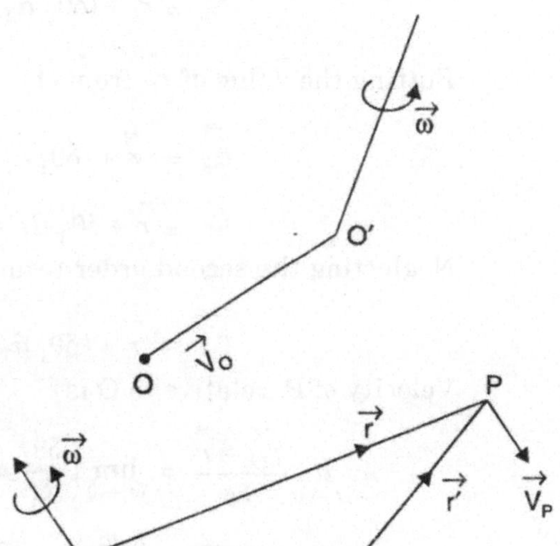

Fig. 5

Suppose P is any other point of the body such that $\vec{OP} = \vec{r}$ and velocity of P is $\vec{V}_P$. Velocity of P relative to O is $\vec{\omega} \times \vec{r}$

$$\text{Thus,} \quad \vec{V}_P = \vec{V}_o + \vec{\omega} \times \vec{r} \qquad \text{...(1)}$$

where V_O is velocity of O.

We now prove that $\vec{\omega}$ is independent of the choice of O.

Proof. Let $\vec{OO'} = \vec{s}$, $\vec{O'P} = \vec{r'}$, then

$$\vec{OP} = \vec{r} = \vec{s} + \vec{r'} \quad ...(2)$$

we know that $$\vec{V_P} = \vec{V_O} + \vec{\omega} \times \vec{r}$$

and also $$\vec{V_O}' = \vec{V_O} + \vec{\omega} \times \vec{s}$$

$\therefore$ $$\vec{V_p} - \vec{V_O}' = \vec{\omega} \times (\vec{r} - \vec{s}) \quad \text{using (2)}$$

or $$\vec{V_P} = \vec{V_O}' + \vec{\omega} \times \vec{r'}$$

Thus, it is independent of the choice of O.

8. Composition of Angular velocities

Let O and P be two fixed point on a rigid body such that $\vec{OP} = \vec{r}$. Let the body takes a small turn $\delta\theta_1$ about an axis $\hat{a}_1$ through O in positive sense.

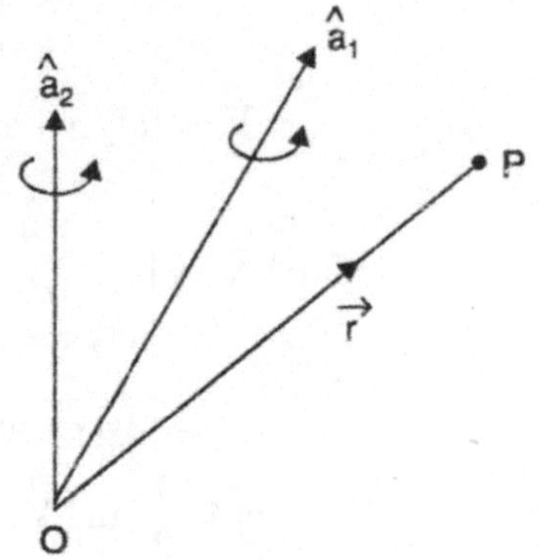

Fig. 6

Then the moment of P is $(\delta\theta_1\ \hat{a}_1) \times \hat{r}$.

$\therefore$ New position of P is

$$\vec{r_1} = \vec{r} + (\delta\theta_1\ \hat{a}_1) \times \bar{r} \quad ...(1)$$

Suppose the body rotates another small angle $\delta\theta_2$ about the axis $\hat{a}_2$ through O.

Hence, the new position of P relative to O is

$$\vec{r_{12}} = \vec{r_1} + (\delta\theta_2\ \hat{a}_2) \times \vec{r_1}$$

Putting the value of $\vec{r_1}$ from (1)

$$\vec{r_{12}} = \vec{r} + (\delta\theta_1\ \hat{a}_1) \times \vec{r} + (\delta\theta_2\ \hat{a}_2) \times \{\vec{r} + (\delta\theta_1\ \hat{a}_1) \times \vec{r}\}$$

$$= \vec{r} + \delta\theta_1\ \hat{a}_1 \times \vec{r} + \delta\theta_2\ \hat{a}_2 \times \vec{r} + \delta\theta_2\ \delta\theta_1\ \hat{a}_2 \times (\hat{a}_1 \times \vec{r})$$

Neglecting the second order terms, we have

$$\vec{r_{12}} = \vec{r} + (\delta\theta_1\ \hat{a}_1 + \delta\theta_2\ \hat{a}_2) \times \vec{r}$$

Velocity of P relative to O is

$$\underset{\delta t \to 0}{lt} \frac{\vec{r_{12}} - \vec{r}}{\delta t} = \lim_{\delta t \to 0} \left(\frac{\delta\theta_1}{\delta t} \hat{a}_1 + \frac{\delta\theta_2}{\delta t} \hat{a}_2 \right) \times \vec{r}$$

$$\vec{V} = \left(\frac{d\theta_1}{dt} \hat{a}_1 + \frac{d\theta_2}{dt} \hat{a}_2 \right) \times \vec{r} = (\vec{\omega_1} + \vec{\omega_2}) \times \vec{r}$$

This result confirms that vector angular velocities about a point of a rigid body are additive in the same way as other vector quantities.

9. Velocity and Acceleration of Moving Axes

Let the cartesian coordinate frame OX, OY, OZ is rotating about an axis through O with angular velocity

$$\vec{\omega} = \omega_1 \hat{i} + \omega_2 \hat{j} + \omega_3 \hat{k}$$

O is considered fixed for a moment.

Let P be a point in space such that

$$\vec{OP} = \vec{r} = x\hat{i} + y\hat{j} + z\hat{k}$$

Then velocity of P is

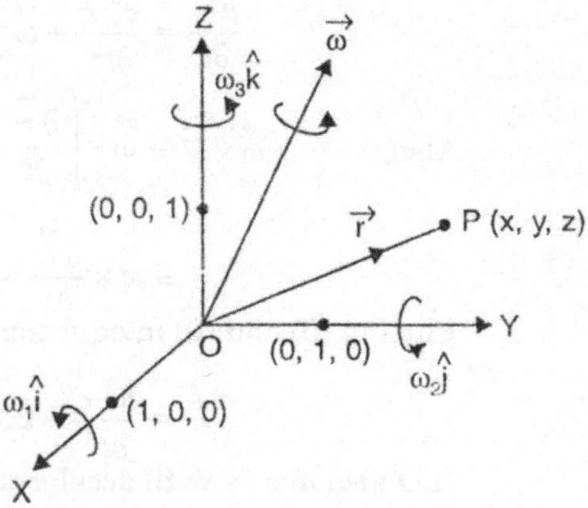

Fig. 7

$$\vec{V} = \frac{d\vec{r}}{dt}$$

$$= \frac{d}{dt}(x\,i + y\,j + z\,\hat{k})$$

$$= \dot{x}\,\hat{i} + x\frac{d\hat{i}}{dt} + \dot{y}\,\hat{j} + y\frac{d\hat{j}}{dt} + \dot{z}\,\hat{k} + z\frac{d\hat{k}}{dt}$$

or $$\vec{V} = (\dot{x}\,\hat{i} + \dot{y}\,\hat{j} + \dot{z}\,\hat{k}) + x\frac{d\hat{i}}{dt} + y\frac{d\hat{j}}{dt} + z\frac{dk}{dt} \qquad ...(1)$$

Since the frame is not fixed, therefore $\frac{d\hat{i}}{dt}, \frac{d\hat{j}}{dt}$ and $\frac{d\hat{k}}{dt}$ are not zero. Now $\frac{d\hat{i}}{dt}$ is the velocity of point $\hat{i}$ (1,0, 0) due to spin which is given by

$$\frac{d\hat{i}}{dt} = \vec{\omega} \times \hat{i} = \begin{vmatrix} i & j & k \\ \omega_1 & \omega_2 & \omega_3 \\ 1 & 0 & 0 \end{vmatrix} = \omega_3\hat{j} - \omega_2\,\hat{k}$$

Similarly, $\frac{d\hat{j}}{dt} = \omega_1\hat{k} - \omega_3\,\hat{i}$ and $\frac{d\hat{k}}{dt} = \omega_2\hat{j} - \omega_1\,\hat{j}$

Putting these values in (1), we get

$$\therefore \quad \vec{V} = \dot{x}\,\hat{i} + \dot{y}\,\hat{j} + \dot{z}\,\hat{k} + x(\omega_3\,\hat{j} - \omega_2\,\hat{k}) + y(\omega_1\,\hat{k} - \omega_3\,\hat{i}) + z(\omega_2\,\hat{i} - \omega_1\,\hat{j})$$

$$= \dot{x}\,\hat{i} + \dot{y}\,\hat{j} + \dot{z}\,\hat{k} + \hat{i}\,(z\omega_2 - y\omega_3) + \hat{j}\,(x\omega_3 - z\omega_1) + \hat{k}\,(y\omega_1 - x\omega_2)$$

$$\vec{V} = \frac{\partial\vec{r}}{\partial t} + \vec{\omega} \times \vec{r} \qquad ...(2)$$

Here, partial derivative means the motion is relative to frame. If we allow the origin to move with velocity v_O, the velocity of V is

$$\vec{V}_P = \vec{v}_O + \frac{\partial\vec{r}}{\partial t} + \vec{\omega} \times \vec{r} \qquad ...(3)$$

Now acceleration of P is f and acceleration of origin is f_0, then

If we first consider O fixed

$$\vec{f} = \frac{d\vec{V}}{dt} = \frac{\partial\vec{V}}{\partial t} + \vec{\omega} \times \vec{V} \qquad ...(4)$$

Differentiating (2)

$$\frac{\partial \vec{V}}{\partial t} = \frac{\partial^2 \vec{r}}{\partial t^2} + \vec{\omega} \times \frac{\partial \vec{r}}{\partial t} + \dot{\vec{\omega}} \times \vec{r} \quad \text{...(5)}$$

Also, $$\vec{\omega} \times \vec{V} = \vec{\omega} \times \left(\frac{\partial \vec{r}}{\partial t} + \vec{\omega} \times \vec{r} \right)$$

$$= \vec{\omega} \times \frac{\partial \vec{r}}{\partial t} + \vec{\omega} \times (\vec{\omega} \times \vec{r}) \quad \text{...(6)}$$

Putting (5) and (6) in equation (4), we get

$$\vec{f} = \frac{\partial^2 \vec{r}}{\partial t^2} + 2\vec{\omega} \times \frac{\partial \vec{r}}{\partial t} + \dot{\vec{\omega}} \times \vec{r} + \vec{\omega} \times (\vec{\omega} \times \vec{r}) \quad \text{...(7)}$$

If O also moves with acceleration f_0,

then $$\vec{f} = \vec{f_0} + \frac{\partial^2 \bar{r}}{\partial t^2} + 2\bar{\omega} \times \frac{\partial \bar{r}}{\partial t} + \dot{\vec{\omega}} \times \vec{r} + \vec{\omega} \times (\vec{\omega} \times \vec{r}) \quad \text{...(8)}$$

10. Acceleration in Terms of Cylinderical Coordinates (M.D.U. 2005)

Let OX, OY, OZ be the cartesian system of axis. P(x, y, z) be cartesian coordinates. From P draw PM perpendicular on XY plane. Let ∠MOX = θ, OM = R.

Let (R, θ, Z) be the coordinates in cylinderical coordinate system.

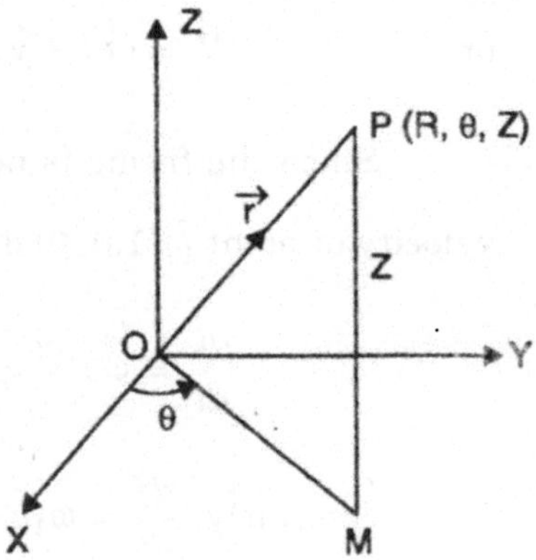

Fig. 8

Let $$\vec{OP} = \vec{r}$$

$$\bar{r} = R\hat{R} + \theta\hat{\theta} + Z\hat{Z} \quad \text{...(1)}$$

where $\hat{R}$, $\hat{\theta}$ and $\hat{Z}$ be the unit vectors in the direction of R, θ, Z increasing. These constitute a right hand system for which $\hat{R} \times \hat{\theta} = \hat{Z}$, $\hat{\theta} \times \hat{Z} = \hat{R}$ and $\hat{Z} \times \hat{R} = \hat{\theta}$. If P is moving with angular velocity ω, then

$$\vec{\omega} = \dot{\theta}\hat{Z}$$

∴ Velocity of P is $$\vec{V} = \frac{\partial \vec{r}}{\partial t} + \vec{\omega} \times \vec{r} \quad \text{...(2)}$$

From (1), $$\frac{\partial \vec{r}}{\partial t} = \dot{R}\hat{R} + \dot{Z}\hat{Z}$$

Putting in equation (2), we have

$$\vec{V} = \dot{R}\hat{R} + \dot{Z}\hat{Z} + \dot{\theta}\hat{Z} \times (R\hat{R} + Z\hat{Z})$$

$$\vec{V} = \dot{R}\hat{R} + \dot{Z}\hat{Z} + \dot{\theta}R\hat{\theta}$$

$$\vec{V} = \dot{R}\hat{R} + R\dot{\theta}\hat{\theta} + \dot{Z}\hat{Z} \quad \text{...(3)}$$

The acceleration of P is

$$\vec{f} = \frac{\partial \vec{V}}{\partial t} + \vec{\omega} \times \vec{V} \qquad \text{...(4)}$$

From (3), $$\frac{d\vec{V}}{dt} = \ddot{R}\,\hat{R} + \ddot{Z}\,\hat{Z} + (R\,\ddot{\theta} + \dot{\theta}\,\dot{R})\,\hat{\theta}$$

Putting in (4)

$$\vec{f} = \ddot{R}\,\hat{R} + (R\,\ddot{\theta} + \dot{R}\,\dot{\theta})\,\hat{\theta} + \ddot{Z}\,\hat{Z} + \dot{\theta}\,\hat{Z} \times (\dot{R}\,\hat{R} + R\,\dot{\theta}\,\hat{\theta} + \dot{Z}\,\hat{Z})$$

$$\vec{f} = \ddot{R}\,\hat{R} + (R\,\ddot{\theta} + \dot{R}\,\dot{\theta})\,\hat{\theta} + \ddot{Z}\,\hat{Z} + \dot{\theta}\,\dot{R}\,\hat{Z} \times \hat{R} + R\dot{\theta}^2\hat{Z} \times \hat{\theta} + 0$$

$$= \ddot{R}\,\hat{R} + (R\,\ddot{\theta} + \dot{R}\,\dot{\theta})\,\hat{\theta} + \ddot{Z}\,\hat{Z} + \dot{\theta}\,\dot{R}\,\hat{\theta} - R\dot{\theta}^2\,\hat{R}$$

$$= (\ddot{R} - R\,\dot{\theta}^2)\;\hat{R} + (R\,\ddot{\theta} + 2\,\dot{R}\,\dot{\theta})\,\hat{\theta} + \ddot{Z}\,\hat{Z} \qquad \text{...(5)}$$

If Z is constant then $\vec{f} = (\ddot{R} - R\dot{\theta}^2)\;\hat{R} + \ddot{\theta} + 2\dot{R}\dot{\theta})\hat{\theta}$ where $\ddot{R} - R\dot{\theta}^2$ and $R\,\ddot{\theta} + 2\,\dot{R}\,\dot{\theta}$ are radial and transverse components in plane.

11. Acceleration in Terms Spherical Coordinates

Let OX, OY, OZ be fixed set of rectangular axis. Let $P(x, y, z)$ be the cartesian coordinates and $P(r, \theta, \phi)$ be spherical coordinates.

$$OP = r,\ \angle POZ = \theta$$

$\angle MOX = \phi$, where M is the foot of $\perp$ from P on XOY plane.

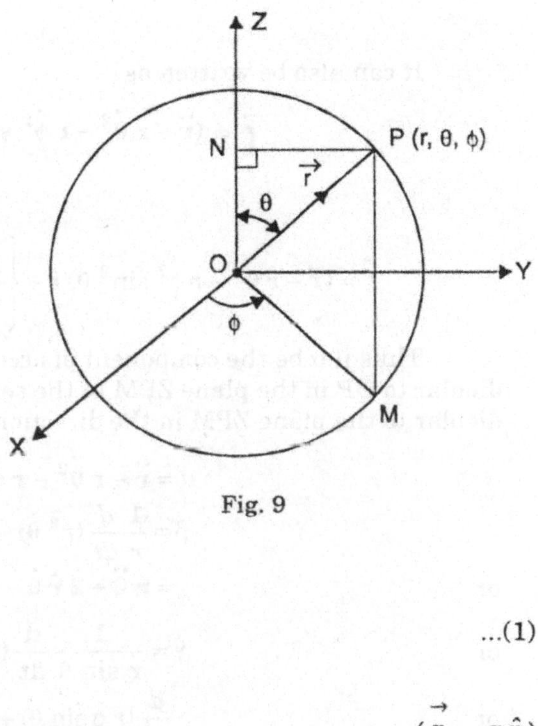

Fig. 9

Let $\hat{r}$, $\hat{\theta}$, $\hat{\phi}$ be unit vectors in the direction of r, θ, ϕ increasing and form a right hand system such that

$\hat{r} \times \hat{\theta} = \hat{\phi}$, $\hat{\theta} \times \hat{\phi} = \hat{r}$ and $\hat{\phi} \times \hat{r} = \hat{\theta}$.

This frame has angular velocity

$$\vec{\omega} = \dot{\theta}\,\hat{\phi} + \dot{\phi}\,\hat{k}$$

$$= \dot{\theta}\,\hat{\phi} + \dot{\phi}\,(\cos\theta\,\hat{r} - \sin\theta\,\hat{\theta})$$

$$\vec{\omega} = \dot{\phi}\cos\theta\,\hat{r} - \dot{\phi}\sin\theta\,\hat{\theta} + \dot{\theta}\,\hat{\phi} \qquad \text{...(1)}$$

Hence, the velocity of P is

$$\vec{V} = \frac{\partial \vec{r}}{\partial t} + \vec{\omega} \times \vec{r} \qquad (\vec{r} = r\,\hat{r})$$

$$= \dot{r}\,\hat{r} + (\dot{\phi}\cos\theta\,\hat{r} - \dot{\phi}\sin\theta\,\hat{\theta} + \dot{\theta}\,\hat{\phi}) \times \vec{r} = \dot{r}\,\hat{r} + 0 + r\,\dot{\phi}\sin\theta\hat{\phi} + \dot{\theta}\,r\,\hat{\theta}$$

$$\vec{V} = \dot{r}\,\hat{r} + r\,\dot{\theta}\,\hat{\theta} + r\,\dot{\phi}\sin\theta\,\hat{\phi} \qquad \text{...(2)}$$

The acceleration of P is

$$\vec{f} = \frac{\partial \vec{V}}{\partial t} + \vec{\omega} \times \vec{V} \quad \text{...(3)}$$

$$\frac{\partial \vec{V}}{\partial t} = \ddot{r}\,\hat{r} + (\dot{r}\,\dot{\theta} + r\,\ddot{\theta})\,\hat{\theta} + (\dot{r}\,\dot{\phi}\sin\theta + r\,\ddot{\phi}\sin\theta + r\dot{\phi}\cos\theta\,\dot{\theta})\,\hat{\phi}$$

$$\vec{\omega} \times \vec{V} = \begin{vmatrix} \hat{r} & \hat{\theta} & \hat{\phi} \\ \dot{\phi}\cos\theta & -\dot{\phi}\sin\theta & \dot{\theta} \\ \dot{r} & r\dot{\theta} & r\dot{\phi}\sin\theta \end{vmatrix}$$

$$= (-r\,\dot{\phi}^2\sin^2\theta - r\dot{\theta}^2)\,\hat{r} + (\dot{r}\,\dot{\theta} - r\,\dot{\phi}^2\sin\theta\cos\theta)\hat{\theta} + (r\,\dot{\theta}\,\dot{\phi}\cos\theta + \dot{r}\,\dot{\phi}\sin\theta)\,\hat{\phi}$$

Putting $\frac{\partial \vec{V}}{\partial t}$ and $\vec{\omega} \times \vec{V}$ in (3), we get

$$\vec{f} = (\ddot{r} - r\,\dot{\theta}^2 - r\,\dot{\phi}^2\sin^2\theta)\,\hat{r} + (r\,\ddot{\theta} + 2\dot{r}\,\dot{\theta} - r\,\dot{\phi}^2\sin\theta\cos\theta)\,\hat{\theta} + (2\dot{r}\,\dot{\phi}\sin\theta + r\,\ddot{\phi}\sin\theta + 2r\,\dot{\theta}\,\dot{\phi}\cos\theta)\,\hat{\phi} \quad \text{...(4)}$$

It can also be written as

$$\mathbf{f} = (\ddot{\mathbf{r}} - \mathbf{r}\,\dot{\theta}^2 - \mathbf{r}\,\dot{\phi}^2\sin^2\theta)\,\hat{\mathbf{r}} + (\mathbf{r}\,\ddot{\theta} + 2\,\dot{\mathbf{r}}\,\dot{\theta} - \mathbf{r}\,\dot{\phi}^2\sin\theta\cos\theta)\,\hat{\theta} + \left(\frac{\mathbf{d}}{\mathbf{dt}}(\mathbf{r}\,\dot{\phi}\sin\theta) + \mathbf{r}\,\dot{\theta}\dot{\phi}\cos\theta + \dot{\mathbf{r}}\,\dot{\phi}\sin\theta\right)\hat{\phi} \quad \text{...(5)}$$

$$\vec{f} = (\ddot{\mathbf{r}} - r\,\dot{\theta}^2 - r\,\dot{\phi}^2\sin^2\theta)\,\hat{r} + \left\{\frac{1}{r}\frac{d}{dt}(r^2\,\dot{\theta}) - r\cos\theta\,\dot{\phi}^2\right\}\hat{\theta} + \frac{1}{r\sin\theta}\frac{d}{dt}\{r^2\sin^2\theta\dot{\phi}\}\,\hat{\phi}$$

Thus if α be the component of acceleration of P along OP, β be the acceleration perpendicular to OP in the plane ZPM in the sense of θ increasing and r be the acceleration perpendicular to the plane ZPM in the direction of ϕ increasing then

$$\alpha = \ddot{\mathbf{r}} - \mathbf{r}\,\dot{\theta}^2 - \mathbf{r}\,\dot{\phi}^2\sin^2\theta$$

$$\beta = \frac{1}{r}\frac{d}{dt}(r^2\,\dot{\theta}) - r\cos\theta\,\dot{\phi}^2$$

or

$$= \mathbf{r}\,\ddot{\theta} + 2\,\dot{\mathbf{r}}\,\dot{\theta} - \mathbf{r}\,\dot{\phi}^2\sin\theta\cos\theta$$

or

$$\gamma = \frac{\mathbf{1}}{\mathbf{r}\sin\theta}\frac{\mathbf{d}}{\mathbf{dt}}(\mathbf{r}^2\sin^2\theta\,\dot{\phi})$$

or

$$\frac{d}{dt}(r\,\dot{\phi}\sin\theta) + r\,\dot{\theta}\,\dot{\phi}\cos\theta + \dot{r}\,\dot{\phi}\sin\theta.$$

Example 1. *A particle is constrained to move along a circular helix so that its coordinates at time t are (a cos θ, a sin θ, aθ tan α) where a, α are constants and $0 < \alpha < \pi/2$. The speed increases linearly with t from zero at t = 0 to V at t = T. Find the acceleration at any time t < T. Motion takes place in the sense of θ increasing and starting from (a, 0, 0).* (M.D.U. 2005)

Sol. $$\vec{r} = a(\cos\theta\,\hat{i} + \sin\theta\,\hat{j} + \theta\tan\alpha\,\hat{k})$$

$$\therefore \quad \frac{d\vec{r}}{dt} = a\left(-\sin\theta\frac{d\theta}{dt}\hat{i} + \cos\theta\frac{d\theta}{dt}\hat{j} + \tan\alpha\frac{d\theta}{dt}\hat{k}\right)$$

$$= \dot{\theta}a\left(-\sin\theta\,\hat{i} + \cos\theta\,\hat{j} + \tan\alpha\,\hat{k}\right)$$

$$\left(\frac{Vt}{T}\right)\hat{t} = a\dot{\theta}\sec\alpha\left[-\sin\theta\cos\alpha\,\hat{i} + \cos\theta\cos\alpha\,\hat{j} + \sin\alpha\,\hat{k}\right]$$

The magnitude of $-\sin\theta\cos\alpha\,\hat{i} + \cos\theta\cos\alpha\,\hat{j} + \sin\alpha\,\hat{k}$ is 1

$$\therefore \quad \frac{Vt}{T} = a\dot{\theta}\sec\alpha \qquad \ldots(1)$$

and $$\hat{t} = -\sin\theta\cos\alpha\,\hat{i} + \cos\theta\cos\alpha\,\hat{j} + \sin\alpha\,\hat{k} \qquad \ldots(2)$$

From (1), $$d\theta = \frac{V}{T}\frac{t\cos\alpha}{a}dt \quad \text{or} \quad \theta = \frac{Vt^2\cos\alpha}{2aT}$$

From (2), $$\frac{d\vec{t}}{ds} = (-\cos\theta\cos\alpha\hat{i} - \sin\theta\sin\alpha\,\hat{j})\frac{d\theta}{ds} \qquad \ldots(3)$$

But $$\frac{d\theta}{ds} = \frac{d\theta}{dt}\Big/\frac{ds}{dt} = \frac{Vt\cos\alpha}{Ta}\Big/\left(\frac{Vt}{T}\right) = \frac{\cos\alpha}{a}$$

Putting in (3)

$$\frac{d\vec{t}}{ds} = (-\cos\theta\hat{i} - \sin\theta\,\hat{j})\frac{\cos^2\alpha}{a}$$

$$\therefore \quad k = \cos^2\alpha/a \quad \text{and} \quad \hat{n} = -\cos\theta\hat{i} - \sin\theta\hat{j}$$

Thus, particle moves with acceleration having component $\frac{V}{T}$ along tangent and $kv^2 = \frac{V^2t^2}{T^2}\frac{\cos^2\alpha}{a}$ along principal normal.

Example 2. *Show that two equal and opposite rotations of a rigid body about distinct parallel axes are equivalent to a translation of the body.*

Sol. Let the body rotate about two parallel axes AB and CD with angular velocities ω, $-\omega$ respectively. Let O_1 and O_2 be chosen on these axes such that $\vec{O_1O_2} = \vec{s}$. P be any point on the rigid body such that $\vec{O_1P} = \vec{r_1}$, $\vec{O_2P} = \vec{r_2}$.

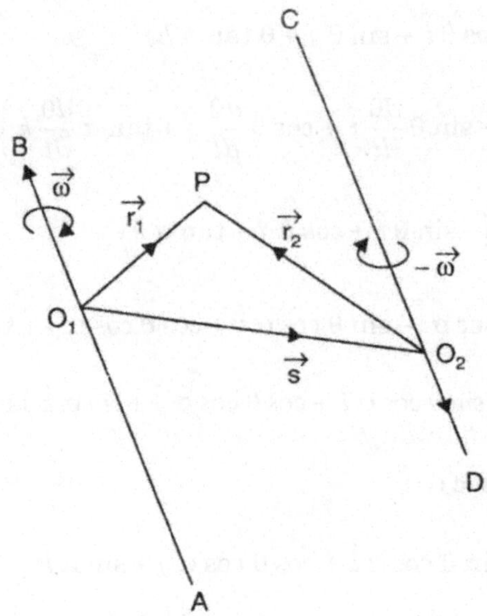

Fig. 10

Velocity of P is $\vec{V} = \vec{\omega} \times \vec{r_1} + (-\vec{\omega}) \times \vec{r_2} = \vec{\omega} \times (\vec{r_1} - \vec{r_2}) = \vec{\omega} \times \vec{s}$

This is independent of the choice of P and perpendicular to the plane of $\vec{\omega}$, $-\vec{\omega}$ and $\vec{s}$. Thus every point moves with the same velocity and hence the motion is translation only.

Example 3. *An aircraft persues a straight course with constant velocity V and is being chased by a guided missile moving with constant speed 2V and the motion is always directed at the target. Initially the missile is at right angle to the course of aircraft and distance R from it. Find the polar equation of the missile's pursuit curve relative to the target, taking the path of the plane as initial line $\theta = 0$, and find the time taken to strike the target.*

Sol. Let M be the position of the missile and T be the position of the target at time t. Let TM = r, $\angle$MTX = θ.

To consider the motion relative to target V is replaced at M in opposite direction. Resolving in radial and transverse direction

$$\dot{r} = (-2V) - V\cos\theta \quad \text{...(1)}$$

$$r\dot{\theta} = V\sin\theta \quad \text{...(2)}$$

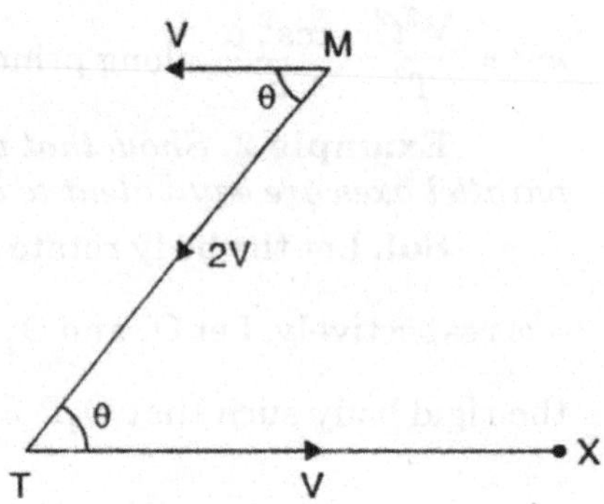

Fig. 11

Dividing (1) by (2)

$$\frac{dr}{r\,d\theta} = \frac{-V(2+\cos\theta)}{V\sin\theta}$$

or

$$\frac{dr}{r} = -\left(\frac{2}{\sin\theta} + \cot\theta\right)d\theta$$

Integrating $-\int \frac{dr}{r} = -\int (2\operatorname{cosec}\theta + \cot\theta)\,d\theta$

$$-\log r + \log c = 2 \log \tan \theta/2 + \log \sin \theta$$

or $$\log \frac{c}{r} = \log (\sin \theta \tan^2 \theta/2)$$

or $$\frac{c}{r} = \sin \theta \tan^2 \theta/2$$

At $$\theta = \pi/2, r = R \Rightarrow c = R . \sin \pi/2 \tan^2 \pi/4 = R.$$

Hence, $$\frac{R}{r} = \sin \theta \tan^2 \theta/2 = \frac{2 \sin^3 \theta/2}{\cos \theta/2} \quad ...(3)$$

which is the required equation of the path.

when $\theta = \pi$, $r = 0$.

From (2) $$\frac{d\theta}{dt} = \frac{V \sin \theta}{r} = \frac{V . 2 \sin \theta/2 \cos \theta/2 (2 \sin^3 \theta/2)}{R \cos \theta/2}$$

or $$\frac{d\theta}{dt} = \frac{4V}{R} \sin^4 \frac{\theta}{2}$$

Time taken to hit the aircraft

$$t = \int_{\pi/2}^{\pi} \frac{R}{4V} \frac{d\theta}{\sin^4 \theta/2} = \frac{R}{4V} \int_{\pi/2}^{\pi} \operatorname{cosec}^4 \theta/2 \, d\theta$$

$$= \frac{R}{4V} \int_{\pi/2}^{\pi} (1 + \cot^2 \theta/2) . \operatorname{cosec}^2 \theta/2 \, d\theta$$

$$= -\frac{R}{4V} \int_{1}^{0} (1 + z^2) \, 2dz$$

$$\cot \theta/2 = Z, \; -\frac{1}{2} \operatorname{cosec}^2 \theta/2 \, d\theta = dz$$

$$= \frac{2R}{4V} \left[z + \frac{z^3}{3} \right]_0^1 = \frac{2R}{4V} . \frac{4}{3} = \frac{2}{3} R/V.$$

EXERCISE 1

1. A particle is constrained to move along the equiangular spiral $r = ae^{h\theta}$. So that the radius vector moves with constant angular velocity ω. Determine the velocity and acceleration components.
2. Obtain the radial and transverse components of acceleration of a particle which describes the plane curve $r = f(\theta)$ in the form

   $\ddot{r} - r\dot{\theta}^2$ and $\frac{1}{r} \frac{d}{dt} (r^2 \dot{\theta})$.

   If the curve is $r = ae^{\theta \cot \alpha}$ and the radius vector of the particle has constant angular velocity, show that the resultant acceleration of the particle makes an angle 2α with the radius vector and is of magnitude v^2/r where v is the speed of the particle.

3. If P is point specified by polar coordinates (r, θ) in a plane and position vector $\overrightarrow{OP}$ at time t is represented by complex number

$Z = re^{i\theta}$ show that

$$\dot{z} = \dot{r}\, e^{i\theta} + r\dot{\theta}\,(i\, e^{i\theta})$$

$$\ddot{z} = (\ddot{r} - r\,\dot{\theta}^2)\, e^{i\theta} + (r\ddot{\theta} + 2\dot{r}\,\dot{\theta})\,(i\, e^{i\theta}).$$

Hence, deduce the radial and transverse velocity and acceleration components of P.

4. An aeroplanes travels at a maximum speed V. It has to travel from A to B and back in the shortest possible time in a cross wind blowing at constant velocity v in a direction making an angle ϕ with $\overrightarrow{BA}$. Find the time taken and determine the directions in which the aircraft must steer on the outward and return journeys.

5. A point moves on the surface given by $x = (c + a \sin\theta) \cos\phi$, $y = (c + a \sin\theta) \sin\phi$, $z = a \cos\theta$. Denoting $\hat{a}, \hat{b}, \hat{c}$ the unit vectors in the normal, meridian and transverse directions at P, so that $\hat{a} \times \hat{b} = \hat{c}$, $\hat{b} \times \hat{c} = \hat{a}$, $\hat{c} \times \hat{a} = \hat{b}$ show that

$$\overrightarrow{OP} = \vec{r} = (a + c \sin\theta)\,\hat{a} + c \cos\theta\,\hat{b} \text{ and } \vec{\omega} = \dot{\phi} \cos\theta\hat{a} - \dot{\phi} \sin\theta\hat{b} + \dot{\theta}\hat{c}$$

Hence, or otherwise deduce the acceleration components of P.

ANSWERS

1. $(b^2 - 1)\,\omega^2 a\, e^{b\omega t}$, $2a\,\omega^2 e^{b\omega t}$.

EXAMINATION PAPERS

M.D.U., ROHTAK

B.A. Part II, Examination, April 2007

Paper III (B.M. 203)

Mechanics

Time : Three Hours] *[Maximum Marks : 30*

Before answering the question, candidate should ensure that they have been supplied the correct and complete question paper. No complaint in this regard, will be entertained after examination.

Note : Attempt **Five** questions in all, selecting at least **One** question from each section. All questions carry equal marks.

SECTION I

1. (*a*) A heavy uniform beam of weight W and length $2l$, rests against a smooth vertical wall upon a peg at a distance b from the wall. If α be inclination of the beam to the vertical then show that $l \sin^3 \alpha = b$.

   (*b*) A ladder of length $2b$ and weight W has its centre of gravity at a distance $\frac{3}{4}b$ way up the ladder, stands on a smooth horizontal plane resting against a smooth vertical wall and middle point is tied to a point in the wall by a horizontal rope of length l. Find the tension in the rope and reactions due to wall and horizontal plane.

2. (*a*) A uniform beam of weight W rests with its ends in contact with two smooth planes inclined at angles β and γ to the horizon. If α be the inclination of the beam to the horizon, show that $2 \tan \alpha = \cot \beta - \cot \gamma$. Also find the reactions at the ends of the beam.

   (*b*) If a system of forces acting at a point produces in it a small displacement, the algebraic sum of the work done by the separate forces is equal to the work done by their resultant force, prove it.

3. (*a*) Four uniform rods each of weight W are jointed to form a parallelogram ABCD, which is suspended from A, it is kept in shape by a string joining A and C. If T is the tension in the string prove that T = 2 W.

   (*b*) A regular hexagon of weight 6 W is composed of six heavy uniform rods jointed together and two opposite edges are connected by a string which is horizontal, one rod being in contact with a horizontal plane, at the middle point of the opposite rod is placed a weight W_1. Find the tension in the string.

SECTION II

4. (*a*) Find the equations of the central axis of any given system of forces acting on a rigid body.

   (*b*) OA, OB, OC are edges of a cube of side a and OO′, AA′, BB′, CC′ are its diagonals. Forces each equal to P, 2P, 3P, 4P act along OB′, O′A, BC, C′A′ respectively. Reduce the system to a force through O and a couple about an axis through O.

5. (*a*) Two forces P and Q act along the lines $x \sin\alpha - y \cos\alpha = 0$, $z = c$ and $x \sin\alpha + y \cos\alpha = 0$, $z = -c$ respectively. Show that their central axis lies on a straight line

   $$x(\mathrm{P} - \mathrm{Q}) \sin\alpha - y(\mathrm{P} + \mathrm{Q}) \cos\alpha = 0,\ z(\mathrm{P}^2 + \mathrm{Q}^2 + 2\mathrm{PQ} \cos 2\alpha) + (\mathrm{Q}^2 - \mathrm{P}^2)\,\mathrm{C} = 0.$$

   (*b*) A solid homogeneous hemisphere of radius r has a solid right circular cone of the same substance constructed on its base; the hemisphere rests on the convex side of a fixed sphere of radius R, the axis of of cone being vertical. Show that the greatest height of the cone consistent with stability for a small rolling displacement is :

   $$\frac{r}{(\mathrm{R}+r)}\left[\sqrt{(3\mathrm{R}+r)(\mathrm{R}-r)} - 2r\right].$$

SECTION III

6. (*a*) To derive the expressions of radial and transverse acceleration for the curve $r = f(\theta)$.

   (*b*) A particle moves along the curve $x = 4t$, $y = 6t - t^2$. Find tangential and normal acceleration at $t = 3$.

7. (*a*) If the displacement of a moving point at any time t is given by equation $x = 3 \cos 2t + 4 \sin 2t$. Show that the motion is S. H. M., find the period, amplitude, maximum velocity and maximum acceleration of motion.

   (*b*) A point executes S.H.M. such that in two of its positions the velocities are v_1, v_2 and the corresponding accelerations are f_1, f_2. Show that the amplitude is :

   $$\frac{\sqrt{(v_1^2 - v_2^2)(f_2^2 v_1^2 - f_1^2 v_2^2)}}{(f_2^2 - f_1^2)}$$

8. (*a*) An elastic string of natural length l and modulus λ is fixed to a point on a smooth horizontal table and the other end is tied to a particle of mass m on the table. The particle is pulled to a distance such that the string becomes equal to twice its natural length and then released. Find the time of one complete oscillation.

   (*b*) A particle of mass 1 unit, is falling under the influence of gravity through a medium whose resistance equal to λ times the velocity. If the particle be released from rest, show that the distance fallen through in time t is :

   $$\frac{8}{\lambda^2}[e^{-\lambda t} + \lambda t - 1]$$

SECTION IV

9. (*a*) **Derive differential equation of central orbit in polar form.**

(*b*) **If law of force is $\frac{5\mu}{r^3} + \frac{8\mu c^2}{r^5}$ and the particle is projected from an apse at a distance c with velocity $\frac{3\sqrt{\mu}}{c}$, prove that the orbit is $r = c \cos\left(\frac{2\theta}{3}\right)$.**

10. (*a*) **Deduce three Kepler's Laws from Newton's Law of gravitation.**

(*b*) **If v_1 and v_2 are maximum and minimum velocities of a planet then prove that $(1 - e)v_1 = (1 + e)v_2$ for an elliptic path.**

M.D.U., ROHTAK

B.Sc. Part II, Examination, April 2007

Paper III (B.M. 203)

Mechanics

Time : Three Hours] *[Maximum Marks : 30*

Before answering the question, candidate should ensure that they have been supplied the correct and complete question paper. No complaint in this regard, will be entertained after examination.

Note : Attempt **Five** questions in all, selecting at least **One** question from each section. All questions carry equal marks.

SECTION I

1. (*a*) A uniform rod AB, movable about a hinge at A rests with one end in contact with a smooth wall. If α be the inclination of the rod to the vertical, prove that the reaction at the hinge is $\frac{1}{2}W\sqrt{3+\sec^2\alpha}$. Also find the direction. 5

   (*b*) Two uniform rods AB, BC rigidly joined at B so that angle ABC is a right angle, hang freely in equilibrium from a fixed pt. A. The length of rods are a & b and their wts are aw and bw. Prove that if AB makes an angle θ with the vertical. 4

$$\tan\theta = \frac{b^2}{a^2 + 2ab}$$

2. (*a*) Prove that the virtual works done by the tension in a virtual extension of a string from length l to $l + \delta l$ is $- T. \delta l$ where T is the tension in the string. 3

   (*b*) Five equal uniform rods, freely jointed at their ends form a regular pentagon. ABCDE and BE is jointed by a light rod. The system is suspended from A in a vertical plane. Prove that thrust in BE is $w \cot\left(\frac{\pi}{10}\right)$ where w is the wt. of each rod. 6

3. (*a*) A uniform beam of length $2a$ rests in equilibrium against a smooth vertical wall and upon a smooth peg at a distance b from the wall. Show that in the position of equilibrium, the rod is inclined to wall at an angle $\sin^{-1}\left(\frac{b}{a}\right)^{1/3}$. 4

   (*b*) A heavy uniform rod of length $2l$ is resting in contact with a smooth sphere of radius r and the lower end of the rod presses against a smooth vertical wall. If the distance of centre of the sphere from the wall is C and the rod makes an angle α with the wall then prove that $l \sin^3 a + r \cos\alpha = C$. 5

SECTION II

4. (*a*) A force P acts along the axis of x and another force nP along a generator of the cylinder $x^2 + y^2 = a^2$. Show that central axis lies on the cylinder : 5

$$n^2 (nx - z)^2 + (1 + n^2)^2 y^2 = n^4 a^2$$

(*b*) Two forces P and Q act along the straight lines whose equations are $y = x$, $z = c$ and $y = -x$, $z = -c$ respectively. Show that central axis lies on the straight line : 4

$$y = x\left(\frac{P-Q}{P+Q}\right) \text{ and } \frac{z}{c} = \frac{P^2 - Q^2}{P^2 + Q^2}$$

5. (*a*) A square lamina rests with its plane perpendicular to a smooth wall; one corner being attached to a point in the wall by a fine string of length equal to side of the square. Find position of equilibrium and show that it is stable. 4

(*b*) A uniform smooth rod passes through a ring at the focus of a fixed parabola whose axis is vertical and vertex below the focus and rests with one end on the parabola. Prove that the rod will be in equilibrium if it makes with the vertical, an angle θ given by the equation $\cos^4 \frac{\theta}{2} = \frac{a}{2c}$ where $4a$ = latus rectum of the parabola and $2c$ = length of the rod. 5

SECTION III

6. (*a*) If a particle moving in a plane curve $r = f(\theta)$. Then find expressions for radial and transverse acceleration. 5

(*b*) Prove that if the tangential and normal accelerations of a particle describing a plane curve be constant through out the motion, the angle ψ through which the direction of motion is given by $\psi = A \log (1 + Bt)$. 4

7. (*a*) A particle rests in equilibrium under the attraction of two centres of force which attract directly as the distance, their intensity being μ, μ'; the particle is displaced slightly towards one of them. Show that time of small oscillation is $\frac{2\pi}{\sqrt{\mu + \mu'}}$. 5

(*b*) A particle moves with S.H.M. in a straight line. In the Ist second after starting from rest it travels a distance a and in the next second, it travels a distance b in the same direction. Prove that amplitude of the motion is $\frac{2a^2}{3a - b}$. 4

8. (*a*) A particle is attached to the middle pt. of a uniform elastic string which is stretched between two pts. A and B on a smooth horizontal table. If the particle is pulled to some point C between A and B and then liberated, show that the motion is simple harmonic and find the period of the motion what is the amplitude. 5

(*b*) An elastic string of natural length l has one end fixed at a pt. O and other end attached to a particle, the wt. of which in equilibrium position would extend the string to a length l_1. Show that if the particle is dropped from rest at 0, it will come to instantaneous rest at a depth $\sqrt{l_1^2 - l^2}$ below the equilibrium position. 4

SECTION IV

9. (*a*) **Prove analytically that when the central acceleration varies as some integral powers of the distance, there are at the most two apsidal distances.** 5

(*b*) **A particle moves with a central acceleration $\frac{\lambda}{(\text{distance})^3}$. Find the path and distinguish the cases.** 4

10. (*a*) **Two inelastic particles of masses 3m and 1m are describing the same ellipse of eccentricity e in opposite direction under a force to the focus. They collide and coalesce at an extrimity of the minor axis. Prove that eccentricity of new orbit is $\frac{1}{4}\sqrt{9+7e^2}$.** 5

(*b*) **A heavy particle moves in a smooth sphere. Show that if the velocity be that due to the level of the centre, the reaction of the surface will vary at the depth below the centre.** 4

K.U., KURUKSHETRA

B.A./B.Sc. Part II, Examination, April 2007

Paper III (B.M. 203)

Mechanics

Time : Three Hours] *[Maximum Marks : B.A.–30*
B.Sc.–45

Note : Attempt Five questions in all, selecting at least One question from each section. All questions carry equal marks.

SECTION I

1. (*a*) Find resultant of a system of Coplanar Forces acting on a Rigid Body. Also write the condition of equilibrium.

   (*b*) One end of a uniform rod is attached to a hinge and the other end is supported by a string attached to the extremity of the rod ; the rod and the string are inclined at the same angle θ to the horizontal. If W be the weight of the rod, show that the reaction at the hinge is $\frac{1W}{4}\sqrt{8+\operatorname{cosec}^2\theta}$ or $\frac{W}{4}\sqrt{9+\cot^2\theta}$.

   Also find the tension of the string.

2. (*a*) A string of length a forms the shorter diagonal of a rhombus formed of four uniform rods each of length *b* and weight W, which are hinged together. If one of the rod is supported in a horizontal position, prove that the tension of the string is

$$\frac{2W(2b^2-a^2)}{b\sqrt{4b^2-a^2}}.$$

   (*b*) A uniform beam of length $2a$ rests in equilibrium against a smooth vertical wall and upon a smooth peg at a distance b from the wall. Show that in the position of equilibrium, the rod is inclined to the wall at an angle $\sin^{-1}\left(\frac{b}{a}\right)^{1/3}$.

3. (*a*) Define Catenary. Also find intrinsic equation of the common catenary.

   (*b*) A Kite is flowing with 600 metres of string from the hand to the kite and a spring balance held in the hand shows a pull equal to the weight of 100 metres of the string inclined at 30° to the horizon. Find the vertical height of the Kite above the hand.

SECTION II

4. (*a*) Equal forces act along two perpendicular diagonals of opposite faces of a cube of side a. Show that they are equivalent to a single force R acting along a line through the centre of the cube and a couple $\frac{1}{2}aR$ with the same line for axis.

(*b*) Show that the minimum distance between two forces which are equivalent to a given system (R, K) and which are inclined at a given angle 2α is $\frac{2K}{R}$ cot α and the forces are then each equal to $\frac{R}{2}$ sec α.

5. (*a*) Find the null point of the plane $ex + my + nz = 1$ for the system of forces.

(*b*) A heavy uniform rod of length $2a$, rests partly within and partly without a fixed smooth hemispherical bowl of radius r, the rim of the bowl is horizontal and one point of the rod is in contact with the rim. If θ be the inclination of the rod to the horizontal, show that $2r \cos 2\theta = a \cos \theta$. Also show that the equilibrium of the rod is stable.

SECTION III

6. (*a*) A particle moves in a circular path of radius a so that its angular velocity about a fixed point in the circumference is constant and equal to w. Show that the resultant acceleration of the particle at every point is of constant magnitude of $4aw^2$.

(*b*) A particle is performing S.H.M. of period T about a centre O and it passes through the position P with velocity v in the direction OP where OP = a. Prove that the time that elapses before it returns to P is $\frac{T}{\pi} \tan^{-1} \frac{vt}{2\pi a}$.

7. (*a*) Prove that the work done against the tension in stretching a light elastic string is equal to the product of its extension and the mean of the initial and final tensions.

(*b*) A body of mass m is projected vertically upwards in a medium for which the resistance is $m\,K^2\,\vartheta^2$. If initial velocity is ϑ_0 ; show that $\vartheta_1^2 = \frac{g\vartheta_0^2}{g + K^2\vartheta_0^2}$.

8. (*a*) A particle of mass m is projected with constant velocity u in a resisting medium whose resistance varies as the velocity. To discuss its motion.

(*b*) A small bead is projected with any velocity along a smooth circular wire under the action of force varying inversely, as the fifth power of distance from a centre of force situated on the circumference. Prove that pressure on the wire is constant.

SECTION IV

9. (*a*) A particle moving under the influence of a central force, describes a circle through the centre of the force. Prove that the force is attractive and inversely proportional to the fifth power of the distance.

(*b*) A particle moving under a constant force from a centre is projected at a distance '*a*' from the centre in a direction perpendicular to the radius vector with velocity acquired in falling to the point of projection from the centre. Show that its path is

$$\left(\frac{a}{r}\right)^3 = \cos^2\left(\frac{3}{2}\theta\right).$$

10. (*a*) If a planet were suddenly stopped in its orbit when at a distance '*a*' from the sun, show that it would fall in the sun in time $\frac{\sqrt{2}\pi a^{3/2}}{4\sqrt{\mu}}$ which is $\frac{\sqrt{2}}{8}$ times the period of the planet's revolution.

(*b*) A heavy particle moves in a smooth sphere. Show that if the velocity be that due to the level of the centre, the reaction of the surface will vary at the depth below the centre.